MATHEMATICAL LOGIC AND FORMAL SYSTEMS

LECTURE NOTES

IN PURE AND APPLIED MATHEMATICS

1. *N. Jacobson,* Exceptional Lie Algebras
2. *L. -Å. Lindahl and F. Poulsen,* Thin Sets in Harmonic Analysis
3. *I. Satake,* Classification Theory of Semi-Simple Algebraic Groups
4. *F. Hirzebruch, W. D. Newmann, and S. S. Koh,* Differentiable Manifolds and Quadratic Forms (out of print)
5. *I. Chavel,* Riemannian Symmetric Spaces of Rank One (out of print)
6. *R. B. Burckel,* Characterization of C(X) Among Its Subalgebras
7. *B. R. McDonald, A. R. Magid, and K. C. Smith,* Ring Theory: Proceedings of the Oklahoma Conference
8. *Y.-T. Siu,* Techniques of Extension on Analytic Objects
9. *S. R. Caradus, W. E. Pfaffenberger, and B. Yood,* Calkin Algebras and Algebras of Operators on Banach Spaces
10. *E. O. Roxin, P.-T. Liu, and R. L. Sternberg,* Differential Games and Control Theory
11. *M. Orzech and C. Small,* The Brauer Group of Commutative Rings
12. *S. Thomeier,* Topology and Its Applications
13. *J. M. Lopez and K. A. Ross,* Sidon Sets
14. *W. W. Comfort and S. Negrepontis,* Continuous Pseudometrics
15. *K. McKennon and J. M. Robertson,* Locally Convex Spaces
16. *M. Carmeli and S. Malin,* Representations of the Rotation and Lorentz Groups: An Introduction
17. *G. B. Seligman,* Rational Methods in Lie Algebras
18. *D. G. de Figueiredo,* Functional Analysis: Proceedings of the Brazilian Mathematical Society Symposium
19. *L. Cesari, R. Kannan, and J. D. Schuur,* Nonlinear Functional Analysis and Differential Equations: Proceedings of the Michigan State University Conference
20. *J. J. Schäffer,* Geometry of Spheres in Normed Spaces
21. *K. Yano and M. Kon,* Anti-Invariant Submanifolds
22. *W. V. Vasconcelos,* The Rings of Dimension Two
23. *R. E. Chandler,* Hausdorff Compactifications
24. *S. P. Franklin and B. V. S. Thomas,* Topology: Proceedings of the Memphis State University Conference
25. *S. K. Jain,* Ring Theory: Proceedings of the Ohio University Conference
26. *B. R. McDonald and R. A. Morris,* Ring Theory II: Proceedings of the Second Oklahoma Conference
27. *R. B. Mura and A. Rhemtulla,* Orderable Groups
28. *J. R. Graef,* Stability of Dynamical Systems: Theory and Applications
29. *H.-C. Wang,* Homogeneous Branch Algebras
30. *E. O. Roxin, P.-T. Liu, and R. L. Sternberg,* Differential Games and Control Theory II
31. *R. D. Porter,* Introduction to Fibre Bundles
32. *M. Altman,* Contractors and Contractor Directions Theory and Applications
33. *J. S. Golan,* Decomposition and Dimension in Module Categories
34. *G. Fairweather,* Finite Element Galerkin Methods for Differential Equations
35. *J. D. Sally,* Numbers of Generators of Ideals in Local Rings
36. *S. S. Miller,* Complex Analysis: Proceedings of the S.U.N.Y. Brockport Conference
37. *R. Gordon,* Representation Theory of Algebras: Proceedings of the Philadelphia Conference
38. *M. Goto and F. D. Grosshans,* Semisimple Lie Algebras
39. *A. I. Arruda, N. C. A. da Costa, and R. Chuaqui,* Mathematical Logic: Proceedings of the First Brazilian Conference

40. *F. Van Oystaeyen,* Ring Theory: Proceedings of the 1977 Antwerp Conference
41. *F. Van Oystaeyen and A. Verschoren,* Reflectors and Localization: Application to Sheaf Theory
42. *M. Satyanarayana,* Positively Ordered Semigroups
43. *D. L. Russell,* Mathematics of Finite-Dimensional Control Systems
44. *P.-T. Liu and E. Roxin,* Differential Games and Control Theory III: Proceedings of the Third Kingston Conference, Part A
45. *A. Geramita and J. Seberry,* Orthogonal Designs: Quadratic Forms and Hadamard Matrices
46. *J. Cigler, V. Losert, and P. Michor,* Banach Modules and Functors on Categories of Banach Spaces
47. *P.-T. Liu and J. G. Sutinen,* Control Theory in Mathematical Economics: Proceedings of the Third Kingston Conference, Part B
48. *C. Byrnes,* Partial Differential Equations and Geometry
49. *G. Klambauer,* Problems and Propositions in Analysis
50. *J. Knopfmacher,* Analytic Arithmetic of Algebraic Function Fields
51. *F. Van Oystaeyen,* Ring Theory: Proceedings of the 1978 Antwerp Conference
52. *B. Kedem,* Binary Time Series
53. *J. Barros-Neto and R. A. Artino,* Hypoelliptic Boundary-Value Problems
54. *R. L. Sternberg, A. J. Kalinowski, and J. S. Papadakis,* Nonlinear Partial Differential Equations in Engineering and Applied Science
55. *B. R. McDonald,* Ring Theory and Algebra III: Proceedings of the Third Oklahoma Conference
56. *J. S. Golan,* Structure Sheaves over a Noncommutative Ring
57. *T. V. Narayana, J. G. Williams, and R. M. Mathsen,* Combinatorics, Representation Theory and Statistical Methods in Groups: YOUNG DAY Proceedings
58. *T. A. Burton,* Modeling and Differential Equations in Biology
59. *K. H. Kim and F. W. Roush,* Introduction to Mathematical Consensus Theory
60. *J. Banas and K. Goebel,* Measures of Noncompactness in Banach Spaces
61. *O. A. Nielson,* Direct Integral Theory
62. *J. E. Smith, G. O. Kenny, and R. N. Ball,* Ordered Groups: Proceedings of the Boise State Conference
63. *J. Cronin,* Mathematics of Cell Electrophysiology
64. *J. W. Brewer,* Power Series Over Commutative Rings
65. *P. K. Kamthan and M. Gupta,* Sequence Spaces and Series
66. *T. G. McLaughlin,* Regressive Sets and the Theory of Isols
67. *T. L. Herdman, S. M. Rankin, III, and H. W. Stech,* Integral and Functional Differential Equations
68. *R. Draper,* Commutative Algebra: Analytic Methods
69. *W. G. McKay and J. Patera,* Tables of Dimensions, Indices, and Branching Rules for Representations of Simple Lie Algebras
70. *R. L. Devaney and Z. H. Nitecki,* Classical Mechanics and Dynamical Systems
71. *J. Van Geel,* Places and Valuations in Noncommutative Ring Theory
72. *C. Faith,* Injective Modules and Injective Quotient Rings
73. *A. Fiacco,* Mathematical Programming with Data Perturbations I
74. *P. Schultz, C. Praeger, and R. Sullivan,* Algebraic Structures and Applications Proceedings of the First Western Australian Conference on Algebra
75. *L. Bican, T. Kepka, and P. Nemec,* Rings, Modules, and Preradicals
76. *D. C. Kay and M. Breen,* Convexity and Related Combinatorial Geometry: Proceedings of the Second University of Oklahoma Conference
77. *P. Fletcher and W. F. Lindgren,* Quasi-Uniform Spaces
78. *C.-C. Yang,* Factorization Theory of Meromorphic Functions
79. *O. Taussky,* Ternary Quadratic Forms and Norms
80. *S. P. Singh and J. H. Burry,* Nonlinear Analysis and Applications
81. *K. B. Hannsgen, T. L. Herdman, H. W. Stech, and R. L. Wheeler,* Volterra and Functional Differential Equations

82. *N. L. Johnson, M. J. Kallaher, and C. T. Long,* Finite Geometries: Proceedings of a Conference in Honor of T. G. Ostrom
83. *G. I. Zapata,* Functional Analysis, Holomorphy, and Approximation Theory
84. *S. Greco and G. Valla,* Commutative Algebra: Proceedings of the Trento Conference
85. *A. V. Fiacco,* Mathematical Programming with Data Perturbations II
86. *J.-B. Hiriart-Urruty, W. Oettli, and J. Stoer,* Optimization: Theory and Algorithms
87. *A. Figa Talamanca and M. A. Picardello,* Harmonic Analysis on Free Groups
88. *M. Harada,* Factor Categories with Applications to Direct Decomposition of Modules
89. *V. I. Istrătescu,* Strict Convexity and Complex Strict Convexity: Theory and Applications
90. *V. Lakshmikantham,* Trends in Theory and Practice of Nonlinear Differential Equations
91. *H. L. Manocha and J. B. Srivastava,* Algebra and Its Applications
92. *D. V. Chudnovsky and G. V. Chudnovsky,* Classical and Quantum Models and Arithmetic Problems
93. *J. W. Longley,* Least Squares Computations Using Orthogonalization Methods
94. *L. P. de Alcantara,* Mathematical Logic and Formal Systems
95. *C. E. Aull,* Rings of Continuous Functions
96. *R. Chuaqui,* Analysis, Geometry, and Probability
97. *L. Fuchs and L. Salce,* Modules Over Valuation Domains

Other Volumes in Preparation

MATHEMATICAL LOGIC AND FORMAL SYSTEMS

A Collection of Papers in honor of Professor Newton C.A. da Costa

edited by

Luiz Paulo de Alcantara

Instituto de Matematica
Universidade Estadual de Campinas
Campinas, São Paulo, Brazil

MARCEL DEKKER, INC. New York and Basel

Library of Congress Cataloging In Publication Data
Main entry under title:

Mathematical logic and formal systems.

(Lecture notes in pure and applied mathematics ; v. 94)
English and French.
Includes index.
1. Logic, Symbolic and mathematical--Addresses, essays, lectures. 2. Costa, Newton C. A. da. I. Costa, Newton C. A. da. II. Alcantara, Luiz Paulo de, [date]. III. Series.
QA9.2.M38 1985 511.3 84-25984
ISBN 0-8247-7330-6

MARCEL DEKKER, INC.
270 Madison Avenue, New York, New York 10016

Current printing (last digit):
10 9 8 7 6 5 4 3 2 1

PRINTED IN THE UNITED STATES OF AMERICA

Professor Newton C. A. da Costa

FOREWORD

The editor of this volume has asked me to write its Foreword. It is an honor and a pleasure for me to do so. An honor, because of my high esteem for the scientific work of Newton da Costa. A pleasure, because of my appreciation of him as a person and a friend.

I do not have to write too much about the importance of his technical work as a logician. There is a detailed analysis of his published work in this volume, written by himself, and the quality and quantity of the papers included in this book attest to the appreciation that scientists from different parts of the world have for him.

I shall only make some remarks of a general character about his scientific work. Most of his papers are of a mathematical nature, but are almost always philosophically inspired. Thus, he has been a mathematician whose work originates, not from mathematics itself, but from other sources. I think that the main difference between a pure and an applied mathematician lies not so much in the direct practical application of the work of the latter, but in the source of his problems. A pure mathematician finds his inspiration in mathematics itself and is not so much interested in the problem to be solved but in the methods to be used. An applied mathematician, on the other hand, derives his work from outside sources and he is interested in the intrinsic importance of the problems, caring less for the particular methods employed. Using this distinction, Newton da Costa is clearly an applied mathematician, that is, his work is applied to philosophy. (I should really say "was an applied mathematician," because lately he would probably prefer to be called a "philosopher who uses mathematics.")

An overview of his work shows the remarkable wealth of new ideas that he has produced. I shall just discuss two fields which I think contain his most fruitful ideas. First, he is one of the originators of paraconsistent logic. He was bold enough to admit the possibility of allowing contradictions in a logical system, and he showed that it was possible to have interesting systems of this sort. When he began working in this area in 1954 it was a very novel idea. Now there are many logicians doing research in the field, not only in Brazil, where Newton formed a school, but in many other countries, such as Poland, the U.S.A., the Soviet Union, and Australia.

The second subject I will undertake to discuss is his current interest: inductive logic and probability. Although he has not yet published very much in this area, he has a wealth of ideas that promise important results in the future. This is the field in which I am most interested and we have discussed the matter frequently. I think I am the person who is best acquainted with his ideas. For me, inductive logic and probability are at the heart of the philosophy of science; I believe that Newton da Costa agrees with me.

What I have just said implies that Newton da Costa is in the middle of his scientific career and that probably some of his most important work is yet to be done. A volume of the character of the present one is usually published in honor of scientists at the end of their careers. This is certainly not the case here. Newton is just 55 years old and he still has much important work to do. Why, then, is this book appearing? Because his work is not only important in itself, but it had a decisive role in the development of science in Brazil and the rest of Latin America. In this respect, his teaching activity has been very important. Almost all logicians in Brazil have been his students at one time or another in their careers. The Brazilian school of logic (yes, there is such a school) owes its existence to him. He has directly advised at least 12 students in their Master's or Doctoral dissertations. Some have completed their studies abroad. In addition to his students, he has influenced almost all other logicians in Brazil. This activity is not just a reflection of his scientific endeavors, but also, and perhaps mainly, of his attractive personality, and of his human values and kindness. I must say that it is very easy to be a friend of Newton and, probably, it is just as agreeable to be his student.

Newton's direct influence is felt, not only in Brazil, but also in the rest of South America. He has visited most of the countries of the region, worked with many logicians there, and made many friends. This last feat is not the least of his accomplishments, since this world full of wars needs friendships.

Newton da Costa is a pioneer not only in the subjects he chooses but also in the development of logic and even mathematics in Brazil. This has shaped part of his career. At first, in his twenties, he was torn by the moral doubt of whether it was proper or not to work in a subject with no immediate application in a country that has many pressing problems such as a large number of poor people. This is a difficult decision not just for individuals, but also for the countries themselves in their budgetary allocations. Da Costa, after wavering a little while working for four years as an engineer, made the right decision: It is more important for a country to develop its creative spirits than to invest in its material development. Without creativity and innovation there is no sustained economic development. Underdevelopment is, in a large measure, of cultural origin.

Newton suffered all the difficulties of being a pioneer and a self-taught man: isolation, misunderstandings, and many others. But he also enjoyed its advantage: the freedom to work in any subject you choose without being entangled by scientific fashions or being forced into the "mainstream" of research in the area. However, this is an advantage that can, and has been, badly misused to do work with no value that, not being understood by anybody in the country, is highly regarded. In order to escape this danger, you have to be exceptionally talented, and scientifically very honest. Fortunately for our region, Newton da Costa has these qualities. Logic and philosophy would not be the same without him.

Rolando Chuaqui

PREFACE

This volume, dedicated to Professor N.C.A. da Costa, contains papers on logic and foundations of mathematics. Though primarily directed to professional logicians, some of the papers will also be useful for graduate students.

As his colleagues and former students, we wish to show our esteem and friendship to Professor da Costa on the occasion of his retirement from the University of São Paulo.

Professor Rolando Chuaqui, in his Foreword, delineates better than we could the significance of Professor da Costa's work and personality for the development of logic in Latin America, especially in Brazil.

I would like to express my sincere gratitude to those whose cooperation made the publication of this book possible. I especially wish to thank my friend J. E. de Paula Assis who proofread the manuscript and enabled me to overcome many difficulties, Marcel Dekker's staff, and M. L. Soares de Silva for her beautiful typing.

I am also grateful to the Fundação de Amparo a Pesquisa do Estado de São Paulo (FAPESP) for its financial support.

Finally my warmest personal thanks are addressed to Professor L. Nachbin for his interest in the project and for his kind assistance in proposing the publication of this volume to Marcel Dekker, Inc.

Luiz Paulo de Alcantara

CONTENTS

CONTRIBUTORS

JORGE E. BOSCH Universidad Centro de Altos Estudios en Ciencias Exactas, Buenos Aires, Argentina

ROLANDO CHUAQUI Facultad de Matemáticas, Pontificia Universidad Católica de Chile, Santiago, Chile

MANUEL CORRADA* Instituto de Matemática e Estatística, Universidade de São Paulo, São Paulo, Brazil

NEWTON C. A. DA COSTA Department of Philosophy, Universidade Estadual de Campinas, Campinas, São Paulo, Brazil

PAUL DEDECKER† Escuela de Física y Matemática, Facultad de Ciencias and Departamento de Matemática y Física, Universidad Central de Venezuela and Instituto Universitario Pedagogico, Caracas, Venezuela

CARLOS A. DI PRISCO Departamento de Matemáticas, Instituto Venezolano de Investigaciones Cientificas, Caracas, Venezuela

ÍTALA MARIA LOFFREDO D'OTTAVIANO Instituto de Matemática, Estatistica e Ciênciá da Computação, Universidade Estadual de Campinas, Campinas, São Paulo, Brazil

Current Affiliations:
*Facultad de Matematicas, Universidad Católica de Chile, Santiago, Chile
†Departmento de Matemática, Universidad de Puerto Rico, Mayagüez, Puerto Rico

ROLAND FRAÏSSÉ Departément de Mathématique, Université de Provence à Marseille, Marseille, France

SIEGFRIED GOTTWALD Department of Philosophy, Karl Marx University, Leipzig, German Democratic Republic

MARCEL GUILLAUME Departément de Mathématiques Pures, Université de Clermont-Ferrand II, Clermont-Ferrand, France

WILLIAM S. HATCHER Department of Mathematics, Laval University, Quebec, Canada

E. G. K. LOPEZ-ESCOBAR* Instituto de Matemática, Estatistica e Ciênciá da Computação, Universidade Estadual de Campinas, Campinas, São Paulo, Brazil

WIKTOR MAREK† Departamento de Matemáticas, Instituto Venezolano de Investigaciones Científicas, Caracas, Venezuela

ROBERT K. MEYER Department of Philosophy, Research School of Social Sciences, Australian National University, Canberra, Australia

CHRIS MORTENSON Philosophy Department, University of Adelaide, Adelaide, Australia

WILLIAM N. REINHARDT Department of Mathematics, University of Colorado, Boulder, Colorado

A. M. SETTE Department of Mathematics, Universidade Estadual de Campinas, Campinas, São Paulo, Brazil

L. W. SZCZERBA‡ Department of Mathematics, Iowa State University, Ames, Iowa

---

Current Affiliations:
*Department of Mathematics, University of Maryland, College Park, Maryland
†Department of Computer Science, University of Kentucky, Lexington, Kentucky
‡Instytut Matematyki, Uniwersytet Warszawski, Warsaw, Poland

MATHEMATICAL LOGIC AND FORMAL SYSTEMS

SCIENTIFIC CURRICULUM VITAE

Newton C. A. da Costa/Department of Philosophy, Universidade Estadual de Campinas, Campinas, São Paulo, Brazil

CONTENTS

BIOGRAPHICAL DATA

Name: Newton Carneiro Affonso da Costa.

Date and place of birth: September 16, 1929 in Curitiba, Paraná, Brazil.

Wife: Neusa Feitosa Affonso da Costa.

Children: Newton C.A. da Costa Júnior, Sylvia Lúcia F.A. da Costa and Marcelo F.A. da Costa.

Present professional address: The State University of Campinas, Department of Philosophy, Campinas, São Paulo, Brazil.

ACADEMIC AND PROFESSIONAL QUALIFICATIONS

Education

1. Civil Engineer, Federal University of Paranã, 1952.
2. Bachelor of Mathematics, Federal University of Paranã, 1955.
3. Licenciate in Mathematics, Federal University of Paranã, 1956.
4. Doctor of Mathematics, same university, 1961.
5. 'Docente Livre' in Mathematical Analysis, same university, 1961.

Academic Positions

1. Instructor of Mathematics, Federal University of Paranã, 1953-1955.
2. Assistant Professor of Mathematics, same university, 1956-1959.
3. Assistant Professor of Mathematics, Instituto Tecnológico de Aeronãutica, São José dos Campos, 1960-1961.
4. Associate Professor of Mathematics, Federal University of Paranã, 1962-1964.
5. Full Professor of Mathematics, same university, 1965-1967.
6. Full Professor of Mathematics, The State University of Campinas, 1968-1969.
7. Full Professor of Mathematics, University of São Paulo, 1970-1981.
8. Full Professor of Philosophy, The State University of Campinas, 1982.

Visiting Positions

Visiting Professor, Visiting Scholar or Occasional Lecturer in several institutions in Latin America, Europe, United States, and Australasia, such as the following: Institut Henri Poincaré, Bulgarian Academy of Sciences, Polish Academy of Sciences, and Universities of Paris, Clermont-Ferrand, Lille, Montpellier, Lyon, Louvain, Warsaw, Katowice, Toruñ, Kracow, Munich, California (Berkeley), Buenos Aires, Bahía Blanca, Cordoba (Argentina), San Marcos, Lima, Chile, F. Santa Maria (Valparaíso), Melbourne, Sidney, Wollongong, Auckland, Los Andes (Colombia), and Brasília; Australian National University, Catholic University of Chile, National University of Colombia, and Federal University of Rio de Janeiro.

HONORARY MEMBERSHIP IN SOCIETIES

1. Honorary Member of the Institute of Philosophy of Peru, 1975.

2. Member of the Academy of Sciences of the State of São Paulo, 1978.
3. Member of the Center of Logic, Epistemology and History of Science of the State University of Campinas, 1978.
4. Honorary Member of the Institute of Philosophical Investigations of the University of Lima, Peru, 1981.
5. Corresponding Member of the Chilean Academy of Sciences, 1982.

SERVICE TO THE SCIENTIFIC COMMUNITY

1. Former Member of the Council of the Association for Symbolic Logic, U.S.A.
2. Chairman of the Advisory Committee on Logic in Latin America of the Association for Symbolic Logic, U.S.A.
3. Former President of the Mathematical Society of Paraná, Brazil.
4. Former President of the Brazilian Society of Logic.
5. Vice-Diretor of the Institut of Mathematics, Statistics and Computation Sciences of the State University of Campinas, 1968-1969.
6. Vice-Diretor of the Institut of Mathematics of the University of São Paulo, 1979-1981.

MAIN AREAS OF RESEARCH

Since the beginning of my career, I have chosen the topics in which I have worked not only in accordance with the satisfaction, aesthetical or otherwise, of solving interesting problems. On the contrary, I have always chosen them taking into account their importance for the system of knowledge as a whole. My basic concern has been to understand the very essence of the act of knowing, in particular the nature of the scientific endeavour. It may be truly said that the major motivation of nearly all my works has invariably been philosophical.

I became interested in logic, mathematics and philosophy early in my life, when I was about fifteen years old. Nonetheless, for personal reasons, my scientific career, properly speaking, began rather late, after I had obtained the dregree of Civil Engineer (in 1952, at the Federal University of Paraná, Brazil) and had worked as an engineer for four years (1953-1956); simultaneously, in this last period, I regularly attended courses at that university, obtaining the degrees of Bachelor of

Mathematics and of Licenciate in Mathematics.

During my undergraduate years, I was very much under the influence of the Protuguese statistician Rémy Freire, who had been teaching mathematics and statistics at the Federal University of Paranã for some years; and, after graduating, the guidance of Professors Edison Farah (University of São Paulo, Brazil) and of Marcel Guillaume (University of Clermont-Ferrand, France) was of inestimable value to me. Notwithstanding, I am essentially a self-taught man, which may explain most of the characteristics of my research activity.

I have been engaged in research in six principal areas, namely: 1) Paraconsistent logic; 2) Set-theoretical foundations of category theory; 3) Lattice theory; 4) The theory of variable binding term operators; 5) Inductive logic and probability; and 6) Philosophy of logic and, in general, philosophy of the formal and empirical sciences.

GENERAL DESCRIPTION OF MY PUBLISHED WORK

1. Paraconsistent logic - Let us suppose that the language of a theory T has the symbol $\neg$ for negation. T is said to be inconsistent if, and only if, there is at least one formula A such that A and $\neg A$ are both theorems of T; otherwise, T is called consistent. T is said to be trivial when all formulas of its language are also theorems of it; in the opposite case, T is called non-trivial. If the logic of T is classical logic or several other kinds of logic, such as, for instance, intuitionistic logic, then T is trivial if, and only if, it is inconsistent. Therefore, in order that one be able to construct inconsistent but non-trivial theories, one must formulate a new class of logics: the so-called paraconsistent logics.

Em 1910, N.A. Vasiliev and J. Łukasiewicz in Russian and Poland, respectively, independently considered the possibility of developing paraconsistent logic. They were the two forerunners in this kind of logic, although some would argue that Aristotle shares this distinction; perhaps this can be said of almost all thinkers who have been engaged in the investigation of the field of the dialectics, since Heraclitus up to some of the present day Marxists.

Nonetheless, S. Jáskowski was the first logician to develop a paraconsistent propositional calculus in 1948, which was baptized by him 'discussive propositional calculus'. Starting a few years later, I independently constructed for the first time hierarchies of paraconsistent propositional calculi, of paraconsistent first-order predicate calculi (with or without equality) and of paraconsistent description calculi (see [10] - [20]). I have applied these calculi to the axiomatization of paraconsistent set theories and higher-order predicate calculi ([15], [17], [28], [53], [78] and [79]).

The semantical analysis of the paraconsistent logics and theories was also undertaken (cf. [48], [51] and [58]). Some paraconsistent semantics constitute an extension of classical semantics as it was formulated by Tarski; in particular, Tarski's definition of truth can be generalized to cope with some paraconsistent systems.

Little by little, a vast literature on paraconsistency emerged, involving for example topics such as the following: algebrization of some paraconsistent systems, paraconsistent model theory, and applications to dialectics and, in general, to philosophy. Several of my papers contain contributions to these themes and to other subjects related to paraconsistency (see, for instance, [3], [5], [23], [24], [37], [39], [44], [56], [58], [61], [62], [68], [69], [70], [74], [75], [77] and [80]).

Today, paraconsistent logic is being studied in several countries, especially in Brazil, Poland, U.S.A., Sovietic Union, and Australia, and it owes very much to logicians of those countries.

Since modal logic is intimately connected with paraconsistent logic, various papers of mine are dedicated to that class of logics (see [47], [53], [54], [58], [67] and [80]). Partly for the same reason, I have two papers on relevant logic ([23] and [59]).

For further observations on the history and applications of paraconsistent logic one way consult, for instance, the following works: A. I. Arruda, 'A survey of paraconsistent logic', in Mathematical Logic in Latin America, A. I. Arruda, R. Chuaqui and N.C.A. da Costa (editors), North-Holland, 1980, pp. 1-41; D. Marconi (editor), La Formalizzazione della Dialettica, Rosenberg & Sellier, 1979; G. Priest and R. Routley (editors), Paraconsistent Logics, Philosophia Verlag, 1985.

2. Set theoretical foundations of category theory - It was clear from its very beginnings that the category theory of Eilenberg and Maclane (1945) could not be based on the usual systems of set theory, such as the systems of Zermelo-Fraenkel, of von Neumann-Bernays-Gödel, and of

Kelley-Morse; thus, several authors tried to strengthen the customary set theories in order that they would function as a foundation for the concept to category. Among these authors, I mention only Sonner, Grothendieck, Dedecker and Feferman.

There are two possible ways to modify axiomatic set theory in order to provide a basis for category theory. One is to introduce in the extant set theories appropriate axioms of infinity, leading to notions such as that of Grothendieck-Sonner-Tarski or Ehresmann-Dedecker universes. The second is to change the main lines of the common formalizations of set theory, leading us to new systems of set theory; Houdebine and Oberschelp have exploited this possibility. In [21], [26], [29], [33], [35], [36] and [42], I presented various reformulations of extant set theories, in which the notion of category can be naturally handled, and in [30], [38] and [40], I studied new types of universes, suggested by Ehresmann and Dedecker.

Some set theories that I have investigated have given rise to interesting by-products: particular models of first-order theories, which I have called α-models (see [42]), and intimately connected with ideas previously introduced by Henkin, Orey, Mostowski and other authors, and inference rules very much stronger than the known infinitary ones, for example Carnap's ω-rule.

3. Lattice theory - In the algebrization of certain logical calculi, the standard method of obtaining the Lindenbaum algebra by passing to the quocient by a congruence relation cannot be applied. This occurs when some of the connectives are not compatible with any significant equivalence relation defined among the formulas of the calculus. Thus, all algebrizations of such a calculus (if any) are not standard. Anyhow, generalizing an idea of Curry, who has considered non-monotone operators in lattices (cf. [27] and Curry's works refered to there), one is able to define a new species of weak algebrization of the logical calculi; in this extended acception, my calculi C_n, $1 \leq n \leq \omega$, for example, may be algebrized, although negation is not compatible here with any non-trivial equivalence relation (as C. Mortensen has shown).

I have referred to the Curry algebra as a lattice, in the sense of Curry (see [27]), in which there is at least one operator that is not compatible with its basic equivalence relation. In some of my publications, I have studied properties of Curry algebras, and applied these structures to the investigation of non-classical logical calculi: various paraconsistent propositional calculi, certain versions of intuitionistic logic and some special modal logics.

It may be worthwhile to observe that the notion of Curry algebra or, better, of Curry lattice, may find applications even in questions relative to classical logic (cf. M. Eytan, 'Tableaux de Smullyan, ensembles de Hintikka et tout ça: un point de vue algèbrique', Math. Sci. Humaines 48 (1975), 21-27).

4. The theory of variable binding term operators - By a variable binding term operator (vbto) is understood an operator which binds variables of formulas to form terms. Examples of vbtos: the description operator, Hilbert's symbols τ and ε, the classifier { : }, and Russell's n-ary abstraction operator ($(\hat{x}_1,\hat{x}_2,\ldots,\hat{x}_n\ F(x_1,x_2,\ldots,x_n)$ is a term which denotes the n-adic relation determined by the formula $F(x_1,x_2,\ldots,x_n)$). Usual vbtos are sometimes introduced by contextual definition, and therefore can be eliminated. Nonetheless, owing to the fact that the elimination of vbtos requires elaborate metatheorems and to the circunstance that there are relevant cases in which vbtos are employed as primitive symbols, a systematic syntactic and semantic treatment of such operators is quite important.

I presented in [46] and [60] the outline of a general syntactic and model-theoretical approach to vbtos, starting from some ideas of Hatcher (see W.S. Hatcher, Foundations of Mathematics, Saunders, 1968); in particular, I have obtained a completeness theorem for standard vbtos in first-order languages independently of, but simultaneously to, Corcoran, Hatcher and Herring (see [60]). It seems convenient to remark that most results of model theory, for instance the theorems of Löwenhein-Skolen, of compactness, of Loś-Tarski and Padoa-Beth, as well as Keisler's model-extension theorem and Ryll-Nardzewski's theorem on categoricity, may be extended to first-order language with vbtos.

My treatment of vbtos may be adapted to deal with higher-order logics, modal logics, and other species of non-classical logics ([60], [67] and [71]). Moreover, it has several applications, including in linguistics (see [71]).

5. Inductive logic and probability - In [72], [73], and [81], I delineated an inductive logic founded on a new form of subjectivistic

theory of probability.

In inductive logic, one investigates inferences which are not deductive. Examples of inductive inferences: induction by simple enumeration, analogy, the statistical syllogism, Mill's methods, the hypothetical-deductive method, classical statistical inference, and (subjectivistic) Bayesian statistical inference.

We may say that inductive logic was created by Aristotle, though perhaps Bacon should be more appropriately considered as the real founder of that branch of logic. During the last century, many people contributed to the development of inductive logic; for instance, Herschel, Whewell, Mill, Stanley Jevons, and Boole. In our century, several thinkers worked in this field, deserving mention among them are Keynes, Jeffreys, and Carnap and his followers (Kemeny, Shimony, Hintikka,...), and the adherents of the subjectivistic theory of probability, specially de Finetti. Notwithstanding, there are philosophers, for instance Popper, who deny the possibility or the convenience of inductive logic.

The common objections to the viability of inductive logic are overcome in my stance: a new variety of probability, of subjectivistic bias, is characterized and employed to reconstruct the standard inductive reasonings. However, induction can only be completely justified by philosophical arguments similar to Kant's transcendental arguments or to Aristotle's elenctic defense of the principle of contradiction (Book Γ of Metaphysics); but I contend that in inductive logic a definite concept of (inductive) correctness may be identified, which distinguishes the sound inductive inferences from the unsound ones. Because the role of Baye's theorem in may exposition is quite important, one may legitimately say that my stand constitutes a kind of Bayesianism; but my Bayesianism is a moderate one, distinct from the position held by the common adherents of subjectivism.

Finally, I remark that in my theory of probability, which I call theory of pragmatic probability, it is possible to base Bayesian decision theory (and perhaps the greather part of extant objectivistic and subjectivistic statistics; but the outstanding open problem is to try to make compatible these two kinds of statistics).

6. <u>Philosophy of logic and, in general, philosophy of the formal and empirical sciences</u> - In [64], I presented my own philosophical conception of logic and mathematics, the so-called formal sciences. I believe that

the formal sciences are historical, i.e., made by men is the course of history; they to not have essences to be discovered, but are the result of a certain type of cultural activity, subject to all uncertainties and limitations inherent to human realizations. The main justification of such an activity can be only a pragmatical one, in the sense that the word 'pragmatical' has in modern semiotics. Nonetheless, some philosophical theses, such as a Platonistic interpretation of logic and of mathematics, may be defended from my point of view, which I would like to call critical pragmatism.

In order to construct a reasonable inductive logic, I had to introduce the notion of quasi-truth (or of pragmatic truth), which is intimately correlated with the conception of pragmatic thinkers such as Peirce and James, and with stances of some sympathizers of this school, for example Vaihinger. Positively speaking, the aim of the empirical sciences is the search of quasi-truth.

With the help of the concept of quasi-truth, it is possible to give a rationale for a theory of science which does incorporate P. Suppes' set-theoretic analysis of the scientific theories, and even naturally encompasses various traits of Sneed and Stegmüller's metatheory of empirical sciences.

I became, in this way, aware of a possible unified conception of the formal and empirical sciences (see [64] and [73]).

PRINCIPAL PUBLICATIONS

1. Research Papers

1956

1. 'Une généralization du théorème de Bouniakowski', An. Soc. Parana. Mat. 3, 12-16.
2. 'Alguns teoremas elementares sobre divisibilidade', An. Soc. Parana. Mat. 3, 60-63.

1958

3. 'Nota sobre o conceito de contradição', An. Soc. Parana. Mat. 1, N.S., 6-8.

4. 'Sobre a teoria lógica da linguagem', Revista Brasileira de Filosofia 8, 58-70.

1959

5. 'Observações sobre o conceito de existência em matemática', An. Soc. Parana. Mat. 2, N.S., 16-19.
6. 'O significado da obra de Kurt Gödel para os fundamentos da matemática' Revista Brasileira de Filosofia 3, 310-319.
7. Espaços Topológicos e Funções Contínuas, Doctor Thesis, Federal University of Paraná, Curitiba.

1960

8. 'Conceptualizatión de la filosofía cientifica', Rev. Fil. Universidad de Costa Rica 11, 363-366.

1962

9. Introdução aos Fundamentos da Matemática, Globo, Porto Alegre.

1963

10. Sistemas Formais Inconsistentes, Thesis, Federal University of Paraná, Curitiba.
11. 'Calculs propositionnels pour les systèmes formels inconsistants', C.R. Acad. Sc. Paris 257, 3790-3793.

1964

12. 'Calculs de prédicats pour les systèmes formels inconsistants', C.R. Acad. Sc. Paris 258, 27-29.
13. 'Calculs de prédicats avec égalité pour les systèmes formels inconsistants', C.R. Acad. Paris 258, 1111-1113.
14. 'Calculs de descriptions pour les systèmes formels inconsistants', C.R. Acad. Sc. Paris 258, 1366-1368.
15. 'Sur un système inconsistant de théorie des ensembles', C.R. Acad. Sc. Paris 258, 3144-3147.
16. 'Sur un théorème de Hilbert et Bernays', C.R. Acad. Sc. Paris 258, 6311-6312 (with A.I. Arruda)
17. 'Sur une hiérarchie de systèmes formels', C.R. Acad. Sc. Paris 259, 2943-2945 (with A.I. Arruda).
18. 'Sur les calculs C_n', Anais Acad. Brasil. Ciências 36, 379-382 (with M. Guillaume).

<u>1965</u>

19. 'Sur les systèmes formels C_i, C_i^*, $C_i^=$, D_i et NF_i', C.R. Acad. Sc. Paris 260, 5427-5430.
20. 'Negations composées et loi de Peirce dans les systèmes C_n' Portugalia Math. 24, 201-210 (with M. Guillaume).
21. 'On two systems of set theory', Proceedings Konenkl. Ak. Wetens. 68, 95-99.

<u>1966</u>

22. 'Transformadas no cálculo restrito de predicados', Anais Acad. Brasil. Ciências 38, 385-390 (with A.I. Arruda).
23. 'O paradoxo de Curry-Moh Schaw Kwei', Bol. Soc. Mat. São Paulo 18, 83-89 (with A.I. Arruda).
24. 'Opérations non monotones dans les treillis', C.R. Acad. Sc. Paris 263, 429-432.

<u>1967</u>

25. 'Filtres et idéaux d'une algèbre C_n', C.R. Acad. Paris 264, 549-552.
26. 'Two formal systems of set theory', Proceedings Konenkl. Ak. Wetens 70, 45-51.
27. Álgebras de Curry, University of São Paulo, São Paulo.
28. 'Une nouvelle hiérarchie de théories inconsistantes,' Publications du Département de Mathématiques, Université de Lyon, 4, 2-8.
29. 'Un nouveau système formel suggéré par Dedecker', C.R. Acad. Sc. Paris 265, 85-88.
30. 'Remarques sur les univers d'Ehresmann-Dedecker', C.R. Acad. Sc. Paris 265, 861-763 (with A. de Caroli).

<u>1968</u>

31. 'Sur la logique discoursive de Jaśkowski', Bull. Acad. Polonaise des Sciences 15, 551-557 (with L. Dubikajtis).

<u>1969</u>

32. 'Les algèbres C_n', C.R. Acad. Sc. Paris 268, 1011-1014 (with A.M. Sette).
33. 'On a set theory suggested by Dedecker and Ehresmann', Proc. Japan Academy of Sciences 45, 880-888.

1970

34. 'Sur un problème de Jaśkowski', C.R. Acad. Sc. Paris 270, 1349-1353 (with I.M. D'Ottaviano).
35. 'On the underlying logic of two systems of set theory', Proceedings Konenkl. Ak. Wetens. 73, 1 - 8.
36. 'Sur les système D* suggéré par Dedecker', C.R. Acad. Sc. Paris 270, 841-844 (witj D. Brignole).
37. 'Sur le schéma de la séparation', Nogoya Math. Journal 38, 71-84 (with A.I. Arruda).

1971

38. 'On supernormal Ehresmann-Dedecker universes', Math. Zeitschrift 122, 342-350 (with D. Brignole).
39. 'Remarques sur les système NF_1' . C.R. Acad. Sc. Paris 272. 1149-1151.

1972

40. 'Modèles et univers de Dedecker', C.R. Acad. Sc. Paris 275, 483-486.

1973

41. 'Sobre o conceito de transformada no cálculo restrito de predicados', IME-USP, Série Mat. 2, 53-57.

1974

42. 'α-models and the systems T and T*', Notre Dame J. of Formal Logic 14, 443-454.
43. 'Remarques sur les calculs C_n, C_n^*, $C_n^=$ et D_n', C.R. Acad. Sc. Paris 278, 819-821.
44. 'Le schéma de la séparations dans les systèmes I_n', Math. Japonicae 19, 183-186 (with A.I. Arruda).
45. 'On the theory of inconsistent formal systems', Notre Dame J. of Formal Logic 15, 497-510.

1975

46. 'Sur les "vbtos" selon M. Hatcher', C.R. Acad. Sc. Paris 281, 741-743 (with I.F. Druck).
47. 'Remarks on Jaśkowski's discussive logic', Reports on Math. Logic 4, 7-16.

1976

48. 'Une sémantique pour le calcul C_1', C.R. Acad. Sc. Paris 283, 729-731 (with E.H. Alves).
49. 'Sur le système D* de théorie des ensembles', C.R. Acad. Sc. Paris 282, 5-7 (with M.F. Dias).
50. 'α-logic and infinitary languages', Zeitschrift Math. Logik u. Grundl. d. Math. 29, 105-112 (with C. Pinter).

1977

51. 'Une sémantique pour le calcul $C_1^=$', C.R. Acad. Sc. Paris 284, 279-282 (with A.I. Arruda).
52. 'A semantical analysis of the calculi C_n', Notre Dame J. of Formal Logic 18, 621-630 (with E.H. Alves).
53. 'On Jaśkowski discussive logic', in Non-Classical Logics, Model Theory and Computability, A.I. Arruda, N.C.A. da Costa and R. Chuaqui, eds., North-Holland, pp. 37-56 (with L. Dubikajtis).
54. 'On some modal logical systems defined in connection with Jaśkowski's problem', ib. pp. 57-73 (with J. Kotas).
55. Non-Classical Logics, Model Theory and Computability, editor, with A.I. Arruda and R. Chuaqui, North-Holland.

1978

56. 'On the problem of Jaśkowski and the logics of Lukasiewicz', in Mathematical Logic: Proceedings of the First Brazilian Conference, A.I. Arruda, N.C.A. da Costa and R. Chuaqui, eds. Marcel Dekker, pp. 127-139 (with J. Kotas).
57. Mathematical Logic: Proceedings of the First Brazilian Conference, editor, with A.I. Arruda and R. Chuaqui, Marcel Dekker.

1979

58. 'A New formulation of discussive logic', Studia Logica 38, 429-445 (with J. Kotas).

1980

59. 'On the relevant systems P and P* and some related systems', Relatório Interno nº 174, IMECC-UNICAMP, 23 pp. (with A.I. Arruda).
60. 'A model-theoretical approach to variable binding term operators', in Mathematical Logic in Latin America, A.I. Arruda, R. Chuaqui and N.C.A. da Costa, eds. North-Holland, pp. 133-162.

61. 'Some problems on logical matrices and valorizations', in Proceedings of the Third Brazilian Conference on Math. Logic., A.I. Arruda, N.C.A. da Costa and A.M. Sette, eds., Soc. Bras. Lógica, 131-146 (with J. Kotas).

62. 'Studies in paraconsistent logic I: the dialectical principle of the unity of opposites', Philosophia 9, 189-217 (with R.G. Wolf).

63. 'El simbolo ε de Hilbert', Lecturas Matemáticas 1, 1-13.

64. Ensaio sobre os Fundamentos da Lógica, Hucitec, São Paulo.

65. Mathematical Logic in Latin America, editor, with A.I. Arruda and R. Chuaqui, North-Holland.

66. Proceedings of the Third Brazilian Conference on Math. Logic, editor, with A.I. Arruda and A.M. Sette, Soc. Bras. Lógica, São Paulo.

1981

67. 'Variable binding term operators in modal logic', to appear in Atas da Soc. Brasileira de Mat. 10, D.G. Figueiredo, A.F. Izé and C.S. Hönig, eds. (with C. Mortensen).

68. 'Problems of modal and discussive logics', Trabalhos de Departamento de Matemática nº 11, IME-USP, 35 pp. (with J. Kotas).

69. 'Relations between paraconsistent logic and many-valued logic', Bull. Sect. Logic, Polish Acad. Sc., 10, 185-191 (with E.H. Alves).

70. 'In the philosophical import of paraconsistent logic', Journal of Non-Classical Logic, 1 (1982), 1-19.

71. 'Notes in the theory of variable binding term operators', to appear in Journal of History and Philosophy of Logic (with C. Mortensen).

72. Lógica Indutiva e Prcbabilidade, IME-USP.

1982

73. 'Ciencia y Verdad', Informe Tecnico, Fac. de Mat., Universidad Catolica de Chile, 15 pp.

74. 'Studies in paraconsistent logic II: quantifiers and the unity of opposites', Pré-publicações do CLE-UNICAMP, nº 2, 12 pp.

75. 'On a paraconsistent calculus', in Collected Papers Dedicated to Professor Edison Farah on the Occasion of His Retirement, O.T. Alas, N.C.A. da Costa and C.S. Hönig, eds., IME-USP, pp. 83-90 (with E.H. Alves), São Paulo.

76. Collected Papers Dedicated to Professor Edison Farah on the Occasion of His Retirement, editor, with O.T. Alas and C.S. Hönig, IME-USP, S.Paulo.

77. 'Logic and Ontology', Pré-publicações do CLE-UNICAMP, nº 3, 20 pp.
78. 'On paraconsistent set theories', Relatório Interno nº 215, IMECC-UNICAMP, 6 pp. (with L.P. de Alcantara).
79. 'A note on type theory', Relatório Interno nº 216, IMECC-UNICAMP, 6 pp. (with L.P. de Alcantara).
80. 'Remarks on higher-order modal logic', Relatório Interno nº 217, IMECC-UNICAMP, 7 pp. (with L.P. de Alcantara).

<u>1983</u>

81. 'The pragmatic theory of truth', to appear (with I. Mikenberg and R. Chuaqui).
82. 'Pragmatic probability', to appear.
83. 'Paraconsistency, paracompleteness, and valuations', to appear (with A. Loparic).

2. Abstracts

<u>1962</u>

I. 'Sobre um subsistema do cálculo proposcional clássico, Resumo das Comunicações, XIV Reunião Anual da Sociedade Brasileira para o Progresso da Ciência, p. 1.

<u>1965</u>

II. 'Some properties of two systems of set theory', Notices Am. Math. Society 12, nº 4, p. 472.

III. 'Non-monotone operations in lattices', Notices Am. Math. Society 12, nº 5, pp. 601-602.

<u>1967</u>

IV. 'On Jaśkowski's propositional calculus', Notices Am. Math. Society 14, nº 5, p. 712 (with L. Dubikajtis).

V. 'Universes of Ehresmann-Dedecker', Notices Am. Math. Society 14, nº 5, p. 2.

<u>1968</u>

VI. 'Nota sobre la teoría de los tipos', Revista de la Unión Matemática Argentina XXIII, nº 4, p. 199 (with A.I. Arruda)

VII. 'On the postulate of separation', Notices Am. Math. Society 15, nº 2, pp. 399-400 (with A.I. Arruda).

VIII. 'Further considerations on the postulate of separation', Notices Am. Math. Society 15, nº 3, p. 555 (with A. I. Arruda).

1969

IX. 'The predicate calculus of order ω as the underlying logic of set theory', Notices Am. Math. Society 16, nº 5, p. 845 (with C.M. de Barros).

1970

X. 'On Ehresmann-Dedecker universes', Notices Am. Math. Society 17, nº 2, pp. 453-454 (with D. Brignole).

XI. 'On supernormal Ehresmann-Dedecker universes', Notices Am. Math. Society 17, nº 5, p. 833 (with D. Brignole).

1971

XII. 'On the systems T and T* ', The Journal of Symbolic Logic 36, nº 3, p. 578.

1978

XIII. 'On Jaśkowski's discussive logic', The Journal of Symbolic Logic 43, nº 2, pp. 354-355 (with L. Dubikajtis).

XIV. 'On some modal logical systems defined in connection with Jaśkowski's problem', The Journal of Symbolic Logic 43, nº 2, p. 354 (with J. Kotas).

XV. 'On Jaśkowski's problem and the logics of Lukasiewicz', Bulletin of the Section of Logic, Polish Academy of Sciences, 7, nº 2, p. 91 (with J. Kotas).

XVI. 'A modal theoretical approach to variable binding term operators', The Journal of Symbolic Logic, 46, nº 1, p. 184.

SUR LES NOMBRES NATURELS ET LES FONDEMENTS DE L'ARITHMÉTIQUE ET LA GÉOMETRIE

Jorge E. Bosch/Universidad Centro de Altos Estudios en Ciencias Exactas, Buenos Aires, Argentina

SOMMAIRE

INTRODUCTION (RÉSUMÉ - "ABSTRACT")

Ce travail a pour but d'introduire une modification dans la philosophie formaliste de la mathemátique, de telle façon que l'essentielle de cette philosophie se conserve et qu'il soit possible d'établir un fondement integré pour l'arithmétique et la géométrie.

Ce programme se développe en introduisant la notion d'espace semi-groupal, qui est une généralisation de celle d'espace vectoriel, strictement adaptée au besoin de donner un fondement de style formaliste à l'idée de nombre naturel.

L'exposé est divisé en cinq chapitres: dans le premier on étudie les généralités sur les espaces semi-groupaux; le deuxième introuduit la notion basique de scalaire naturel associé à un espace semi-groupal, comme fonction de cet espace dans lui même satisfaisant un axiome d'induction; dans le troisième chapitre on étudie les espaces semi-groupaux simplifiable et sans torsion, dont les scalaires naturels associés sont précisément les nombres naturels correspondants; le chapitre quatrieme est dédié à certains compléments théoriques , notamment quelques considerations catégorielles; le chapitre cinquième, enfin, est consacré a l'exposé des thèses philosophiques, en particulier les fondements formalistes de l'arithmétique et un programme concernant les ensembles hypothétiques.

CHAPITRE 1 - LA NOTION D'ESPACE SEMI-GROUPAL

1.1. DÉFINITION ET GÉNERALITÉS

1.1.1. Motivation Vectorielle

Nous voulons donner une version purement algébrique et abstraite de l'idée de ensemble de vecteurs fixes du plan (ou de l'espace), muni de l'opération d'addition, celle-ci définie seulement pour le cas des vecteurs ayant l'origine commune. On remarque que, si nous appelons V cet ensemble de vecteurs, alors pour chaque point 0 du plan (ou de l'espace) il existe un sous-ensemble de V, que nous appellerons V_o, formé par tous les vecteurs fixes d'origine 0, qui est un groupe abélien par rapport à l'addition de vecteurs. La classe de sous-ensembles du type V_o est une partition de l'ensemble V. On définit la relation de sommabilité dans V comme celle qui a lieu entre le vecteur $\underline{u}$ et le vecteur $\underline{v}$ si set seulement s'il existe la somme $\underline{u}+\underline{v}$; c'est à dire, si et seulement si $\underline{u}$ et $\underline{v}$ ont même origine. Cette relation est une relation d'èquivalence dans V et ses classes d'équivalence sont les ensembles de type V_o que nous venons de décrire. Dans ces conditions, l'ensembles V et l'opération d'addition (partiellement définie) constituent un espace de groupes ou espace groupal.

Cependant, dans ce travail nous ne ferons pas usage de toute la structure de groupe des ensembles V_o, mais seulement d'une partie: celle concernant l'associativité et l'existence d'élement neutre. Nous nous passerons de l'axiome d'existence d'élements symétriques (ou opposés). Si l'on "efface" cet axiome on obtient la structure de semi-groupe, qui est la base sur laquelle on arrive a définir les nombres naturels.

Nous appellerons espace groupal ou espace de groupes tout couple (E,+), où E est un ensemble et + est une opération binaire (addition), partiellement définie dans E, de telle façon que les conditions suivantes soient vérifiées: (i) La relation de sommabilité, qui a lieu entre x et y si et seulement s'il existe la somme x + y, est une relation d'équivalence dans E; (ii) Chaque classe d'équivalence correspondant à cette relation est un groupe abélien par rapport à la restriction de l'operation de somme à cette classe.

1.1.2. Semi-Groupes et Espaces Semi-Groupaux

Un semi-groupe es un couple (G,.) tel que: (i) G est un ensemble et le point dénote une opération binaire (globale) associative G; (ii) Il existe un élément e, appartenent à G, appelé elément neutre, tel que pour tout x appartenent à G soit x.e = e.x = x .

On démontre facilement que l'élément neutre est unique. Si, en plus, l'opération est commutative, le semi-groupe est dit commutatif ou abélien; dans ce cas il est habituel d'adopter les conventions suivantes: l'opération s'appelle somme et se désigné par "+"; l'élément neutre s'appelle zéro et se désigne par "0". Il est évident que tout groupe est un semi-groupe et que tout groupe abélien est un semi-groupe abélien.

On définit la notion d'espace semi-groupal ou espace de semi-groupes en substituant le mot "semi-groupe" au mot "groupe" dans la définition d'espace groupal donnée dans 1.1.1. Par abus de notation, l'espace semi-groupal (E,+) pourra être noté simplement E.

1.1.3. Note sur les Classes Structurales

Chaque classe d'équivalence dont il s'agit dans la condition (ii) d'espace groupal (1.1.1) ou de la définition correspondante d'espace semi-groupal (1.1.2) est caracterisée par son respectif élément neutre. Si a est un élément neutre, la classe d'équivalence correspondante sera denotée par E_a. Cette notation, appliquée au cas des vecteurs décrit dans 1.1.1, éxigerait la notation $V_{\vec{oo}}$ au lieu de V_o, mais il est préférable d'adopter la notation V_o. Nous désignerous aussi comme d'habitude (bien qu'il ne soit pas strictement rigoureux) par E_a le semi-groupe abélien dont la classe sous-jacente est E_a. Tous ces semi-groupes seront appelés classes structurales ou semi-groupes structuraux de l'espace semi-groupal E.

1.1.4. Cas Particuliers

Nous avons déjà dit que tout espace groupal est un espace semi-groupal. Aussi tout semi-groupe est un espace semi-groupal (moyennant une identification canonique), et l'ensemble de tous les vecteurs du plan (ou de l'espace) peut être muni d'une façon évidente d'une structure d'espace semi-groupal.

1.2. TRANSFORMATIONS COHÉRENTES ET MORPHISMES D'ESPACES SEMI-GROUPAUX

1.2.1. Définition de Transformation Cohérente

Si (E,+) et (F,+) sont des espaces semi-groupaux (en utilisant, par abus de notation, le même signe pour les deux opérations de somme), une transformation cohérente du premier dans le second est une fonction $f : E \longrightarrow F$ qui respecte les semi-groupes structuraux; plus précisément, ceci veut dire que si x et y appartiennent à un même semi-groupe structural de E, alors $f(x)$ et $f(y)$ appartiennent aussi à un même structural de F. Ce dernier sera dit correspondant du premier par la fonction f.

1.2.2. Définition de Morphisme

Un morphisme de l'espace semi-groupal (E,+) dans l'espace semi-groupal (F,+) est transformation cohérente du premier dans le second tel que: (1) Si x e y appartiennent à un même semi-groupe structural de E, alors $f(x + y) = f(x) + f(y)$; (2) Si e est un élément neutre de E, $f(e)$ est un élément neutre de F.

Remarque. S'il s'agissait de groupes il n'y aurait pas besoin d'établir en forme indépendente la condition (2), car elle serait une conséquente immédiate de (1). Mais dans le cas de semi-groupes la condition (2) est vraiment indépendente, comme le montre l'exemple suivant: Soit $S = \{0;1\}$ et définissons la somme ainsi: $0 + 0 = 0$, $0 + 1 = 1 + 0 = 1 + 1 = 1$. On vérifie que (S,+) est un semi-groupe et que 0 en est l'élement neutre.

Par conséquent (S,+) est aussi un espace semi-groupal (d'après 1.1.4). Définissons maintenant la fonction $\underline{f} : S \to S$ par $\underline{f}(\underline{x}) = 1$ (constante). Or cette fonction $\underline{f}$ satisfait la condition (1) mais non point la condition (2).

1.2.3. Restriction et Recollement de Morphismes

Un morphisme (ou homomorphisme) de semi-groupes se définit aussi par les conditions (1) et (2) de 1.2.2. Il est clair que si E_a est un semi-groupe structural de E et F_b est le semi-groupe structural de F correspondant par le morphisme $\underline{f}$, alors la restriction de f à E_a et F_b est un morphisme de semi-groupes abéliens. Et réciproquement: si pour chaque semi-groupe structural de E l'on donne un morphisme de semi-groupes dans quelque semi-groupe structural de F, le recollement de tous ces morphismes est un morphisme d'espaces semi-groupaux de E dans F. Le recollement est la fonction de E dans F dont la restriction à chaque semi-groupe structural de E est le morphisme de semi-groupes donné.

1.2.4. Propriétés Catégorielles des Morphismes

Les morphismes d'espaces semi-groupaux se composent comme les fonctions. Nous dirons que le morphisme $\underline{f}$ est composable avec le morphisme $\underline{g}$ si la fonction composée $\underline{g} \circ \underline{f}$ existe. Cette opération de composition (partiellement définie) entre morphismes, jouit des propriétés suivantes, dites catégorielles:

(a) Composabilité et associativité. Si $\underline{f},\underline{g}$ et $\underline{h}$ sont morphismes d'espace semi-groupaux et si $\underline{f}$ est composable avec $\underline{g}$ et celui-ci l'est avec $\underline{h}$, alors les fonctions $\underline{g} \circ \underline{f}$ et $\underline{h} \circ \underline{g}$ sont des morphismes d'espaces semi-groupaux, $\underline{g} \circ \underline{f}$ est composable avec $\underline{h}$, $\underline{f}$ est composable avec $\underline{h} \circ \underline{g}$ et en plus: $\underline{h} \circ (\underline{g} \circ \underline{f}) = (\underline{h} \circ \underline{g}) \circ \underline{f}$.

(b) Existence d'élément neutre (ou morphisme neutre). Pour chaque espace semi-groupal E la fonction identité 1_E est un morphisme d'espaces semi-groupaux, qui est caracterisé par les conditions suivantes:

(bi) Pour tout morphisme $f : E \to F$ il vient $\underline{f} \circ 1_E = \underline{f}$

(bii) Pour tout morphisme $\underline{g} : G \to E$ il vient $1_E \circ \underline{g} = \underline{g}$

1.2.5. Propriétés Catégorielles des Transformations Cohérentes

Les transformations cohérentes entre espaces semi-groupaux comprennent les morphismes comme des cas particuliers, et elles aussi possèdent des propriétés catégorielles strictement analogues à celles de 1.2.4.

1.2.6. Isomorphismes

Un isomorphisme d'espaces semi-groupaux est un morphisme $f : E \longrightarrow F$ pour lequel il existe un morphisme $\underline{g} : f \longrightarrow E$ tel que $\underline{g} \circ \underline{f} = 1_E$ et $\underline{f} \circ \underline{g} = 1_F$. Dans ce cas le morphisme $\underline{g}$ est dit inverse du morphisme $\underline{f}$. L'espace semi-groupal E est isomorphe a l'espace semi-groupal F s'il existe un isomorphisme $\underline{f} : E \longrightarrow F$. On démontre facilement les propriétés suivantes:

(a) Un morphisme d'espaces semi-groupaux est un isomorphisme si et seulement s'il est bijectif.

(b) Le morphisme inverse de l'isomorphisme $\underline{f}$ est unique et il est aussi un isomorphisme; on le dénote par $\underline{f}^{-1}$.

(c) L'inverse de l'inverse est l'isomorphisme donné, c'est a dire: $(\underline{f}^{-1}) = \underline{f}$.

(d) Le composé de deux isomorphismes est un isomorphisme et son inverse est le composé des inverses des isomorphismes donnés, mais pris dans l'ordre inverse; c'est à dire: $(\underline{g} \circ \underline{f})^{-1} = \underline{f}^{-1} \circ \underline{g}^{-1}$.

(e) Les morphismes identiques sont les isomorphismes que coïncident avec leurs respectifs inverses, soit: $1_E^{-1} = 1_E$.

(f) La relation d'isomorphie d'espaces semi-groupaux est une relation d'équivalence.

(g) Si l'espace semi-groupal E est isomorphe à l'espace semi-groupal F, tout isomorphisme $\underline{f} : E \longrightarrow F$ induit une fonction bijective entre les ensembles respectifs de semi-groupes structuraux, et en plus il induit un isomorphisme de chaque semi-groupe structural de E sur le correspondant semi-groupe structural de F.

CHAPITRE 2 - LES SCALAIRES NATURELS D'UN ESPACE SEMI-GROUPAL

2.1. GÉNÉRALITÉS ET INDUCTION

2.1.1. Définition de Somme de Fonctions à Valeurs dans un Espace Semi-Groupal

Soient E un espace semi-groupal et A un ensemble. Les fonctions $\underline{f},\underline{g} : A \longrightarrow E$ seront dites <u>sommables dans</u> E, si pour tout $x \in A$ les valeurs $\underline{f}(x)$ et $\underline{g}(x)$ appartiennent à un même semi-groupe structural de E. Dans ce cas, on appelle <u>somme</u> de $\underline{f}$ et $\underline{g}$ la fonction $\underline{f}+\underline{g} : A \longrightarrow E$ donnée par $(\underline{f} + \underline{g})(x) = \underline{f}(x) + \underline{g}(x)$.

2.1.2. Théorème

L'ensemble de toutes les fonctions de l'ensemble A dans l'espace semi-groupal E, est un espace semi-groupal par rapport à la somme définie dans 2.1.1, dont les éléments neutres sont les fonctions constantes qui prennent comme valeur un élément neutre de E. Si E est un semi-groupe cet ensemble de fonctions l'est aussi.

La démonstration est évidente.

2.1.3. Définition de Système Inductif

On appelle <u>système inductif associé</u> à l'espace semi-groupal E, tout ensemble F de fonctions de E dans E tel que:

(i) Pour tout $\underline{f} \in F$ et tout $\underline{x} \in E$, l'élément $\underline{f}(\underline{x})$ appartient au même semi-groupe structural que $\underline{x}$.

(ii) <u>La fonction nulle</u> sur E, qui à chaque élément de E assigne l'élément neutre de son propre semi-groupe structural, appartient à F.

(iii) Si $f \in F$ alors $f + 1_E \in F$. (Compte tenu que, d'après la condition (<u>i</u>), la fonction f est sommable avec l'identité 1_E).

2.1.4. Définition de Scalaires Naturels

On appelle <u>système de scalaires naturels associé</u> à l'espace semi-groupal E, l'ensemble intersection de tous les systèmes inductifs associés à E.

Si N_E est le système de scalaires naturels de E et S_E est l'ensemble de tous les systèmes inductifs associés à E, on a alors par définition:

$$N_E = \bigcap_{F \in S_E} F .$$

2.1.5. Théorème

L'intersection de tout ensemble de systèmes inductifs associés à E est aussi un système inductif associé à E. En particulier, N_E est un système inductif associé à E.

La démonstration est évidente.

2.1.6. Quelques Éléments de N_E

Si l'on développe la démonstration précédente on est conduit à mettre en évidence la fonction nulle $0_E : E \longrightarrow E$, donnée par $0_E(x) = e_G$, celui-ci étant l'élément neutre du semi-groupe structural G à qui l'élément $\underline{x}$ appartient. Et l'on voit sans peine que le scalaire 0_E est élément neutre pour l'addition dans N_E. Et puisque N_E est un système inductif, on a $0_E + 1_E \in N_E$, donc $1_E \in N_E$. Ce scalaire s'appelle le <u>scalaire Un</u>, ou <u>le scalaire unité</u>, de N_E, ou simplement de E. On définit de façon analoque les scalaires successifs de N_E, ou de E : 2_E, 3_E, $4_E, \ldots, n_E$, ...

2.1.7. Notation Multiplicative

On définit la multiplication des scalaires de E par les éléments de E, comme la fonction de $N_E \times E$ dans E qui à chaque couple $(\underline{n}, \underline{x})$ fait correspondre l'élément $n(x)$, noté $\underline{n} . \underline{x}$.

2.1.8. Le Principle D'Induction Finie

<u>Théorème</u>. Soit $C \subset N_E$ tel que: $(\underline{a})$ $0_E \in C$; $(\underline{b})$ $n_E \in C \Longrightarrow n_E + 1_E \in C$. Alors $C = N_E$.

La démonstration est évidente.

On peut établir le même principe avec le langage des propriétés, en remplaçant l'ensemble C par une propriété P et la formule "$x \in C$" par la formule "x a la propriété P".

2.2. SOMME DE SCALAIRES NATURELS

2.2.1. Théorème

Les scalaires $\underline{n}$ et $\underline{m}$ de l'espace semi-groupal E sont toujours sommables.

En effet, il suffit d'appliquer la définition 2.1.1, compte tenu que $\underline{n}$ et $\underline{m}$ ont la propriété (<u>i</u>) de 2.1.3, puisque E est un système inductif par 2.1.5.

2.2.2. Théorème

Si m_E et n_E sont des scalaires naturels de E, leur somme $m_E + n_E$ l'est aussi.

La démonstration consiste en la construction, pour chaque m_E, de l'ensemble

$$C_m = \{\underline{k} \mid k \in N_E \quad \text{et} \quad m_E + k \in N_E\} ,$$

qui est égal a N_E par induction.

2.2.3. N_E Comme Semi-Groupe Abélien

<u>Théorème 1</u>. Pour tout espace semi-groupal E, l'ensemble N_E, des scalaires naturels de E, est un semi-groupe abélien par rapport à la somme.

C'est évident en vertu de 2.2.2, 2.1.2 et 2.1.6.

2.2.4. Distributivité de la Multiplication de Scalaire Naturel par Élement de E, par Rapport à la Somme de Scalaires Naturels

$$(m+n) \cdot x = m \cdot x + n \cdot x \; . \tag{1}$$

La démonstration est triviale.

2.2.5. Distributivité de la Multiplication de Scalaire Naturel par Élement de E, par Rapport à la Somme dans E

$$m . (x + y) = m . x + m . y . \qquad (2)$$

La démonstration est moins évidente et se fait par induction.

2.3. LA MULTIPLICATION DE SCALAIRES NATURELS

2.3.1. Idée Intuitive

L'idée basique de ce paragraphe consiste en la définition de la multiplication des scalaires naturels comme composition de fonctions. Mais pour que cette idée puisse être développée on doit résoudre deux question: (a) Deux scalaires naturels, peuvent toujours se composer? (b) En cas affirmatif, est toujours cette composition un scalaire naturel?

2.3.2. Lemme

Si $\underline{f}$, $\underline{g}$, $\underline{h}$, sont des fonctions de E dans E et si $\underline{g}$ et $\underline{h}$ sont sommables dans E, alors la distributivité $(g+h)_o f = g_o f + h_o f$ est vérifiée.

La démonstration est immédiate. L'autre distributivité ne se vérifie pas en général, mais elle est valable pour scalaires naturels, comme on verra dans 2.3.7.

2.3.3. Théorème

Si m_E et n_E sont des scalaires naturels associés à l'espace semi-groupal E, il existe la fonction composée $n_E {}_o m_E$, qui est à son tour un scalaire naturel de E.

La démonstration se fait par induction, en considérant l'ensemble associé à m_E:

$$C_m = \{\underline{k} \mid \underline{k} \in N_E \text{ et } k_o m_E \in N_E\}$$

qui résulte égal à N_E .

2.3.4. Définition de Multiplication de Scalaires Naturels

Si $\underline{m}$ et $\underline{n}$ sont des scalaires naturels de E, on appelle produit $\underline{m} . \underline{n}$

le scalaire naturel qui résulte de la composition de $\underline{m}$ avec $\underline{n}$, c'est à dire: $m.n = n_o m$. Donc:

$$(m \, . \, n)(x) = (n_o m)(\underline{x}) \quad \text{pour} \quad \underline{x} \in E.$$

En vertu de 2.3.3, N_E est un ensemble fermé par rapport à la multiplication.

2.3.5. N_E Comme Semi-Groupe par Rapport à la Multiplication

Nous venons de voir que N_E est fermé par rapport à la multiplication. D'ailleurs la composition de fonctions est associative, et finalement le scalaire 1_E est élément neutre pour la multiplication. Donc, en vertu de 1.1.2, N_E est un semi-groupe par rapport à la multiplication.

2.3.6. Théorème

Chaque scalaire naturel $\underline{n}$ (de l'espace semi-groupal E) est un morphisme d'espaces semi-groupaux, de E dans E.

En effet: chaque élément $\underline{n}$ de N_E est cohérent dans le sens de 1.2.1, car N_E est un système inductif et la condition $(\underline{i})$ de la définition de système inductif (2.1.3) implique que les éléments de N_E sont cohérents. La propriété additive,

$$n \, . \, (x + y) = n \, . \, x + n \, . \, y \; ,$$

et la conservation des éléments neutres, $n.e_G = e_G$, se démontrent par induction.

2.3.7. Semi-Anneau Unitaire et Morphismes

Rappelons que:

$\underline{a)}$ <u>Un semi-anneau</u> est un triplet (ordonné), (A,+,.), tel que (A,+) soit un semi-groupe abélien et le point désigne une opération binaire dans A (multiplication) qui soit distributive à droite et à gauche par rapport à la somme:

$$k \, . \, (m + n) = k.m + k.n \quad , \qquad (m+n).k = m.k + n.k \; .$$

La première est une conséquence immédiate du lemme 2.3.2, compte tenu que l'ordre utilisé pour la notation de la composition des fonctions est opposé à celui utilisé pour la notation de la multiplication. La deuxième distributivité ne se vérifie pas en général pour toutes les fonctions, mais elle est valable pour le cas spécifique des scalaires naturels. La démonstration est simple. On prouve ainsi que $(N_E,+,.)$ est un semi-anneau. Or, on dit que le semi-anneau (A,+,.) est unitaire si (A,.) est un semi-groupe. On voit alors que $(N_E,+,.)$ est un semi-anneau unitaire. Bien entendu, tout semi-anneau trivial (avec un seul élément) est aussi unitaire, puisque dans ce cas il vient $0 = 1$. Mais il y a des semi-anneaux non triviaux pour lesquels se vérifie aussi $0 = 1$.

b) Morphismes. Un morphisme de semi-anneaux unitaires, de (A,+,.) dans (B,+,.) où par commodité on a désigné avec les mêmes notations les opérations dans A et dans B, est une fonction $f : A \longrightarrow B$ qui respecte les sommes, les multiplications et les éléments neutres: $f(a+b)=f(a)+f(b)$, $f(a.b) = f(a).f(b)$, $f(0) = 0$, $f(1) = 1$. On démontre sans peine:

b1) La fonction identité sur chaque semi-anneau unitaire est un morphisme;

b2) La composition de morphismes (quand elle existe) est un morphisme.

On peut remarquer que, pour tout morphisme $f : A \longrightarrow B$, de semi-anneaux, si A est unitaire l'élément f(1) est idempotent, c'est à dire: $f(1).f(1) = f(1)$. Ceci est une restriction importante quand on veut définir un morphisme de semi-anneaux qui ne soit pas un morphisme de semi-anneaux unitaires, dans le cas où A et B sont unitaires.

c) Composition et isomorphismes. Ou prouve facilement les propriétés categorielles:

c1) La composition de morphismes de semi-anneaux (respectivement, de morphismes de semi-anneaux unitaires) et un morphisme de semi-anneaux (respectvement, morphisme de semi-anneaux unitaires);

c2) La fonction identique sur un semi-anneau (respectvement, unitaire) est un morphisme de semi-anneaux (respectivement, unitaires).

Dans les deux cas (semi-anneaux et semi-anneaux unitaires) les isomorphismes sont définis comme d'habitude: le morphisme $f : A \longrightarrow B$ est un isomorphisme si et seulement s'il existe un morphisme $g : B \longrightarrow A$ tel que $g_o f = 1_A$ et $f_o g = 1_B$. Condition nécessaire et suffisante pour que le morphisme f soit un isomorphisme et qu'il soit bijectif. La relation d'isomorphie entre semi-anneaux (respectivement, unitaires), est définie comme d'habitude. On remarque que si A est un semi-anneau unitaire et B est un semi-anneau, et si $f: A \longrightarrow B$ est un isomorphisme de semi-anneaux, alors B est aussi semi-anneau unitaire avec unité f(1), et f est un isomorphisme d'anneaux unitaires.

2.3.8. Commutativité de la Multiplication

Bien que la première des distributivités vues en 2.3.7 pourrait être employée pour démontrer la deuxième moyennant la commutativité, celle-ci ne peut se démontrer (dans cette présentation) qu'en utilisant précisément la deuxième distributivité.

La démonstration se fait par induction.

2.3.9. Associativité Combinée

Nous appelons ansi la formule

$(n.m).x = n.(m.x)$, pour $n,m \in N_E$, $x \in E$.

Elle se démontre en utilisant la commutativité vue en 2.3.8.

2.3.10. Résumé: Structure de N_E

Structure interne: N_E est un semi-anneau unitaire (2.3.7) et commutatif (2.3.8).

Structure externe: La multiplication d'éléments de N_E par les éléments de E est une fonction de $N_E \times E$ dans E qui jouit des propriétés suivantes: (a) Distributivité à gauche (2.2.4): $(m+n).x = m.x + n.x$. (b) Distributivité à droite (2.2.5): $m.(x+y) = m.x + m.y$. (c) Associativité combinée (2.3.9): $(m.n).x = m.(n.x)$. (d) Identité (démonstration triviale: voir 2.1.6): $1.x = x$.

En plus, comme proprieté interne importante on peut remarquer la loi d'absorption du zéro: $n.0 = 0$.

CHAPITRE 3 - LES ESPACES SEMI-GROUPAUX SIMPLIFIABLES SANS TORSION ET LES NOMBRES NATURELS

3.1. ESPACES SEMI-GROUPAUX SIMPLIFIABLES SANS TORSION

3.1.1. Introduction

On ne peut pas définir de la manière habituelle la propriété d'être sans torsion parce que pour cela on aurait besoin d'employer la notion de nombre naturel, qui est précisément celle qu'on veut justifier dans cet article. On doit offrir donc une autre définition qui accomplisse la même tâche: celle d'éviter la "circularité", c'est à dire d'éviter "le retour à zéro". Pour éviter d'autres "pathologies" on introduira aussi un postulat de "simplification". On remarque que les axiomes qu'on introduira concernent les espaces semi-groupaux mais le but est celui d'éliminer certaines "pathologies" dans le semi-anneau des scalaires naturels correspondant, pour le transformer ainsi en semi-anneau de nombres naturels.

3.1.2. Définition

Nous dirons que l'espace semi-groupal E est simplifiable si pour x, y, z quelconques dans E il vient:

$$x + y = x + z \Longrightarrow y = z.$$

On voit facilement que, pour un tel espace, il ne peut pas se présenter la "pathologie" suivante: qu'il existe un scalaire non nul n tel que n = n+1. On voit aussi qu'il ne peut pas se présenter cette "pathologie": qu'il existe des scalaires non nuls n et p tel que n+p = n. D'ailleurs, si E est simplifiable, N_E l'est aussi.

3.1.3. Définition

Nous dirons que l'élement x, de l'espace semi-groupal E, est un élément de torsion, s'il existe un scalaire non nul n, appartenant à N_E, tel que (n+1).x soit un élément neutre (le zéro du semi-groupe structural à qui x appartient). Et nous dirons que l'espace semi-groupal E n'a pas de torsion (ou qu'il est sans torsion) si les éléments neutres sont les seuls éléments de torsion de E.

On voit facilement que dans un espace sans torsion il ne peut pas se présenter la "pathologie" du "retour a zéro", c'est à dire qu'il existe un scalaire non nul n tel que n+1 = 0.

Tout comme dans le cas de la propriété d'être simplifiable, celle d'être sans torsion est héréditaire dans le sens suivant: si E est sans torsion, N_E l'est aussi.

3.2. LES NOMBRES NATURELS

3.2.1. Définition

Si E est un espace semi-groupal simplifiable et sans torsion, les scalaires correspondants seront appelés nombres naturels associés à E, ou nombres naturels de E.

De cette façon il n'y a pas un ensemble unique de nombres naturels, mais un ensemble pour chaque espace semi-groupal simplifiable et sans torsion. Tous ces ensembles ont même structure, comme nous verrons tout de suite.

3.2.2. Théorème

Pour tout espace semi-groupal simplifiable et sans torsion E, son ensemble de nombres naturels N_E satisfait les "axiomes de Peano": (i) $0 \in N_E$; (ii) Si $n \in N_E$ alors $n+1 \in N_E$; (iii) Si $n \in N_E$ alors $n+1 \neq 0$; (iv) Si $n \neq m$ alors $n+1 \neq m+1$; (v) Si les trois conditions suivantes sont vérifiées, $C \subset N_E$, $0 \in C$, $n \in C \Longrightarrow n+1 \in C$, alors $C = N_E$.

3.2.3. Propriétés des Nombres Naturels

D'après ce théorème on peut considérer que toute l'arithmétique ordinaire est établie; en particulier, les opérations de somme et de multiplication que nous avons déjè défini pour N_E sont les mêmes qui résultent en appliquant les définitions habituelles par induction.

3.2.4. Isomorphie des Ensembles de Nombres Naturels

Théorème. Deux semi-anneaux quelconques de nombres naturels sont canoniquement isomorphes. Plus précisément: si E, F, sont des espaces semi-groupaux simplifiables et sans torsion, il existe un isomorphisme canonique $f_{FE} : N_E \longrightarrow N_F$.

L'isomorphisme en question est défini par induction de la manière évidente: $f_{FE}(0) = 0$, $f_{FE}(n+1) = f_{FE}(n) + 1$. La canonicité est contenue dans les deux propriétés suivantes: Première propriété catégorielle: Pour des espaces semi-groupaux simplifiables et sans torsion E, F, G, on a:

$f_{GF} \circ f_{FE} = f_{GE}$.

Deuxième propriété catégorielle: Pour tout espace semi-groupal simplificable et sans torsion E, on a: $f_{EE} = 1_{N_E}$.

CHAPITRE 4 - COMPLÉMENTS CATÉGORIELS

4.1. CONNEXION AFFINE DANS LES ESPACES SEMI-GROUPAUX

4.1.1. Introduction

Le modèle que nous avons en vue est encore l'ensemble V, des vecteurs fixes du plan (ou de l'espace), tel que nous l'avons présenté en 1.1.1. D'après ce que nous venons de voir, la structure d'espace semi-groupal simplificable et sans torsion suffit pour fonder l'introduction des nombres naturels. Mais cette structure ne suffit pas pour développer une étude proprement géométrique; il semble, en effet, que l'ingrédiant spécifiquement géométrique est la notion d'équipolence de vecteurs fixes, car elle permet de relier l'algèbre vectorielle à la notion de parallélisme, qui est à son tour le concept basique de la géométrie affine. Or, la relation d'équipolence entre vecteurs fixes permet de définir un isomorphisme naturel entre les groupes abéliens de vecteurs V_A (vecteurs d'origine A) et V_B (vecteurs d'origine B). Ceci se fait de la manière suivante: on définie $j_{BA} : V_A \longrightarrow V_B$ en établissant que $j_{BA}(\underline{u})$ est le vecteur (unique) d'origine B équipolent a $\underline{u}$. On démontre facilement que j_{BA} est un homomorphisme de groupes, et qu'il est bijectif; il est donc un isomorphisme. Ces isomorphismes ont deux propriétés categorielles:

a) Pour des points A, B, C, quelconques, il vient: $j_{CB} \circ j_{BA} = j_{CA}$;

b) Pour tout point A on a: $j_{AA} = 1_{V_A}$.

Or, si nous appelons $\mathcal{V}$ la partition de V (ensemble de tous les vecteurs fixes du plan ou de l'espace) dont les éléments sont les classes V_A, V_B, etc., on peut définir une fonction $\underline{j}$ de la manière suivante:

$$j : \mathcal{V} \times \mathcal{V} \longrightarrow \mathcal{I} \quad \text{tel que} \quad j((V_A ; V_B)) = j_{BA} ,$$

$\mathcal{I}$ étant l'ensemble de tous les isomorphismes du type j_{BA}. On reconnaît en cette fonction $\underline{j}$ une conexion affine dans le plan (ou dans l'espace).

4.1.2. Définition

Soit E un espace semi-groupal, tel que deux de ses semi-groupes structuraux (d'après 1.1.3), quelconques, soient isomorphes; soit $\mathcal{V}_E$ l'ensemble de tous les semi-groupes structuraux de E, et soit $\mathcal{I}_E$ l'ensemble de tous les isomorphismes entre éléments de $\mathcal{V}_E$. Alors on appelle connexion affine dans E toute fonction

$$\underline{j} : \mathcal{V}_E \times \mathcal{V}_E \longrightarrow \mathcal{I}_E$$

qui possède les deux propriétés catégorielles suivantes:

a) Si 0_1, 0_2, 0_3, sont des éléments neutres de E, et si E_1, E_2, E_3, sont leurs respectifs semi-groupes structuraux, on a:

$$\underline{j}(E_2,E_3) \circ \underline{j}((E_1,E_2)) = \underline{j}((E_1,E_3));$$

b) Pour tout élément neutre 0 de E, si E_o est son respectif semi-groupe structural, on a:

$$\underline{j}((E_o,E_o)) = 1_{E_o} .$$

4.1.3. Notation

Pour abréger la notation on écrira $\underline{j}_{ik}$ au lieu $\underline{j}((E_k,E_i))$; alors:

$$\underline{j}_{32} \circ \underline{j}_{21} = \underline{j}_{31} \quad ; \qquad \underline{j}_{ii} = 1_{E_i} .$$

On a aussi, pour des éléments neutres 0_1 , 0_2 , quelconques:

$$\underline{j}_{12} = \underline{j}_{21}^{-1} .$$

4.1.4. Définition

Nous appellerons espace semi-groupal affine tout couple $(E,\underline{j})$, où E est un espace semi-groupal dont les semi-groupes structuraux sont isomorphes deux à deux, et $\underline{j}$ est une connexion affine sur E. On dira dans ce cas que E est un espace semi-groupal affine.

4.1.5. Définition

Si E est un espace semi-groupal affine et si $\underline{j}$ est sa connexion affine, soient 0_1, 0_2 des éléments neutres de E, et E_1, E_2 leurs respectifs semi-groupes structuraux. Soit $\underline{x} \in E_1$. Nous appellerons <u>déplacé de</u> $\underline{x}$ à l'origine 0_2, ou bien <u>déplacé de</u> $\underline{x}$ au semi-groupe structural E_2, l'élément $\underline{j}_{21}(\underline{x})$.

4.1.6. Théorème de Conservation du Produit de Scalaire Naturel par Élément de E

Soit $(E,\underline{j})$ un espace semi-groupal affine. Pour des éléments neutres 0_1, 0_2, de E, pour $x \in E_1$, et pour tout scalaire naturel $\underline{n}$ de E, on a: $\underline{j}_{21}(\underline{n}\cdot\underline{x}) = \underline{n}\cdot\underline{j}_{21}(\underline{x})$.

4.1.7. Corollaire

Si $x \in E_i$ et si $\underline{n}\cdot\underline{x}$ est nul (respectivement non nul), alors, pour n'importe quel semi-groupe structural E_k, on a que $\underline{n}\cdot\underline{j}_{ki}(\underline{x})$ est nul (respectivement non nul).

4.2. LE FONCTEUR NOMBRES NATURELS

Dans ce paragraphe nous utiliserons librement des notions et des résultats élémentaires de la théorie générale des catégories et foncteurs.

4.2.1. Le Groupoïde Naturel

Rappelons que l'on appelle groupoïde toute catégorie dans laquelle tous les morphismes sont des isomorphismes. Nous appellerons <u>groupoïde naturel</u> la catégorie dont les objets sont les espaces semi-groupaux simplifiables sans torsion et dont les morphismes sont les couples de tels espaces, la composition étant la composition usuelle de couples (de gauche à droite): $(a;b).(b;c) = (a;c)$. Ce groupoïde sera noté *ESG*

On appelle semi-anneau simplifiable sans torsion tout semi-anneau que, considéré comme espace semi-groupal (en faisant abstraction de la multiplication), soit simplifiable sans torsion.

4.2.2. Théorème

Si à chaque objet E de *ESG* on fait correspondre le semi-anneau N_E de ses nombres naturels, et à chaque morphisme de *ESG* l'isomorphisme canonique entre les deux semi-anneaux de nombres naturels respectifs (voir 3.2.4), on obtient un foncteur

N : *ESG* ⟶ *SA*

SA étant la catégorie dont les objets sont les semi-anneaux unitaires simplifiables sans torsion, et dont les morphismes sont les morphismes de semi-anneaux unitaires avec la composition usuelle.

Il est évident que N_E, considéré comme espace semi-groupal, est simplifiable sans torsion (voir la fin de 3.1.2 et de 3.1.4).

4.2.3. Définition

Le foncteur N : *ESG* ⟶ *SA*, donné par le théorème précédant, sera appelé <u>foncteur nombres naturels</u>.

4.2.4. Itération

Le foncteur nombres naturels peut être itéré, pourvu que l'on fasse une certaine convention <u>ad-hoc</u> sur les structures de semi-anneau et de semi-groupe. Si N_E et le semi-anneau unitaire de nombres naturels de l'espace E, nous ferons la convention (pratiquement déjà adoptée auparavant) de désigner par le même symbole N_E le semi-groupe sous-jacent (celui que l'on obtient en se passant de la multiplication entre nombres naturels). Alors N(E) peut être considéré à son tour comme un semi-groupe, lequel, comme nous l'avons déjà remarqué, est un espace semi-groupal simplifiable sans torsion, c'est à dire un objet de *ESG*. Donc nous pouvons considérer le semi-anneau unitaire N(N(E)), qui est canoniquement isomorphe au semi-anneau unitaire N(E).

Pour exprimer avec précision cette idée nous avons besoin du foncteur de sous-jacence S : *SA* ⟶ *ESG* qui à chaque semi-anneau simplifiable sans torsion fait correspondre l'espace semi-groupal sous-jacent (en se passant de la multiplication), et que à chaque morphisme de *SA* fait correspondre le couple formé par les espaces semi-groupaux respectifs. Alors ce que nous venons de noter (N(N(E)) doit être en toute rigueur N(S(N(E))); et ce semi-anneau est canoniquement isomorphe à N(E). Avec les notations utilisant indices, ces semi-anneaux sont, respectivement: $N_{S(N_E)}$ et N_E .

CHAPITRE 5 - LES THÈSES PHILOSOPHIQUES ET LES ENSEMBLES HYPOTHÉTIQUES

5.1. L'INTUITION FORMALISTE BASIQUE

5.1.1. Manque d'Unicité

L'Introduction des nombres naturels que nous venons de réaliser dans ce texte présente un désavantage apparent qui, d'un point de vue philosophique, est néanmoins un avantage: nous parlons du manque d'unicité; il n'y a pas un ensemble de nombres naturels uniques, mais un tel ensemble pour chaque espace semi-groupal simplifiable sans torsion. Mais il ne s'agit pas ici de la multiplicité de nombres qui se trouve, par exemple, dans la théorie de types de Russell-Withehead, laquelle est en effet une complication peu désirable. Voyons de plus près ce détail.

Quand on étudie un espace semi-groupal simplifiable sans torsion bien determiné, comme par exemple les vecteurs fixes du plan affine, on sait avec précision qui sont les nombres naturels pour un tel sujet d'étude. Il n'y a pas d'inconvénient non plus pour utiliser ces nombres naturels avec sa fonction usuelle de "mesure de la cardinalité". Appelons T la théorie d'un certain espace semi-groupal E (simplifiable sans torsion); nous dirons que l'ensemble C, définissable dans la théorie T, a le cardinal $\underline{n}$, s'il existe dans T une bijection de C sur l'intervalle [[1;n]] (ensemble des nombres naturels $\underline{x}$, appartenant a N_E, tel que $1 \leqq x \leqq n$).

Si T est une théorie dans laquelle on étudie plusieurs espaces semi-groupaux simplifiables sans torsion (définis au préalable dans T), l'introduction de la cardinalité n'offre non plus de difficulté parce que les isomorphismes canoniques vus en 3.2.4 permettent de passer d'un ensemble N_E à un autre N_F sans inconvéniant théorique.

5.1.2. Les Nombres Naturels el L'Intuition Formaliste

L'idée que nous développons ici peut se concevoir comme un perfectionnement de la théorie formaliste de Hilbert, tout au moins dans la partie qui regarde les fondements de la notion de nombre naturel. En effet, le formalisme hilbertien se développe sans peine une fois que l'on est arrivé à identifier les nombres naturels. Mais c'est précisément cette identification qui pose les problèmes philosophiques les plus ardus. Hilbert établie que le nombre 1 est un trait (par exemple: |), que le nombre 2 est obtenu en ajoutant un trait à la droite du nombre 1 (par exemple, ||), et, en général, que le nombre n+1 est obtenu en ajoutant un trait à la droite du nombre n.

Avec ceci Hilbert prétend d'atteindre plusieurs objectifs:

(i) Donner aux nombres naturels en fondement empririque clair;

(ii) Développer l'arithmétique élémentaire par des méthodes finitistes;

(iii) Montrer le principe d'induction comme auto-évident.

À leur tour ces objectifs serviraient à obtenir un résultat plus général et qui occupe un lieu éminent dans le formalisme hilbertient, à savoir:

(iv) La consistence de l'arithmetique elémentaire.

La réussite de l'objectif (iv) à partir de trois précédants est basée sur une idée d'origine kantienne: que les processus réels sont exempts de contradiction.

La critique a déjà indiquè les trois aspects les plus faibles de cette formulation:

(a) Le concept de trait n'est pas du tout clair, et l'est moins encore celui d'ensemble de traits;

(b) Comme les traits que l'on peut produire sont infinis (au moins en potence), il y aurait une multiplicité de nombres 1 (ainsi que de nombres 2, etc.), ce qui fait perdre identité aux nombres naturels;

(c) Ce qui apparaît dans la construction de l'arithmétique n'est pas seulement une collection d'objets empiriques mais aussi une capacité (de l'esprit?) pour itérer indéfinimment le processus d'ajouter un trait. Le status philosophique de cette capacité n'est pas clair non plus.

Bien que le formalisme hibertien se présente comme une théorie rivale de l'intuituonnisme browverien, il n'y a pas de doute que - au moins dans les fondements de l'arithmétique - le premier admet certaines intuitions: celle de trait, celle d'ensemble de traits, et celle de capacité d'ajouter un trait. Ces intuitions (que l'on pourrait appeler "intuitions formalistes") ont été soumises à des critiques assez sévères.

Pour essayer de surmonter les critiques que nous venons d'esquisser, nous introduirons une certaine modification dans les fondementes du formalisme hilbertien, tout en conservant les caractères philosophiques de celui-ci.

5.2. MODIFICATION DES FONDEMENTS DE L'ARITHMÉTIQUE FORMALISTE

5.2.1. La Notion de "Ajouter un Objet"

Pour mantenir l'essentiel du formalisme (c'est à dire, la conception de la mathématique comme science des systèmes formels), ainsi que pour mantenir la possibilité d'expliquer les rapports entre la mathématique pure et l'appliquée, il faudra modifier les fondements de l'arithmétique élémentaire en renonçant de façon définitive aux "traits hilbertiens". Nous parcourirons maintenant ce chemin, mais tout en conservant aussi les aspects structurels de la théorie hilbertienne des traits.

Nous avons vu que l'essentiel des traits et des nombres-traits c'est la capacité pour ajounter toujours un trait. Dans un autre langage, l'essentiel est la capacité pour réaliser la somme (ou l'addition) d'un objet à une certaine classe pre-établie. Comme l'idée de "ajouter un trait" est en elle même obscure, et elle se fait plus obscure quand on tâche de la clarifier (par exemple, en disant que le nouveau trait doit être ajouté "a droite", qu'il doit "être semblable aux autres", etc.), il est préférable partir d'une situation dans laquelle on accepte déjà comme indubitable le sens de la phrase "additionner un objet".

5.2.2. La Base Formaliste de l'Arithmétique et de la Géométrie

Au lieu de chercher une élusive fondation empirique pour les systèmes formels, nous tâcherons d'invertir le procèssus en cherchant une fondation formaliste pour l'arithmétique empirique. Après nous discuterons les problèmes philosophiques que cette inversion entraîne.

Comme l'idée basique de l'arithmétique élémentaire est celle d'additionner un objet, il semble naturel partir d'un système formel qui comprenne un minimum des notions capable d'esquisser ce que d'habitude on appelle "addition". Ce système formel minimum est l'espace semi-groupal introduit dans 1.1.2.

Si l'on accepte cette notion basique (après nous verrons les problèmes philosophiques correspondants) on a une base solide pour établir avec précision la notion de scalaires naturels associés à un espace semi-groupal E, tout en reproduisant - mais avec netteté - la procédure hilbertienne des traits.

Le passage des scalaires naturels aux nombres naturels est fondé sur deux propriétés formelles douées d'un haut degré d'intelligibilité: on introduit ainsi les espaces semi-groupaux simplifiables et sans torsion. (Voir 3.1).

D'ailleurs - et comme il a été dit clairement dans 1.1.1 et 4.1 - les espaces semi-groupaux constituent la base de la géométrie vectorielle. Par conséquent, cette structure permet d'offrir une fondation intégrée de l'arithmétique et la géométrie. Cette idée sera developpée dans ce qui suit.

Comme les scalaires naturels (et par conséquent les nombres naturels) sont des fonctions, et comme les fonctions ont été introduites dans un cadre nettement ensembliste, il est nécessaire de formuler quelques remarques sur les ensembles.

5.3. PROGRAMME POUR UNE THÉORIE D'ENSEMBLES HYPOTHÉTIQUES

5.3.1. Sur les Deux Formes de la Mathematique Pure

Si l'on veut utiliser la mathématique ensembliste pour fonder les applications de la mathématique à la réalité, on doit accepter des très forts postulats logiques et philosophiques. En particulier, on doit accepter une ontologie d'ensembles, de classes d'ensembles, de classes de classes d'ensembles, etc. En plus cette ontologie est restreinte par quelques postulats logiques ayant pour but celui d'éviter les paradoxes. Tout cela pose des problèmes philosophiques tellement ardus qu'il semble plus sage reduire au minimum l'ontologie ensembliste de base. Par exemple, il n'est pas convenable de postuler une ontologie infinie; ce qui ne signifie pas que l'on doit accepter le postulat contraire, d'après lequel il y aurait seulement une quantité finie d'entités; il s'agit plutôt de laisser la question ouverte, sans postulat dans un sens ou dans un autre.

Le dificultés les plus graves ne surgissent pas d'accepter une ontologie de certains ensemble déjà donnés, mais d'accepter un procédé logique pour la "création" d'ensembles au cours de la théorie; en effet, ce qui a conduit à presque tous les paradoxes de la théorie des ensembles est l'idée d'associer un ensemble à chaque fonction propositionnelle. C'est cette idée qui a été restreinte par les diverses solutions que l'on a proposé: théorie de types, axiomatique de Zermelo-Fraenkel, etc. En prenant, par exemple, l'axiomatique de Zermel-Fraenkel telle que l'expose Jean-Louis Krivine, c'est l'axiome de substitution (ou de remplacement) qui garantit l'existence d'un ensemble associé a un énoncé dans certaines conditions. Remarquons en passant que cette axiome est le seul qui fait référence à des énoncés, car les autres ne font référence qu'à des ensembles; donc, pour une interprétation sémantique (que nous ferons par la suite) cet axiome a un statut logique et sémantique très particulier. Alors, pour éviter et le danger des paradoxes et la référence à des enoncés dans les axiomes, nous prendrons un point de vue à la fois sémantique et hypoyhétique, ce que nous expliquerons dans ce paragraphe et dans 5.3.2.

Pour le moment disons qu'il y a deux formes de concevoir la mathématique pure: la forme syntactique et la forme sémantique. La première jouit de la préférence des logiciens; la seconde est celle qu'adoptent en pratique les mathématiciens professionnels et les enseignants, car les définitions qui se trouvent dans les livres de texte et dans les articles de recherche ont toujours un sens référentiel.

Nous prenons ici, du départ, un point de vue sémantique, semblable à celui de Krivine dan l'oeuvre citée, mais non tout à fait pareil. En effet, cet auteur considère comme établi le mot "collection" dans le mâta-langage, et après il définit la notion de "Univers" comme collection munie d'une relation binaire appelée <u>d'appertenance</u>, de telle façon que les axiomes de Zermelo-Fraenkel soient vérifiés. C'est bien un point de vue sémantique, semblable (mais non identique) à celui qui est habituel quand on définit "groupe", "espace vectoriel", etc., car si l'on dit "<u>x</u> est un ensemble", ou "<u>x</u> est dans la collection U", il est entendu que la lettre "<u>x</u>" est utilisée comme le non d'un objet à qui elle fait référence; tandis que, pour le point de vue syntactique, la lettre "<u>x</u>" ne fait référence à aucun objet.

Nous proposons maintenant une légère variation, à savoir: nous n'introduisons pas dans le méta-langage le mot "collection", mais nous dirons tout simplement que le mot "ensemble" est un prédicat qui s'applique à des objets, et "appertenance" indique une relation binaire, de telle façon que les axiomes de Zermelo-Fraenkel (d'après Krivine) sauf celui de substitution, soient vérifiés, en plus de certaines propriétés particulièrement intéressantes qui découlent de l'axiome de substitution, et que nous devons établir explicitement (axiomes de la paire, du complément relatif, du produit cartésien, de l'ensemble vide, etc.). Mais ce n'est pas tout.

5.3.2. Théorie Basique et Existence Hypothpétique

La théorie des ensembles telle que nous venons de l'esquisser sera appelée Théorie Basique (en bref: TB). Elle ne suffit pas à développer la mathématique courante. Pour ce faire il faut introduire d'autres ensembles, quelques uns, peut être, d'une façon *ad-hoc*. La règle (dans le meta-langage) que nous adoptons est la suivante:

Règle Générale. Chaque fois que nous introduirons un nouvel ensemble (tout en démontrant son unicité par le moyen, disons, de l'axiome d'extensionnalité) nous dirons que cet ensemble a une existence hypothétique. La mathématique ensembliste consiste en la rédaction de textes où l'on accepte TB plus qualques ensembles d'existence hypothétique. On dit qu'un texte de mathématique ensembliste n'est pas valable quand ce texte entraîne une contradiction.

Par exemple, un texte de mathématique ensembliste où l'on introduit l'ensemble des ensembles x tel que $x \notin x$, est un texte qui n'est pas valable. De même pour un texte avec ensemble universel, etc.

Il y aura certainement plusieurs textes desquels nous ne saurons pas s'ils sont valables ou non. Cette situation n'est pas moins commode que celle des théories axiomatiques dont nous ne savons pas si elles sont consistentes.

Cette conception de l'activité mathématique comme activité créatrice de textes, est plus cohérente-pensons nous - que les conceptions classiques d'après lesquelles toute la théorie des ensembles est donnée d'emblée, et l'on est obligé de reconnaître *a priori* l'existence d'une foule d'ensembles dont on n'a pas besoin et qui sont, par dessus le marché, assez encombrants. Au contraire, il est plus naturel d'introduire dans chaque texte les ensembles dont on a vraiment besoin. Si l'on a la mauvaise fortune d'introduire un ensemble dont l'existence hypothétique mène à une contradiction, tout ce que l'on doit faire est de déclarer ce texte non valable et rédiger (si l'on peut) un autre texte qui ne fasse pas usage de cet ensemble et qui soit aussi intéressant.

Dans cette théorie les opérateurs logiques existentiel et universel ont un sens sémantique, mais ils ne font référence qu'aux ensembles acceptés dans le texte où ils se trouvent. Par exemple, on peut rédiger un texte on l'on introduise l'ensemble des tous les textes valables, ce qui pose le problème de l'auto-référence; mais on n'est pas obligé de le faire. On peut même développer toute la mathématique connue en se passant de l'ensemble de tous les textes valables.

Ces indications ne forment qu'une esquisse d'une théorie qui sera exposée formellement ailleurs. À titre d'exemple nous verrons maintenant comment on peut interpréter dans le cadre de ces idées la théorie des espaces semi-groupaux, ainsi que les fondements de l'arithmétique et de la géométrie.

5.3.3. Les Fondements de l'Arithmétique et de la Géométrie

Le point de départ est TB. On définit les relations binaires et les fonctions comme d'habitude, par le moyen des couples. Avec ceci on peut déjà accepter les définitions d'espace semi-groupal, de transformation cohérente et de morphisme (voir 1.1.2, 1.2.1 et 1.2.2). La définition de système de scalaires naturels associé à l'espace semi-groupal E (voir 2.1.4) requiert l'introduction hypothétique des ensembles de fonctions de E dans E, ainsi que de certain cas d'ensemble intersection, ce qui ne pose pas de problème. On a besoin aussi, bien entendu, d'introduire un axiome relatif à l'existence de l'ensemble N_E.

La définition des "nombres naturels" associés à E n'a besoin d'introduire au préalable l'axiome de l'infini; en effet, il y a des nombres naturels si et seulement si il y a des espaces semi-groupaux simplifiables et sans torsion. Donc on peut développer la théorie des nombres naturels (donc l'arithmétique) d'une façon hypothétique: si l'on accepte que E soit un espace simplifiable et sans torsion alors N_E est un système de nombres naturels.

Avec ceci l'arithmétique de Peano est fondée sur des bases sémantiques et hypothétiques. Mais en même temps nous avons établi les fondements de la géométrie élémentaire, car les espaces semi-groupaux sont le germe de la théorie des vecteurs, comme on l'a esquissé dans 4.1.

Si en plus nous introduisons de manière hypothétique l'ensemble de tous les espaces semi-groupaux simplifiables et sans torsion, on peut développer la théorie catégorielle commencée dans 4.2. On ne rencontrera pas de problème pourvu qu'on n'arrive à introduire un ensemble pathologique. Mais encore dans ce cas on ne doit pas abandonner la mathématique ensembliste développée jusque là, mais seulement on est obligé d'abandonner l'ensemble pathologique et chercher d'autres voies mathématiques.

HOW TO DECIDE BETWEEN DIFFERENT METHODS OF STATISTICAL INFERENCE

Rolando Chuaqui/Facultad de Matemáticas, Pontificia Universidad Católica de Chile, Santiago, Chile

INTRODUCTION

The philosophy of a science is important its own right because we, as human beings, need to understand what we do. This is especially so for those who work in a particular science. It is important for them to understand the meaning of the terms they are using, to realize the significance of its methods and their limitations, to visualize the formal structure of the science, etc. This is a basic belief that I share with Newton da Costa. More than this, both his work and mine has always had philosophical inspiration and tried to give a better understanding of science.

Probability is extensively used in many sciences, not only in statistics, but also in physics, biology, medicine, etc. For the reason stated above, the foundations of probability should be of interest for all these different types of scientists. Thus, anyone interested in the philosophy of science is naturally led into the study of the foundations of probability. This is what has happened to Newton da Costa and me.

Since we met many years ago, we have also discussed the subject.

These conversations with him have have influenced my thought greatly.

There are many instances of philosophical ideas influencing the development of a science. But, Statistics is the only case I know of where philosophical ideas not only are relevant for the development of the science but also have practical implications. It is well-known that the different interpretations of the concept of probability lead to different statistical methods. For instance, those who favor a subjectivist or logical interpretation are Bayesian; on the other hand, the partisans of a frequency interpretation use classical (i.e. Neyman-Pearson) methods. Although it is not infrequent that the application of the different methods may lead us to similar results, it is also not infrequent that they may yield mutually contradictory consequences both in statistical inference and in decision theory.

An attitude, espoused by some, is to accept one method or another, according to the situation involved. Thus, is a given situation there would be no conflict. The statistician with his experience must decide which is the best method for each situation. This experience is such that it cannot be expressed in words. The statistician, just as a physician in front of a patient, knows which method to apply without being able to say why.

This comparison of a statistian with a physician is not adequate. If two physicians with similar reputations are studying a patient, it is almost sure that both will reach the same conclusion. However, among statisticians, this is certainly not the case. There are many instances where statisticians of the best reputations disagree completely on the methods to be applied. These disagreements are not only on the mathematical techniques, but also on ways of collecting the data. This does not mean that the contention that different techniques work for different problems is false, but it does show, or at least suggest, that it is necessary a theory to characterize in some way the different types of problems, and to explain why the different methods are appropriate to the different problems. My work is precisely an attempt to give such a theory.

It is also possible to think that, just as in the natural sciences, experience will enable us to decide between the different statistical methods. It has been usual in other sciences that competing theories exist side by side until by experiments or observations one theory is accepted and the other rejected. Experience might enable us to decide which methods work or work in particular situations. Thus, finally, the techniques that work well in general or in each situation will be retained and others abandoned or restricted in their use. The difficulty here is to understand what we mean when we say that a certain method _works_. As a first approximation to this, I shall consider an explanation appearing in Kyburg 1974, pp. 148-149:

"Let us suppose (what is no doubt contrary to fact) that we have a clear idea of what is for a statistical technique to work in a given instance. Quite clearly it is not in the nature of a statistical technique to work all the time, or to work in the same degree all the time. It is conceivable, however, that we could collect a number of instances i of the application of a statistical technique, and associate with each instance a number $D(i)$ representing the degree to which that technique worked in that instance. This provide a picture of how well that technique has worked in the past. But when we talk about how well a statistical technique works, we are concerned not merely with how well it has worked in instances of its application in the past, but how well it works _in general_ : that is, what concerns us is not the sample of values of the random quantity D that we have happened to observe in the past, but the general distribution of D. How do we arrive at knowledge of this distribution? Why, by statistical inference, of course!"

Let us analyze more carefully Kyburg's description. In order to apply statistical techniques to the our problem, we should assume that there is a collection of possible methods T, and that for each $t \in T$ we have a random quantity D_t as was introduced by Kyburg. Suppose that applying any method t in T the estimation of D_{t_o} , for a certain t_o, is preferable according to t, to all others. Then clearly, we should conclude that t_o is the best technique. Although for this situation to obtain is extremely unlikely (because, among other reasons, methods differ not only in the mathematics required, but also in other aspects, such us the sampling method used), it would be a case where a decision by experience would be possible. However, there is another problem that I believe is unsurmountable for a strictly empirical procedure: How are we going to choose our set T of techniques? I think that is clearly impossible to take T to be the set of all logically possible methods, since it is always possible to design a silly technique that only approves of itself. Thus, we must select a reasonable T. But on what empirical basis can be do this? We would again have to consider a set F reasonable T's and apply statistical methods to it. Thus, we arrive either to an infinite regress or we must make an initial choice for reasons other than empirical.

We can conclude with Kyburg 1974, pp. 149-150:

"This supports the intuition that what is at issue in conflicts among approaches to statistics is a profound and pervasive conception of the nature of statistical inference and the nature of probability ... We are offered what appear to be ultimate principles - principles so basic that it is difficult to see how to defend them or argue about them".

Thus, we arrive at the basic difficulty. It seems that in dealing with statistical inference and probability we are confronted with fundamental epistemological questions, in particular with the problem of induction. We are dealing here with our basic intuitions concerning the fundamental principles of probabilistic arguments, that is, what may be called the "logic of probability".

It is obvious that statistical inference has practical importance. Thus, we are in the presence of a philosophical problem with immediate practical implications: how to decide between alternative statistical techniques.

I believe that each conception of probability has its roots in a particular epistemological position. In order to support this claim, I would have to examine the different interpretations of probability and show from where they originate. I shall not do this here; I shall not even state completely the philosophical position that has led me do my version of probability. Instead, I shall just make some remarks which are relevant to my view of probability and statistical inference, without attempting to thoroughly justify all of them.

I believe that we can get to know different types of propositions, including logical and mathematical propositions, laws of nature, and, certainly, singular propositions about physical objects. I think that it is possible (and desirable) to justify the possibility of knowing each type of proposition, although this justification could not be of a logical kind. As I said before, I shall not attempt here such a justification.

To know a proposition implies to accept it to be true. Since

I believe that truth is prior to probability, I must hold, as I said before, that we must be able to get to know some propositions. The following argument speaks in favor of the priority of truth over probability, i.e. of the claim that the assignment of probabilities depend on the acceptance of some contingent propositions as true. If probabilities were always based on other probabilities, in order to avoid an infinite regress, we would have to accept a first assignment of probability. This first assignment, either it is not justified at all but only irrationally accepted (as in de Finetti 1937) or justified on purely logical grounds (as in Carnap 1950). The first alternative seems to me too irrational to accept, especially since it has to involve an arbitrary assignment of probabilities to all propositions (each and every one of them), whatever "all propositions" may mean. On the other hand, all attempts to justify the assignment of probabilities on logical grounds seem to depend crucially on the language involved (see Carnap 1950, 1971), and have not succeeded in providing criteria for assigning the probabilities uniquely.

The problem of having to talk about "all propositions" or "all events" seems to me to be the main stumbling block of these positions. The probabilities assigned to particular propositions always depend on the totality of propositions that we are considering (in Carnap's version, it depends on the language chosen). This is not true of deductive logic, where the fact that a proposition is deducible from others depends just on the propositions involved. In order to define logical probabilities, we have to decide on the class of propositions that we are considering, and this cannot always be done probabilistically, if we are to avoid an infinite regress. Hence, the first selection of the class of all propositions is either arbitrary or based on rational but not probabilistic grounds, i.e. on what we accept as true.

PROBABILITY AND TRUTH

Similarly as logical inference, statistical inference gives rational rules for starting from some propositions accepted as true, arriving at other propositions that are to be accepted as true. The paradigmatic case of the application of statistical inference is the evaluation of scientific hypotheses. Decision Theory, on the other hand, gives rules of rationality for guiding our actions in the face of uncertainty. The fundamental difference between the evaluation of scientific hypotheses and decision theory is that in the latter we are forced to decide on an action, while in the first one we are only interested in the truth or validity of the hypothesis and, if there is not sufficient evidence, we just say we do not know and suspend judgement.

In decision theory we must act, although our evidence may not be enough to ascertain truth or falsity. We must be based just on our degree of belief. Thus, in decision theory we pass from degree of belief to degree of belief, while in statistical inference proper, from truth to truth. However, in order to avoid an infinite regress, even in decision theory our degrees of belief should be based on propositions accepted as true.

Some people have argued that statistical inference is not essentially different from decision theory, since they consider accepting or rejecting a hypothesis as an action. I think that this stretches too much the meaning of the word "action" and prevents us from giving a good account of inference. Besides this, although truth is a guide to action, it is not its only function.

I will now briefly describe my conception of probability and its relation to truth (for a fuller technical description see Chuaqui 1983, 1984). Probability models or structures are built for reality These structures incorporate two basic elements: a notion of equiprobability based on a refined version of the Principle of Insuficient Reason, and a notion of stochastic dependence (or independence). These are considered the primitive notions from which the probability measure for the given structure is defined. The events (or propositions) of the model are also determined by the given structure.

That which is true or false are the propositions. A proposition is true or false according to that obtains in the real world. Logic studies what is true independently of our knowdledge of the real world. Thus, we relativize truth to a possible state of the world, formally, a relational structure. When we want to apply logic, we assume a certain possible state $\mathcal{A}$. Now, if in the real world a proposition obtains that is false in $\mathcal{A}$, then $\mathcal{A}$ is not considered to be an accurate description of the real world, and another is sought after. On the other hand, if many propositions obtain that are true in $\mathcal{A}$, it is, at least provisionally, accepted.

I think of probability as a sort of "degree of truthfulness". Thus, it should be applied to propositions, just as truth is. A proposition might have a certain degree of truthfulness according as to what obtains and is possible in the real world. Just as in logic, we relativize probability to a possible world K, that in this case is a probability structure which in general determines not the truth of propositions, but just their probabilities. Again, in order to apply the theory, we assume K to be an accurate description of the world, and calculate the probabilities of propositions (or events) according to K. Since only some propositions are either true or false in K, and most have only probabilities assigned to them, we cannot have the same principles of rejection and acceptance as before. The main purpose of statistical inference is to provide and justify principles of this sort for probability structures.

When we have to decide on a certain action, we must accept a possible world K as the real world, calculate the probabilitites in K, and act accordingly. Thus, in decision theory, the question of the fitness of K to reality does not appear. Since we must act, we have to accept a certain K, although our reasons for accepting it may not be strong.

CLASSICAL STATISTICAL INFERENCE

As I mentioned before, I believe that the main purpose of classical statistical inference is to give rules for the rejection or acceptance of hypotheses. I this paper, I shall concentrate on rules for rejection.

We consider a possible model of the world K. The only thing we need to know about the structure of K, at this stage, is that K determines a probability distribution over the propositions (or events) we are interested in (for details see Chuaqui 1983, 1984). A natural general rule would be to reject a model K if the next best thing to an event false in K obtains, namely, an unlikely event relative to K occurs. But, what is an unlikely event? An unlikely event should be one that is approximately false. Thus, we are tempted to say that a proposition φ is unlikely relative to K, if φ has low probability in K. Hence, we would assume that if an event with low probability in K occured, we should reject K, and accept it provisionally otherwise. However, this rule does not work. Suppose that the experiment consists of tossing a coin $\underline{n}$ times. We assume a model K for this experiment that gives the distribution for a fair coin. The probability of any particular sequence of heads and tails is low, and all sequences have the same probability in K. But it is clear that intuitively some would count as favorable to K, while others not.

We are, hence, forced to consider classes of alternative hypotheses $K_{\underline{i}}$, for each $\underline{i}$ in a certain index set $\underline{I}$. These are all the hypotheses we consider possible under the situation involved. How to decide on which class of hypotheses to consider is not determined, in general, by statistical methods. There may be general natural laws or particular facts relevant for this decision. Anyway, we must accept as true that the class $\{K_{\underline{i}} : \underline{i} \in \underline{I}\}$ contains all the possible models (i.e. hypotheses). Now, we can improve the definition of an unlikely event, by saying that a proposition φ is unlikely if we accept $K_{\underline{i}_o}$, relative to the set $\{K_{\underline{i}} : \underline{i} \in \underline{I}\}$, if φ has low probability in $K_{\underline{i}_o}$, and high probability in $K_{\underline{i}}$, for a certain $\underline{i} \in \underline{I}$.

It is not, in general, possible to take $\{K_i : i \in I\}$ as containing all logically possible hypotheses because for any event there is a hypothesis under which it has high probability. As an example[(2)] of this, let us suppose that we toss a coin ten times. A sequence X = 0 1 0 0 1 0 1 1 1 0, counts in favor of the hypothesis K of the coin being fair, if the alternatives are the coin being biased with different degrees of biasedness. However, it counts against K, if one of the possibilities is: "the sequence is generated by a machine that produces X".

The simple consideration of alternative hypotheses, as above, is not enough. The rule of rejection that we get (and which has been described above) is what has been called the <u>likelihood principle</u>. There are examples where it does not quite work. One of these examples is due to Birnbaum (I give it as quoted in Kyburg 1974). Let $\Theta=(\mu,\sigma)$. σ may have the value 0 or the value 100, representing the precision of a measurement, and μ may have any integer value between -10^{10} and 10^{10}, representing the value of a quantity to be measured. Suppose that $\underline{X}$ is any observation in the interval $(10^{-8}, 10^{8})$. An easy calculation shows that the probability of $\underline{X}$ under the hypothesis $(\underline{X},0)$ (what is called <u>the likelihood of (X,0) given X</u>) is at least 100 times the likelihood of any others of the alternative hypotheses. Thus, $\underline{X}$ should be counted as strong evidence in favor of $\sigma = 0$ against $\sigma = 100$. However, it seems unnatural that with one observation we would be able to draw such a strong conclusion. Thus, we should amend our rule for rejection.

The intuitive idea behind our intended rule is to reject a hypothesis K, if it is "approximately" false. That is, if we can get as close to falsehood (which is approximately probability zero) as we want. Now, I describe more formally how this kind of rule can be formulated.

We consider the field of events as the algebra of propositions, which may be represented formally by the Lindenbaum algebra of a formal language. This field of events is determined by a probability structure K (see Chuaqui 1983,1984). We shall denote the field of events determined by K, $\mathcal{A}_K$. The structure K also determines a probability measure P_K over $\mathcal{A}_K$.

For K any probability structure let ${}^{\underline{n}}K$ he the set of sequences of n elements of K (where $\underline{n}$ is a natural number) or ω, the set of all such numbers ${}^{\underline{n}}K$ models (see Chuaqui 1983) $\underline{n}$ independent trials of the experiment modelled by K. We assume, as in Chuaqui 1983, that there is a probability P_K and that $P_{{}^{\underline{n}}K}$ is the product measure.

We shall consider an ideal rule of rejection and a semi-practical rule.

(I) Ideal rule of rejection of $K_{\underline{i}_o}$, given $\{K_{\underline{i}} : \underline{i} \in \underline{I}\}$.

Suppose $\underline{A} \in \mathcal{A}_{{}^{\omega}K}$. Reject $K_{\underline{i}_o}$ if:

(i) $\underline{A}$ obtains,

(ii) $P_{{}^{\omega}K_{\underline{i}_o}}(\underline{A}) = 0$,

(iii) there is an $\underline{i} \in \underline{I}$ such that $P_{{}^{\omega}K_{\underline{i}}}(\underline{A}) = 1$.

(II) Semi-practical rule of rejection of $K_{\underline{i}_o}$, given $\{K_i : i \in I\}$.

Reject $K_{\underline{i}_o}$, if for every natural number $\underline{m}$, we can effectively find a natural number $\underline{n}$, such that there is an event $\underline{A}_{\underline{m}} \in \mathcal{A}_{{}^{\underline{n}}K_{\underline{i}_o}}$ satisfying:

(i) $\underline{A}_{\underline{m}}$ obtains,

(ii) $P_{{}^{\underline{n}}K_{\underline{i}_o}}(\underline{A}_{\underline{m}}) < \frac{1}{\underline{m}}$,

(iii) We can effectively find an $\underline{i} \in \underline{I}$ such that $P_{\underline{n}_{K_{\underline{i}}}}(A_{\underline{m}}) \geq \frac{1}{2}$.

(II) is a sort of dialectical rule. It says that if I am given any natural number $\underline{m}$, I can provide (i.e. find effectively) the $\underline{n}$ and $\underline{A}_{\underline{m}}$. Thus if I want to convince somebody that $K_{\underline{i}_o}$ is false, I ask him for an $\underline{m}$ and find the corresponding $A_{\underline{m}}$ and $\underline{n}$.

I shall now explain how to apply this last rule using the relative frequency of the occurence of an event in a finite sequence of repetitions. This will permit me to clarify the effectiveness requirement of the rule. Let K be the hypothesis to be tested and assume that $\underline{B}$ is an event in A_K with $P_K(\underline{B}) = \underline{p}$. Let $\underline{y} \in {}^{\underline{k}}K$ with $\underline{n} \leq k < \omega$, and let $F_{r_{\underline{n}}}(\underline{y},\underline{B})$ be the relative frequency of the occurrence of $\underline{B}$ in the first $\underline{n}$ members of $\underline{y}$. Then, by the law of large numbers, for each $\underline{m}$, there is an $\underline{n}$ such that

$$P_{\underline{m}_K}(\{y : |F_{r_{\underline{n}}}(\underline{y},\underline{B}) - \underline{p}| \geq \underline{f}(\underline{m},\underline{n})\}) < \frac{1}{\underline{m}}$$

for some function $\underline{f}$.

Suppose that $K_{\underline{q}}$ is a probability structure such that $P_{K_{\underline{q}}}(\underline{B}) = \underline{q}$; let $K_{\underline{p}} = K$. Then

K should be rejected with respect to $\{K_{\underline{q}} : \underline{q} \in (0,1)\}$ if the following conditions are satisfied

(i) The function $\underline{f}$ is computable (i.e. recursive in $\underline{m}$, $\underline{n}$)

(ii) There is a recursive function $\underline{g}$ and a sequence $\underline{y} \in {}^{\underline{g(m)}}K$ such that y occurs and

$$|Fr_{\underline{g(m)}}(y,B) - p| \geq \underline{f(m,g(m))}$$

This rule is derived from our semi-practical rule as follows. Take $\underline{A_m} = \{y : |\underline{Fr_{g(m)}}(\underline{y},\underline{B}) - \underline{p}| \geq \underline{f(m,g(m)\}}$. Then $P_{\underline{g(m)}_K}(A_m) < \frac{1}{\underline{m}}$.

Let $\underline{X}$ be the sequence actually observed and let $\underline{q} = \underline{Fr_{g(m)}}(\underline{X},\underline{B})$. Suppose, first, that $Fr_{\underline{g(m)}}(\underline{X},\underline{B}) > \underline{p}$. Then, since $\underline{X} \in \underline{A_m}$, $\underline{Fr_{g(m)}}(\underline{X},\underline{B}) - \underline{p} \geq \underline{f(m,g(m))}$. But $P_{\underline{g(m)}_{K_{\underline{q}}}}(\{\underline{y} : \underline{Fr_{g(m)}}(\underline{y},\underline{B}) \geq q\}) = \frac{1}{2}$.

Also, $Fr_{\underline{g(m)}}(\underline{y},\underline{B}) \geq \underline{q}$ implies $Fr_{\underline{g(m)}}(\underline{y},\underline{B}) - p \geq \underline{f(m,g(m))}$. Hence $P_{\underline{g(m)}_{K_{\underline{q}}}}(\underline{A_m}) \geq \frac{1}{2}$. By a similar argument we obtain the same conclusion for the case when $Fr_{\underline{g(m)}}(\underline{X},\underline{B}) \leq \underline{p}$.

BAYESIAN METHODS

In order to apply Bayesian methods, we have to accept that reality behaves as if it obeyed a model of the following kind. We start with a probability model K_o that determines a probability distribution μ_o on an algebra of events A_{K_o}. Each $A \in K_o$ determines a probability model K_A with its corresponding probability distribution μ_A on the algebra of events A_{K_A}. The compound events are sets of pairs (A,L) where $A \in K_o$ and $L \in K_A$. The elementary event $\langle A,L \rangle$ represents the occurrence of A, followed by the occurrence of an event in K_A, in this cases, L. (For a slightly different and more complete formulation see Chuaqui 1983). The measure μ on these sets is given by

$$\mu(\underline{A}) = \int_{\mathcal{D} \circ A} \mu_A(\underline{A} * \{A\}) d\mu_o$$

where $D \circ \underline{A} = \{A : \exists L(\langle A,L \rangle \in \underline{A})$ and $\underline{A}*\{A\} = \{L : \langle A,L \rangle \in \underline{A}\}$.

Here the notion of dependence is implicit in the requirement that events consist of pairs $\langle A,L \rangle$ with $L \in K_A$, i.e. L belongs to the probability structure determined by A. Thus, the second component of the pair depends stochastically on the first.

I believe that in order to apply Bayesian techniques we have to accept as true to reality a probability model of such a form. We do not apply these methods for testing the model itself, but use the model for computing probability in accordance with it. The reasons for accepting the model lie outside of the Bayesian techniques used by applying it.

As an example, I will discuss here the model already analyzed by me in Chuaqui, 1980, pp. 110-111 for medical diagnosis (this model was studied in Lusted 1968). In this case , we suppose that a physician has to decide what treatment he should give to a certain patient. We assume that the physician knows (or accepts as true) that the patient has one of the diseases $E_o, \ldots, E_{n-1}$, that

K_o models the distribution P_o of these diseases in the population, and that each disease E_i determines a distribution P_{E_i} of the set of symptoms S among the patients with E_i, given by the model K_{E_i}. Then by Bayes' theorem, the probability that the patient has E_i, given that he presents S, $P(E_i \mid S)$, is obtained by the formula:

$$P(E_i \mid S) = \frac{P(E_i) \cdot P(S \mid E_i)}{\sum_{j<n} P(E_j) \cdot P(S \mid E_j)}$$

Here, $P(A|B) = \frac{P(A \cap B)}{P(B)}$, as usual, and P is the measure obtained from P_o and the P_{E_i}'s, as μ was obtained from μ_o and the μ_A's in the general model presented above.

$P(E_i)$, which is the same as $P_o(E_i)$, can be estimated from the frequency of E_i in the population that consults the physician; and $P(S \mid E_i)$ $(= P_{E_i}(S))$ from the frequency of the set of symptoms among the patients with E_i. In order to use, all this information (in particulat $P(E_i)$) we must use Bayesian techniques. The general model, however is not validated by these techniques, but by medical science, which accepts that the symptoms are determined by the diseases.

Notice that in this case we are essentially in front of a decision--theoretic problem. The physician is interested in curing the patient, and the only reason to inquire about which is the disease he suffers, is in order to treat him. Actually the problem should be framed in decision-theoretic terms by defining also a loss function.

It is doubtful whether the Bayesian model is applicable or not when we need to know the disease of the patient for reasons other than his treatment. For instance, let us suppose that the physician is testing a drug to see whether it is effective for the treatment of a disease or not. Then he really needs to <u>know</u> if the patient has the disease. I believe that in this case the frequency of the disease in the population is generally irrelevant, and, hence Bayesian methods cannot be applied. Another example makes this fact even more evident. Suppose that it is not a physician acting, but a judge(3). He has to decide on the guilt of a person, given some evidence of the crime (the evidence represents the symptoms , for this case). Is he justified in using the frequency of guilt among the accused? I believe that the answer is certainly not. Why? Because what we need to know is the truth, and not just decide on an action.

Bayesian methods tend to be used even when there is less evidence than in the medical case for the adequacy of the probability model. There is the danger that in the natural model for the situation involved, no probability measure exists. (I do not believe that it is always possible to find such a measure.) In the decision-theoretic framework, it may be necessary to take such risks, because an action is imperative. But it is also convenient to keep in mind this possibility.

NOTES

1. Research on this paper was partially supported by the Organization of American States through its Regional Scientific and Technological Development Program and DIUC of the Catholic University of Chile. I especially thank the comments on the paper by N. C. A. da Costa, L. Bertosi, and W. N. Reinhardt.
2. This example is due to W. N. Reinhardt.
3. This application arose in discussions with J. I. Malitz.

REFERENCES

1. R. Carnap, Logical Foundations of Probability, Chicago U. Press, 1950.
2. R. Carnap and R. Jeffreys, Studies in Inductive Logic I, U. of California Press, Berkeley and Los Angeles, 1971.
3. R. Chuaqui, Foundations of statistical methods using a semantical definition of probability, in Mathematical Logic in Latin America, Arruda, Chuaqui and da Costa (eds), North-Holland Pub. Co., Amsterdam, 1980, pp. 103-120.
4. R. Chuaqui, Factual and cognitive probability, to appear in the Proceedings of the V Latin American Logic Symposium, to be published by Marcel Dekker, New York, 1983.
5. R. Chuaqui, Models for probability, to apper in the Proceedings of the First Chilean Symposium on Analysis, Geometry, and Probability, to be published by Marcel Dekker, New York, 1984.
6. B. de Finetti, La prévision: ses lois logiques, ses sources subjectives, Annales de l'Institut Henri Poincaré, vol. 7, 1937, pp. 1-68.

7. H.E. Kyburg, The logical foundations of statistical inference, D. Reidel Pub. Co., Dordrecht and Boston, 1974.

8. L.B. Lusted, Introduction to medical decision making, Charles C. Thomas, Springfield, 1968.

Added in the proof to the item
CLASSICAL STATISTICAL INFERENCE:

Working with frequencies we can, under favorable circumstances, use the most natural set of alternative hypotheses $\{K_i : i \in I\}$. If it is assumed that each K_i represents a possible hypothesis for the explanation of an experiment that is repeated independently a large number of times, then we can take the set $\{K_i : i \in I\}$ as consisting of all probability structures that have a fixed algebra of propositions as their field of events. Then we can choose an event B for this field of events such that $P_K(B) = p$ and proceed as above.

THE SET THEORY OF A CLASS SYSTEM

Manuel Corrada*/Instituto de Matemática e Estatística, Universidade de São Paulo, São Paulo, Brazil

The set theory of a class system is, loosely speaking, the body of theorems concerning only sets which can be proved in a class system. Thus this theory is subjected to a proof theoretical description. To characterize which formulae refer to sets we introduce the notion of universe translation of a formula, i.e. the formula obtained relativizing its variables to those that satisfy a previously given formula with one free variable. If this formula with one free variable defines a universal class, i.e. a class to which all sets belong, then a formula refers only to sets if it is equivalent to the universe translation of some formula, the translation relativizing variables to the formula defining the universal class. In this paper we give a short survey of the principal syntactical and semantical properties of this theory. Some notions are presented with great generality. It is conceived in order they can be applied to a wider class of class systems.

To develop axiomatic systems of class systems the choice of a two sorted formalism L is useful. It is the two sorted predicate logic with identity. In this formalism there are two kinds of variables: u,v,w,x,z, ... called set variables and class variables X,Y,Z,... The

*Current Affiliation: Facultad de Matemáticas, Universidad Católica de Chile, Santiago, Chile

only non logical constant is the membership binary predicate whose metalogical designation is $\in$. The expressions $x = y$, $X = Y$, $x = Y$, $x \in Y$, $x \in y$ one called atomic formulae. From these, by means of the sentential connectives $\sim$ (negation symbol), $\vee$ (disjunction symbol) and the existential quantifier, in symbols $\exists$, we construct the set of formulae denoted by Φ.

It is desirable to distinguish some set theoretical symbols between mathematical systems in which they are introduced and the metasystem. In practice for set theoretical symbols mentioned in the discussion without mention explicitly if they are used properly in mathematical systems to which the discussion refers, or in the metasystem, we suppose the reader knows which one it should be. Also we assume that for every system discussed an infinite sequence of terms $\Delta_o, \Delta_1, \ldots, \Delta_n, \ldots$ containing no variables is available and a correspondence between formulae and natural numbers has been established and that it satisfies the usual conditions concerning recursiveness. We denote by Δ_φ the term Δ_m where m is the natural number that corresponds to φ under the above correspondence. If Σ is a finite set of formulae Δ_Σ means Δ_φ where φ is the conjunction of formulae in Σ.

Ordinals are represented by the letters "α", "β". As usual ω is the least infinite ordinal. The symbol $\lim_{>\beta}(\alpha)$ is read "α is a limit ordinal greater than β". R_β denotes the set of all sets whose rank is smaller than the ordinal β. The power set of a set X is denoted by Px.

Since we are particularly interested in the development of formal systems of set theory, relational structures in our discussion will be set-theoretical realizations of L, i.e. structures $\mathbf{A} = \langle A, A; E \rangle$ where A, A the set and class universes in $\mathbf{A}$, symbolically $\mathbf{A}$ and $\mathbf{A}$, are non empty sets, and E is the restriction of the relation $\in$ to members of $A \cup A$. As a kind of compromise if an arbitrary set theoretical realization of L is denoted by $\mathbf{A}$, the corresponding roman letters A, A will denote the class universe and set universe of $\mathbf{A}$. ST will denote the class of set-theoretical realizations of L.

Given $\mathbf{A} = \langle A, A; E \rangle$, $\mathbf{A} \in$ ST, and non-empty sets $B \subseteq A$, $B \subseteq A$, by the restriction of $\mathbf{A}$ to (B, B), in symbols $\mathbf{A} \upharpoonright (B, B)$, we understand the structure $\mathbf{B} = \langle B, B; F \rangle$ where $F = E \cap ((B \cup B) \times (B \cup B))$.

We assume it will be understood what is respectively meant by the statement that a formula φ is true in a relational structure $\mathbf{A}$, or $\mathbf{A}$ is a model of $\mathfrak{J}$ when $\mathfrak{J}$ is an arbitrary axiomatic system formalized in $\mathbf{B}$. These notions are defined semantically in terms of satisfaction and can be defined in the Zermelo-Fraenkel system of set theory. For every system $\mathfrak{J}$ we denote the class of all models of $\mathfrak{J}$ by $\mathrm{Md}\,\mathfrak{J}$. If Σ is a set of formulae $\mathrm{Md}\,\Sigma$ will denote the class of all models of Σ. It is convenient to introduce the proof-theoretically notion of derivability, symbolically $\vdash$, in terms of rules of inference in such a way that for any system $\mathfrak{J}$ with non logical axioms $A\xi[\mathfrak{J}]$ the classes $\mathrm{Md}\,\mathfrak{J}$ and $\mathrm{Md}\,A\xi[\mathfrak{J}]$ coincides. As is customary we suppose it has been done in this way.

Two relational structures $\mathbf{A}$ an $\mathbf{B}$ one called elementary equivalent, $\mathbf{A} \equiv \mathbf{B}$, if they satisfy the same sentences. The related relation of <u>weak elementary substructure</u>, $<$, is defined in terms of satisfaction as follows: $\mathbf{A} < \mathbf{B}$ iff for every $\varphi \in \Phi$ and $a_1 \dots a_n \in S\mathbf{A}$ if $\mathbf{A} \models \varphi(a_1 \dots a_n)$ then $\mathbf{B} \models \varphi(a_1 \dots a_n)$ (i.e. only set parameters are allowed). The relation defined in a relational structure $\mathbf{A}$ by a formula φ is denoted by $\varphi^{\mathbf{A}}$.

Let $\mathbf{A} \in ST$ and let φ,ψ be formulae with one free variable. We define a function $\Gamma_{\langle\varphi,\psi\rangle}$ from ST to ST, called <u>universe interpretation with universes $\langle\varphi,\psi\rangle$</u> by

$$\Gamma_{\langle\varphi,\psi\rangle}\,\mathbf{A} = \mathbf{A}\restriction(\varphi^{\mathbf{A}},\psi^{\mathbf{A}}).$$

The corresponding function mapping formulae to formulae is called <u>universe translation with universes $\langle\varphi,\psi\rangle$</u>, $F_{\langle\varphi,\psi\rangle}$ and is defined inductively as follows:

$$F_{\langle\varphi,\psi\rangle}(x = y) \approx x = y \wedge \varphi(x) \wedge \varphi(y)$$

$$F_{\langle\varphi,\psi\rangle}(X = Y) \approx X = Y \wedge \psi(X) \wedge \psi(Y)$$

$$F_{\langle\varphi,\psi\rangle}(x = Y) \approx x = Y \wedge \varphi(x) \wedge \psi(Y)$$

$$F_{\langle\varphi,\psi\rangle}(x \in y) \approx x \in y \wedge \varphi(x) \wedge \varphi(y)$$

$$F_{\langle\varphi,\psi\rangle}(x \in Y) \approx x \in Y \wedge \varphi(x) \wedge \psi(Y)$$

$$F_{\langle\varphi,\psi\rangle}(\neg\theta) \approx \neg F_{\varphi,\psi}(\theta)$$

$$F_{\langle\varphi,\psi\rangle}(\theta \vee \xi) \approx F_{\langle\varphi,\psi\rangle}(\theta) \vee F_{\langle\varphi,\psi\rangle}(\xi)$$

$F_{\langle\varphi,\psi\rangle}(\exists x\theta) \approx \exists x(F_{\langle\varphi,\psi\rangle}(\theta) \wedge \varphi(x))$

$F_{\langle\varphi,\psi\rangle}(\exists x\theta) \approx \exists X(F_{\langle\varphi,\psi\rangle}(\theta) \wedge \psi(x))$.

We put

$\Phi_{\mathfrak{J}}^{\langle\varphi,\psi\rangle} = \{\theta : \theta \in \Phi$ and there exists $\xi \in \Phi$, $\theta \equiv_{\mathfrak{J}} F_{\langle\varphi,\psi\rangle}\xi\}$

and

$ST^{\langle\varphi,\psi\rangle} = \{\mathfrak{B} : \mathfrak{B} \in ST$ and there exists $\mathfrak{C} \in Md\,\mathfrak{J}$, $\mathfrak{B} \equiv \Gamma_{\langle\varphi,\psi\rangle}\mathfrak{C}\}$

Given a system $\mathfrak{J}$ formalized in L and $F_{\langle\varphi,\psi\rangle}$ a universe translation with universes $\langle\varphi,\psi\rangle$, the <u>$\langle\varphi,\psi\rangle$ theory of $\mathfrak{J}$</u>, $\mathfrak{J}^{\langle\varphi,\psi\rangle}$, is $\{\theta : \theta \in \Phi$ and $\mathfrak{J} \vdash F_{\langle\varphi,\psi\rangle}\theta\}$. $\mathfrak{J}^{\langle\varphi,\psi\rangle}$ is simply the set of formulae whose $\langle\varphi,\psi\rangle$ universe translations are provable in $\mathfrak{J}$.

Some consequences of this definition are:

If $\mathfrak{J}$ is recursive then $\mathfrak{J}^{\langle\varphi,\psi\rangle}$ is recursive

If $\mathfrak{J}$ is recursively enumerable, then $\mathfrak{J}^{\langle\varphi,\psi\rangle}$ is recursively enumerable. By Craig 53 we have

If $\mathfrak{J}$ is decidable then $\mathfrak{J}^{\langle\varphi,\psi\rangle}$ is decidable

If $\mathfrak{J}$ is axiomatizable then $\mathfrak{J}^{\langle\varphi,\psi\rangle}$ is axiomatizable.

The following theorem obtained by Ratajczyk 80 using an improvement of the axiomatization in one scheme given in Vaught 67., provides a set of axioms for $\mathfrak{J}^{\langle\varphi,\psi\rangle}$.

Assume that (i) $\mathfrak{J}$ is a recursively enumerable system formalized in L, (ii) $A\xi[\mathfrak{J}]$ is a recursive set of axioms of $\mathfrak{J}$, (iii) $\Gamma_{\langle\varphi,\psi\rangle}$ is a universe interpretation with universes $\langle\varphi,\psi\rangle$, (iv) j is an interpretation of Peano's arithmetic, P, in $\mathfrak{J}$, and (v) for every $\theta \in \Phi^{\langle\varphi,\psi\rangle}$, $\mathfrak{J} \vdash \theta \rightarrow \mathrm{Con}\, j(\Delta_\theta + \Delta_\Sigma)$ for each finite Σ, $\Sigma \subset A\xi[\mathfrak{J}]$. Under these assumptions the set of all instances of all universal closures of the formulae

$j(\xi)$ for each axiom ξ of P

$\theta \rightarrow \mathrm{Con}_j(\Delta_\theta + \Delta_\Sigma)$ for every $\theta \in \Phi^{\langle\varphi,\psi\rangle}$ for each finite Σ, $\Sigma \subseteq A\xi[\mathfrak{J}]$

form a set of axioms for $\mathfrak{J}^{\langle\varphi,\psi\rangle}$.

Concerning models of $\mathfrak{J}^{\langle\varphi,\psi\rangle}$ the following result is relevant for our purposes:

$Md\,\mathfrak{J}^{\langle\varphi,\psi\rangle} = ST_{\mathfrak{J}}^{\langle\varphi,\psi\rangle}$

Hence we can consider models of $\mathfrak{J}^{\langle \varphi,\psi \rangle}$ as universe restrictions to the universes defined by φ and ψ in models of $\mathfrak{J}$.

We must deal with class systems and with certain classes of models for set systems such that when appropriate sets are added to these universes they yield models for the class systems considered.

When speaking of set theoretical systems we have in mind the system of Zermelo-Fraenkel, ZF. A characteristic feature of the formalization in a two-sorted vocabulary is the inclusion in it axiom set of the sentence $\forall X_\xi \exists v_\mu (v_\mu = X_\xi)$, i.e. the proper classes are excluded of the system.

As a class system we consider a system of the impredicative theory of classes, IC. The most characteristic structural property of this system is the presence of the comprehension schema, or the scheme of class specification:

$$\exists X \forall y (y \in X \leftrightarrow \theta) \quad \text{with} \quad \theta \in \Phi \quad \text{and} \quad X \notin F_v\ \theta.$$

Replacing the previous scheme by weaker versions, class systems known in the literature are obtained. Of course, there are numerous results concerning these systems similar to those in the present discussion. However, to simplify the exposition we disregard then.

From the point of view of set theoretical realizations for ZF and IC, because their structural properties models of ZF are set theoretical realizations $\mathbf{A}$ where $S\mathbf{A} = C\mathbf{A}$, while in models of IC their class universes must be subsets of the power set of the set universe union the set universe. Regarding those structural properties we can discuss those models of ZF on which adding certain subsets of its universes to its class universes union its set universes lead to models of the impredicative theory of classes. In view of this we put

$$E(ZF,IC) = \{\mathbf{A} : \mathbf{A} \in \mathrm{Md}ZF \text{ and there exists } A \subseteq PA, \langle A \cup A, A \rangle \in \mathrm{Md}\ IC\}$$

The following question arises in a natural way. Let $\mathbf{A}$ be a set theoretical realization belonging to E(ZF,IC), what is the power of the set $\{A : A \subseteq PA \text{ and } \langle A \cup A, A\rangle \in \mathrm{Md}\ IC\}$? In general, as far the author knows, the answer to this question is an outstanding problem. Ratajczyk 80a has shown that in case $\mathbf{A}$ is denumerable and transitive it has the power of the continumm.

The inequality $E(ZF,IC) \subseteq \mathrm{Md}\ ZF$ cannot be replaced by an equality. For instance, the least transitive model of Zermelo-Fraenkel set system plus local choice axiom does not belong to E(ZF,IC). Moreover there is no set of formulae Σ for which the following holds:

$$Md(ZF + \Sigma) = E(ZF,IC).$$

This is a direct consequence of the fact due to Krajèwski as reported in Marek & Mostowski 75 that E(ZF,IC) is not an elementary class. It is a pseudo-elementary class. Bielinski 77 has given a nonelementary characterization of E(ZF,IC).

In IC the universal class, V, is introduced in such a way that the following is provable:

$$v_\xi \in V \leftrightarrow \exists v_\mu (v_\xi \in v_\mu).$$

In the subsequent discussion we will write $\mathcal{J}^{\langle V \rangle}$ instead of $\mathcal{J}^{\langle \varphi, \varphi \rangle}$ where φ is the sentence $\exists v\, (v_\xi \in v_\mu)$. Obviously $v_\xi \in V$ expresses the fact that the object represented by X is a set.

s$\mathcal{J}$, <u>the set theory of the class system</u> $\mathcal{J}$ is $\mathcal{J}^{\langle V \rangle}$, it is simply the set of all statements concerning only sets provable in system $\mathcal{J}$.

In case of the impredicative theory of classes we can state the following identity

$$sIC = \cap \{Th\ \mathbf{A} : \mathbf{A} \in E(ZF,IC)\}.$$

Hence the set theory of the impredicative theory of classes is the set of sentences true in all models of ZF that can be extended adding subsets of its set universes union its set universes to models of the impredicative class system. This set seems independent of the Zermelo-Fraenkel system of set theory. However we mention here that for any subsystem Z of ZF formalized in L whose realizations are also set theoretical realizations the following identity holds:

$$\cap \{Th\ \mathbf{A} : \mathbf{A} \in E(ZF,IC)\} = \cap \{Th\ \mathbf{A} : \mathbf{A} \in E(ZF,IC)\}.$$

where E(ZF,IC) has the obvious meaning. This fact is closely related to the proof theoretical definition of sIC.

On the other hand, IC cannot be replaced by one of its subsystems. We can easily exhibit formulae belonging to sIC but not to $\cap \{Th\ \mathbf{A} : \mathbf{A} \in E(ZF,T)\}$ for certain subsystems T of IC.

We state here some properties of the set theory of the impredicative theory of classes

sIC is an undecidable system

sIC and IC are equiconsistent systems
sIC is axiomatizable
sIC is not finitely axiomatizable
sIC is not finitely axiomatizable over ZF
IC is not finitely axiomatizable over sIC.

With regard to the strenght of system sIC we have the following results:

IC — sIC $\neq 0$

sIC — ZF $\neq 0$

In connection with the last result the following improvement of it is due to H. Friedman.

For this purpose we need to introduce some topological notions. If X and Y are topological spaces and $F : X \to Y$, F is called Borel if for every open set $U \subseteq Y$, $F^{-1}(U)$ is a Borel set (the class of Borel sets in X is the smallest class containing all open sets, closed under complementations and countable intersections, i.e. the least σ-algebra of subsets of X generated by the open subsets of X). $[0,1]^{\omega}$, denoted by Q, is called the Hilbert cube. It is $\{f : f : \omega \to [0,1]\}$, so we can think Q as consisting of all infinite sequences $\vec{a} = \langle a_o..a_n... \rangle$ where $a_i \in [0,1]$. $[0,1]$ is provided with the usual topology and impose the infinite product topology on $[0,1]^{\omega}$. Thus a typical basic open set in Q has the form $U_o \times ... \times U_{n-1} \times [0,1] \times ... \times [0,1] = \{f \in Q : \forall i < n(f(i) \subset U_i)\}$ where $U_o .. U_{n-1}$ are open in $[0,1]$.

Let H be the group of all permutations σ of ω which are the identity map except at finitely many places. If $\sigma_1,\sigma_2,\sigma_3,\sigma_4 \in H$ then $F : Q^4 \to Q$ is called σ_1-invariant if $F(y_1 \circ \sigma_1..y_4 \circ \sigma_4) = F(y_1,..,y_4) \circ \sigma_1$. If $F : Q^5 \to Q$ and $\vec{x} \in Q$ define $F_{\vec{X}} : Q^4 \to Q$ by $F_{\vec{X}}(y_1,..y_4) = F(X,y_1,..,y_4)$.

Let θ be the following sentence

Whenever $F : Q^5 \to Q$ is a Borel function such that for every $\vec{X} \in Q$, $F_{\vec{X}}$ is σ_1-invariant then there is a sequence $\vec{X}_o,\vec{X}_1, ... \vec{X}_{m-1}$ $(m < \omega)$ such that the following conditions hold:

(a) For any $i_o,i_1 \in \omega$ there is $q(i_o,i_1) \in Q$ such that

$$F(\vec{X}_{i_o},..\vec{X}_{i_4}) = q(i_o,i_1) \text{ whenever } i_o < ... < i_4,\ i_o,...,i_4 \in \omega$$

(b) $q(i_o,i_1) \in \{X_{i_{1_n}} : n \in \omega\}$

Friedman 1981 proved that

$\theta \in sIC$ — ZF

and moreover

$\theta \in sIC$ — (ZF + V = L)

So, this constitutes the first sentence concerning sets of mathematical character that is independent of ZF and, more surprising, independent of ZF constructibility and that is provable in IC.

As an immediate consequence of the definition of sIC we have

$$\text{Md } sIC = ST^{V}_{\langle IC \rangle}$$

and recalling the definition of $ST^{V}_{\langle IC \rangle}$

$$\text{Md } sIC = \{\mathfrak{B} : \mathfrak{B} \in ST \text{ and there exists } \mathfrak{C} \in \text{Md} IC\ ,\ \mathfrak{B} \equiv \langle C,C \rangle\}.$$

Thus a formula is true in a model of the set theory of the impredicative theory of classes if and only if it is true in the structure formed by the set universe of some model of IC.

Althought sIC is axiomatizable and previously we have presented a set of axioms depending on certain hyotheses which are satisfied by IC, this set of axioms hardly seems a natural choice of basis for a certain set-theoretical system. A much more convenient and simpler axiomatization obtained in joint work by W. Reinhardt and the author, is the set of all universal closures of the following formulae:
the axioms of ZF

$$\exists\alpha\ \exists\beta(\lim_{>\omega}(\alpha) \wedge \lim_{>\alpha}(\beta) \wedge \langle R_\alpha, L_{\delta^n_\alpha}(R_\alpha)\rangle \prec \langle R_\beta, L_{\delta^n_\beta}(R_\beta)\rangle \wedge \theta^{R_\alpha} \leftrightarrow \theta)$$

for each $\theta \in \Phi^{\langle V \rangle}_{IC}$ and each $n \in \omega$

where $L_K(R_\beta)$ is relative constructibility over R_β and δ^n_K is the first β such that $(R_K, L_\beta(R_K)) \models \Delta'_n$ - comprehension (by relative constructibility over R_β we mean the subsets of R_β constructed until a certain level, so that $L_K(R_\beta) \subseteq R_{\beta+1}$) .

We would of course fully dispose all the discussion involved if we decided to present the set theory of the impredicative theory of classes by giving it a set of axioms. Such a procedure puts the theory sIC on the same status of all set theories known from the literature used in mathematical practice.

NOTE

This article was partially written when the author was at the Universidade de São Paulo, São Paulo, Brazil, thanks to the kind invitation of Newton da Costa. I am grateful to Newton da Costa for several valuable conversations.

ACKNOWLEDGMENTS

1. K. Bielinski, Extendability of structures as infinitary property, in Set Theory and Hierarchy Theory V. Lecture Notes in Mathematics, vol. 619, Springer Verlag, Berlin, 1977, pp. 75-93.

2. W. Craig, On axiomatizability within a system, J. Symbolic Logic 18 (1953), pp. 30-32.

3. M. Friedman, On the necessary use of abstract set theory, Advances in Mathematics 41 (1981), pp. 209-280.

4. W. Marek & A. Mostowski, On extendability of models of ZF set theory to models of KM theory of classes, in Logic Colloquium 1974, Kiel. Lecture Notes in Mathematics 499, Springer Verlag, 1975.

5. Z. Ratajcyk, On sentences provable in impredicative extensions of theories, Dissertations Mathematical, vol. CLXXVIII, Warsaw 1980.

6. Z. Ratajcyk, On the number of expansions of the models of ZFC - set theory to models of KM - theory of classes, Logic Colloquium'78, Boffer, Van Dalen, Mc Aloon eds. North-Holland, Amsterdam, 1979, (1980a), pp. 317-333.

7. R.L. Vaught, Axiomatizability by a schema, J. Symbolic Logic 32 (1967), pp. 473-479.

CATEGORY THEORY AND FOUNDATIONS OF MATHEMATICS

Paul Dedecker*/Escuela de Física y Matemática, Facultad de Ciencias and Departamento de Matemática y Física, Universidad Central de Venezuela and Instituto Universitario Pedagogico, Caracas, Venezuela

SUMMARY

The problem of the elaboration of a satisfactory logical language in which to develop freely category theory seems to remain open. A proposal to that effect which I made at a "Centre Belge de Reherches Mathématiques" conference in 1958 at Louvain was the origin of researches of da Costa and others on what they called the "Ehresmann-Dedecker" universes. My purpose was a modification, of a rather revolutionary or exotic character, of the von Neumann-Bernays-Gödel logic (NBG) which is revived hereafter, without, however, reaching a final solution which will require additional efforts. A similar program, in pure set theory and independent of ours and of category theory, was proposed in 1971 by Alonso Church [2].

Our proposed system allows for a development of classical mathematics, including "small" categories, in its lower part which is essentially ZF (the Zermelo-Fraenkel logic), while "large" categories (like that of <u>all</u> groups, ...) require the upper part. These two stages, combined together, constitute a modification of NBG build so as to allow for certain (restricted) classes of classes. In other words, we propose a two-type logic (sets and classes) avoiding the unpopular and uneasy higher types,

---

*Current Affiliation: Departamento de Matemática, Universidad de Puerto Rico, Mayagüez, Puerto Rico

thanks to exotic and anti-conformist rules governing the calculus of classes. This logic legitimates as well (or aims at legitimizing) "very large" categories, including a good standing universal category CAT of all categories (small, large and very large) but no "Russell subcategory", (namely a category of categories, the object of which would be exactly those categories which are not an object of themselves). However, CAT is an object of itself. The technical part of this program remains to be completed but we produce a model of a similar set theory, derived from ZF (and thus equiconsistent), that includes a universal class U which is a member of itself. That universal class (or the universal category) thus appears like a picture reproducing itself in one of its parts, this part reproducing the entire picture and thus itself again, etc ... and similarly the other way around in the opposite direction ... (in the manner of what one sees in two parallel facing mirrors).

INTRODUCTION

The study of Foundation of Mathematics got one of its initial impulses from the work of G. Cantor (1845 - 1918) on set theory and transfinite arithmetic during the last quarter of the 19th century. The systematic use of the concept of set led both to remarkable results and to doubts about the correctness of the method. Objections were raised questioning the legitimacy of Cantor's methods and abstract constructions involving "very large" sets. The discussion culminated with the apparition of logical antinomies essentially linked to the "set of all sets" and its cardinal, contradicting Cantor's theorem to the effect that there is no maximum cardinal. This is essentially the same difficulty as Russell's "impossible set" R consisting exactly of those sets s which to do not belong to themselves: In symbols, this "impossible set" R is defined by

$s \in R$ means $s \notin s$

so that the alternative

$R \in R$ means $R \notin R$ (a contradiction)

and the only other alternative.

$R \notin R$ means $R \in R$ (also a contradiction).

(Indeed, if the "set of all sets" U exists, it must coincide with its power set $\mathcal{P}U$ so that identity $U = \mathcal{P}U$ is a bijection. Then Cantor's diagonal method, to the effect of showing that such a bijection does not exist, precisely yields the "impossible Russell set" R. The argument of Alonzo Church [2] (page 298, lines 30-33) coincides with ours in [6] to the effect that it is not the "set of all sets" but the "Russell set" which is antinomic. The solution consists in saying, as we did in [6], that the bijection exists and that R does not.)

Various systems of axiomatic set theory emerged during the first half of the 20-th century so as to avoid the contradictions and produce an adequate foundation of mathematics. Among them the most popular are the systems ZF of Zermelo-Fraenkel and NBG of von Neumann-Bernays-Gödel which are the natural frames for the Gödel's and Cohen's theorems on incompleteness and the independence of the (generalized) continuum hypothesis or of its negation and about the axiom of choice.

Until recently, it became more and more generally admitted that these axiomatic approaches and abstract constructions both belonged to mathematical reality. They made it possible to develop "to the perfection" the entire body of mathematics. In particular, transfinite induction became a classical tool in algebra and topology, for example in proving fundamental elementary theorems (e.g. the one saying that every subgroup of a free group is free). This, at least, was quite convenient as long as one wanted to limit one's considerations to individual mathematical objects or to limited collections of such objects.

However, a new mathematical device and method of mathematical investigation appeared in the period 1942-45 with the introduction of <u>categories</u>, <u>functors</u> and <u>natural transformations</u> by S. Eilenberg and S. MacLane. The essential idea of this step was the analysis of the concepts of "natural equivalence", "natural transformation" and "canonical association" between mathematical objects. Indeed, such analysis had become urgent due to the development of algebra and algebraic topology. Thinking of the relation between a vector space V and its dual V^* or its bidual V^{**}, further of the "canonical" mapping $V \longrightarrow V^{**}$ of V into its bidual, it was no longer possible to limit oneself to consider only <u>limited collections</u> of mathematical objects of a certain kind or "category". It then became necessary to consider seriously the <u>totality</u> not only of these objects but also the totality of the "morphisms" between them.

For example, we have to consider the "category" TOP consisting of <u>all</u> the topological spaces and <u>all</u> continuous mappings (morphisms) between them; or the category GRP (resp. AB) of <u>all</u> groups (resp. commutative groups) and of <u>all</u> homomorphisms between them; or the category SET of <u>all</u> sets and <u>all</u> mappings between them. We want also to consider the category TOP_* of <u>all</u> pointed topological spaces (i.e. pairs (X,x), (Y,y), ... consisting of topological spaces $X,Y,\ldots$ and points x in X, y in Y, ...) and <u>all</u> morphisms from one (X,x) to another (Y,y), namely continuous functions $X \longrightarrow Y$ sending x to y. In between categories are defined <u>functors</u>. For example, we have the first <u>homotopy</u> group functor π_1 from TOP_* to GRP which associates to every pointed topological space (X,x) its first homotopy group $\pi_1(X,x)$ and to every morphism $f : (X,x) \longrightarrow (Y,y)$ the corresponding homomorphism

$$\pi_1(f) : \pi_1(X,x) \longrightarrow \pi_1(Y,y). \tag{1}$$

We also have the first <u>homology</u> functor H_1 (with integral coefficients) from TOP or TOP_* to AB which associates to (X,x) the abelian group $H_1(X) = H_1(X,x)$ and to f the homomorphism

$$H_1(f) : H_1(X,x) \longrightarrow H_1(Y,y). \tag{2}$$

Then, to every (X,x), (Y,y), ... is associated a homomorphism

$$\alpha(X,x) : \pi_1(X,x) \longrightarrow H_1(x,x),\ \alpha(Y,y) : \pi_1(Y,y) \longrightarrow H_1(Y,y),\ \ldots \tag{3}$$

so that the following diagrams are always commutative:

$$\begin{array}{ccc}
\pi_1(X,x) & \xrightarrow{\quad \pi_1(f) \quad} & \pi_1(Y,y) \\
{\scriptstyle \alpha(X,x)} \downarrow & & \downarrow {\scriptstyle \alpha(Y,y)} \\
H_1(X,x) & \xrightarrow[\quad H_1(f) \quad]{} & H_1(Y,y) .
\end{array}$$

The latter property means that the function α associating to every (X,x), (Y,y), ... the homomorphisms (3) is a <u>natural transformation</u> from the first homotopy functor to the first homology functor.

ROLE OF THE NBG SYSTEM

Because of the necessity of this "global" approach considering "simultaneously" all objects and all morphisms, it was stated from the beginning by Eilenberg and MacLane that the most appropriate logical system for their considerations was NBG which, contrary to ZF, admits sets and classes. In this system, every set is a class but not conversely and every class which belongs to something (set or class) is by necessity a set; a class which is not a set is called a proper class. The intuitive difference between set and class is that the latter are collections much bigger than sets and that the comprehension principle (stating that the objects x satisfying a property P(x) expressible in the logical language form a set) cannot apply to the bigger collections in order to avoid antinomies. It is only admitted that the sets x such that P(x) is true form a class C_P which is only exceptionally a set. This logical system thus allows for the class of all topological spaces, the class of all groups, the class of all abelian groups, the class of all sets, the class of all pointed topological spaces, etc..., as well as for the classes of all morphisms between them. But none of these classes is a set, just for the reason that every set can be endowed with (in general) several structures of topological spaces or of group. Therefore the class of all topological spaces, for example, must be, if anything, larger than the class of all sets. It is however permitted in NBG logic to speak of the "large" categories TOP, TOP_*, GRP, AB, ... They are proper classes. This implies that they constitute mathematical entities of a new type situated outside of classical mathematics. Similarly, it seems reasonable in NBG to speak of a functor from one "large" category A to another B, or of two such functors plus a natural transformation between them. However, either a functor or a natural transformation of this type constitutes a proper class, which makes it illegitimate in NBG to speak of the category B^A or HOM(A,B) of functors from category A to category B: indeed (a) the objects of this category-to-be are the functors F, G, ... from A to B (and these functors are in general proper classes) and (b) its morphisms are the natural transformations between such functors (and therefore are themselves proper classes). A "category" HOM(A,B) is therefore in general illegitimate in NBG because it would be a class containing proper classes. Moreover, as a natural transformation between two such functors F,G already is a class, it is a fortiori illegitimate in NBG to speak of the class of all these classes.

Similarly, there is not in NBG a legitimate category CAT of all categories of which the objects are all categories and the morphisms are all the functors between them.

For any of the above categories, TOP, TOP_*, GRP, AB,..., all morphisms from one object X to another Y form a set HOM(X,Y). For this reason these categories are said to be <u>locally small</u>.

There are of course categories the classes of objects and morphisms of which constitute sets: these are just ordinary mathematical objects (i.e. sets endowed with a structure of a certain kind) and they are called <u>small</u> categories. If we consider two small categories A, B then $Hom(A,B) = A^B$ always exists and is again a small category (objects: functors from A to B; morphisms: natural transformations between them). We can also consider the locally small category <u>Cat</u> of which the objects are the small categories and the morphisms, the functors between them. A category which is not small is said to be <u>large</u>.

HIGHER DIMENSIONAL STRUCTURES

Now, if A, B, C are three categories (large or small) with functors $F, F' : A \longrightarrow B$; $G,G' : B \longrightarrow C$ and natural transformations $\alpha : F \Longrightarrow F'$, $\beta : G \Longrightarrow G'$:

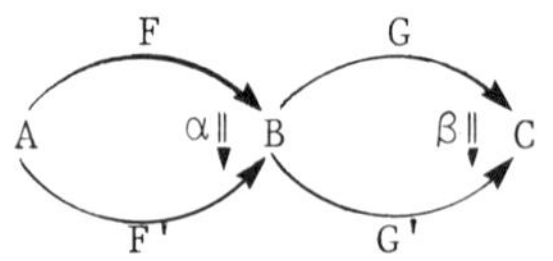

one defines a composite natural transformation

$$\gamma = \beta * \alpha : G \circ F \longrightarrow G' \circ F'$$

which associates to every object X of A the diagonal morphism $\gamma(X)$ in the commutative diagram

$$\begin{array}{ccc} G \circ F(X) & \xrightarrow{\beta(F(X))} & G' \circ F(X) \\ {\scriptstyle G(\alpha(X))}\downarrow & & \downarrow{\scriptstyle G'(\alpha(X))} \\ G \circ F'(X) & \xrightarrow[\beta(F'(X))]{} & G' \circ F'(X) \end{array} .$$

This defines the *two-dimensional* structure or *2-category* structure of category theory. In case of two other functors $F'' : A \longrightarrow B$, $G'' : B \longrightarrow C$ and natural transformations $\alpha' : F' \Longrightarrow F''$, $\beta' : G' \Longrightarrow G''$, the two-dimensional situation

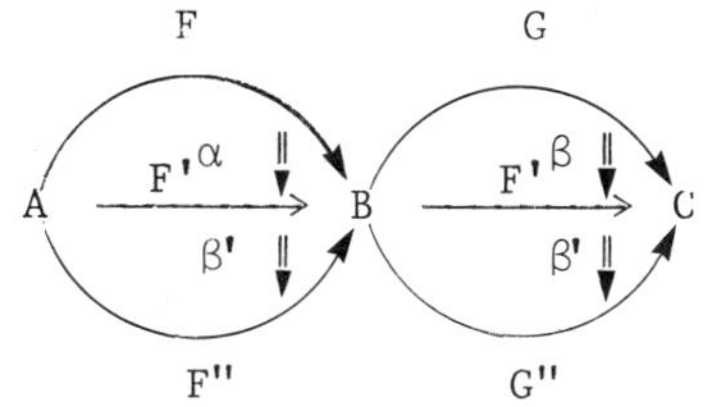

yields the formula

$$(\beta' * \alpha') \cdot (\beta * \alpha) = (\beta' \cdot \beta) * (\alpha' \cdot \alpha) \qquad (4)$$

If we consider a *set* S of small categories A, B, C, ..., then their disjoint union is a mathematical object endowed with a "two-dimensional" structure involving: (a) three stages, namely objects or *0-arrows* , functors or *1-arrows* and natural transformations or *2-arrows*; (b) three non everywhere defined compositions (o on 1-arrows; · and * on 2-arrows) verifying some elementary axioms, including formula (4). Such an object is called a *small 2-category* and belongs to classical mathematics. One then easily defines the "large category of all small 2-categories" to be denoted 2-*Cat*. This is a "large" 2-category, endowed with a possibly "three-dimensional" structure.

One also sees that the large category *Cat* of all small categories can be endowed with an extra structure making of it the "large 2-category of small categories" which we shall denote *Cat* .

Now doubt, both 2-*Cat* and *Cat* are legitimate objects in NBG. But they are classes so that 2-*Cat* (or *Cat*) cannot be an object in itself.

NECESSITY AND UTILITY OF MODIFYING NBG

We would like to modify NBG so as to be able to consider functor categories $HOM(A,B) = B^A$ when A and B are large categories and a large (or very large) category CAT of all large categories and a corresponding large or very large 2-category *CAT*. However, the higher dimensional structure does not seem to involve new logical problems so that we will not take care of it in the rest of this article.

Should we not make the distinction between classes and sets and should we admit the existence of the categories TOP, TOP_* , GRP, AB, ... and their set nature; should we moreover admit a similar existence of a "universal" category CAT; then we would automatically reproduce Russell's paradox by considering some subcategory, say RUSSELL, of CAT retaining as objects exactly those categories which are not an object of themselves. The antinomy then reappears from the question: is then RUSSELL an object of RUSSELL?

An obviously unacceptable, insane situation!

However, as just pointed out: if the distinction is made between classes and sets, and if we want to remain in the respectable NBG system, then we can consider categories like TOP, TOP_*, GRP, AB, SET, etc. ... (as non-classical mathematical objects), but there is no legitimate room for functor categories HOM(A,B) nor for a "universal" categoty CAT. One way out of this difficulty would be to admit the existence of superclasses but we would then be led to the unpopular and uneasy theory of types.

As is well known, a model of NBG in ZF is obtained if one assumes the existence (in ZF) of a Grothendieck-Sonner universe U. To that effect one calls "set" an element of U and "class" a subset of U (possibly U itself). In this interpretation, the "small categories" are elements of U, the "large categories" correspond to subsets of U, while the functor categories HOM(A,B) and the category $\underline{CAT}$ of large and small categories are elements of a larger universe U' of which U is an element. (This $\underline{CAT}$ is however not an object of itself). This way, the "superclasses" reappear through an increasing sequence of universes or "types" and $\underline{CAT}$ is an object of a larger category $\underline{C\ A\ T}$. This is essentially the approach used by the classical categorists as developed, for example by Saunders MacLane [15], from a pragmatical point of view. Indeed, the fundamental theorems of category theory (as Yoneda lemma, adjoint functor and Kan extensions theorems) apply to situations involving only small and large categories and are dominated by smallness assumptions. In this pragmatical theory, the word "universal" in expressions like "universal element" , "universal problem" refers to "representable functors" not at all to a "universal set" or "universal class". If this theory admits to speak of TOP, of GRP and similar categories, possibly of $\underline{CAT}$, everything is done to avoid mentioning non-conformist concepts and, above all, dangerous or subversive concepts like CAT or RUSSELL, considered as tabooed improper utterances.

It is therefore clear that, with category theory, reappears the problem of foundations of mathematics with the peculiarity that neither ZF nor NBG are satisfactory for a full development. For a long time professional logicians considered with disdain this fundamental foundational problem and still recently published books or recent logical colloquia do not mention it. Is, for example, the problem explicitly mentioned at some place in the 500 pages of the Tarski Symposium? The considerations of S. Feferman and G. Kreisel [8] do not seem to solve the problem and it was only recently, in 1968, that a book on the foundations of mathematics included a thorough and careful discussion of the foundational problem of category theory [10]. This book, which has just been reedited and completed [12], 1982. It tries (among other things) to define a logical frame in which category theory can be developed: it concludes that the development of foundations lies ahead of us, not behind. One needs also mention the recent monograph [9], 1979, the purpose of which is rather the description of logic through categories. Lawvere's [14] gives an axiomatic description of the "category C of 'all' categories" but it is not clear what he means by "all", and eventually, SET does not belong to C as an object, but as a subcategory of C. Goldblatt, Lawvere and other "toposophers" want to develop mathematics from the corner-stone consisting of the concept of category (object plus arrows), more particularly from the concept of topos, rather than form classical logic (sets and $\in$-relation). This is no doubt an essential and new idea which, when reaching full success and conclusion, should also clarify the set theoretical foundation of category theory instead of relegating it to antiquary. In the same way the invention of category theory helped to a better understanding of many classical mathematical subjects, including group or module theory, algebraic topology, homological algebra, etc... , and did not relegate them into a museum for the respectable past.

I want to advocate for the feeling or hope, shared by many mathematicians, that a suitable modification of the ZF or NBG systems should provide a satisfactory logical basis for a theory of categories in which it would be legitimate to speak not only of the large categories SET, TOP, TOP_*, GRP, AB, etc. ... but also of the very large one like the categories of functors and even a universal category CAT. If classical mathematics fits into the system ZF of Zermelo-Fraenkel, however, the consideration of those large categories, with the necessary set theoretic restriction or conditions in fundamental theorems of category theory (e.g. completeness or co-completeness), require the set and class distinction of the (more of less "equivalent") system NBG of von Neumann-Bernays-Gödel. It seems moreover that only a more exotic logical system, still to be fully and technically elaborated, can thoroughly legitimate the "very large categories" like CAT. For several distinguished categorists or "toposophers", this "really does not matter" (e.g. P.T. Johnstone [13], p. xix, line 22). (There might, thus, be a disdain of mathematicians versus a disdain of logicians.) The necessity and usefulness of a new axiomatic system was expressed in the following terms by Ch. Ehresmann [7] in 1957, in a kind of profession of faith, similar to the ones of mine in [6] and in the present article:

"We distinguish between sets and classes; a set is also a class. The class of all sets is not a set. We admit the same operation for classes as we do for sets. We do not refuse either the construction of certain classes of sub-classes of a class. Should contradictions be produced by these conceptual operations, it would always be possible to introduce restrictions so as to remain within the limits of the theory of sets; however, this would necessitate a more cautious and complicated theory"[1].

In his quoted article [2], Alonzo Church goes along advocating for a "program" (p. 303) of "exploration of the axiomatic possibilities" (p. 299, lines 11, 12), "based on the usefulness of the theory for mathematical purpose or for the purpose of some special branch of mathematics" (p. 307, last three lines). (Certainly a mention of category theory would have been both nice and in order at that point.)

Essential in Ehrenmann's statement is the acceptance of <u>certain</u> "classes of subclasses of a class", which is clearly in contradiction with NBG's axiom to the effect of which only a set can be a member of a class. But this possibility refers to only <u>certain</u> such classes, so as to avoid antinomies. In order words, this statement advocates for a logical language as liberal as possible with however some non-explicited restricting rules so as to avoid contradictions. At the same time, it is a profession of faith to the effect of which, should a contradiction appear, it would always be possible to find a recipe so as to avoid it.

However, no general restricting rule is explicitly given. (In the language of Church, no exploration has been done by Ehresmann and mine in [6] has not gone far enough. One should also remind at this point of John Isbell's warning against possible inconsistency as well as of necessity "for deep and various further work", in his analysis of Lawvere's [14] in "Mathematical Reviews", vol. 32, # 7332, 1967).

I want to plead for the same optimistic and liberal attitude. All systems explicited so far avoid the contradictions of "antinomic sets" through rather strong restrictions. NBG's axiom to the effect of which a proper class cannot be member of a class is one example and this restriction induces more inconvenience in category theory than in "classical" set theory. Quine's system which admits a universal class, avoid contradiction thanks to the notion of "stratified" formula and the very strong axiom to the effect of which the comprehension principle applies only to stratified predicates. This has, for example, the consequence of forbidding the consideration of pairs $(x,\{x\})$; therefore, in this system (excellently described in [10]), the map $x \longrightarrow \{x\}$ from a set A to its power set P(A) (or to its unitary power set PU(A)) does not exist or is illegitimate. Such a consequence seems desperately too far from our everyday intuition so that Quine's system cannot easily become popular.

AN EARLY EXPLORATION (1958)

On the other hand, I believe that mathematics and logic are natural sciences (together with the lesser abstract ones like biology, chemistry, etc. ...) and aim at a better understanding of the world around us. The principles of logic are therefore projections or interpretations in our mind of natural phenomena (including those related to our way of thinking). A logical or mathematical contradiction is not more dangerous than a deviation between a theoretical forecast and a physical observation; such contradiction does not make life to stop : it can only be the result of an incorrect interpretation in our mind of the real physical world around us.

In 1958, I suggested in [6] a modification of the NBG logic so as to admit certain classes of proper classes . I considered, as in NBG, a two-stage logic, which be a theory T in the sense of Bourbaki, with classes and sets, every set being also a class, but not conversely.

In this theory or language, one can form predicates $P(x)$ relative to a variable x and the fundamental question is whether or not such predicate P is collectivizing: in other words, does there exist a class C_P such that

"$P(x)$ true" is equivalent to "$x \in C_P$",

x being a set or a proper class. Such a system could be formed by adjoining to the symbols of NBG new ones representing new classes, each of which containing at least an element which is a proper class. It would also be reasonable to consider (e.g. admit as definition) that one of these new classes which is equipotent to a set, or is the image of a set by a map, be also a set (therefore, with respect to the initial NBG theory, an additional set of which at least one element is a proper class).

Among the predicates which we would like to be collectivizing are the following. To a class X associate the predicate P_X

1°) $P_X(x) \equiv$ "$x \subset X$"

(x is a subclass of X) defining, for every class X, its power class $\mathfrak{P}X$.

To a class E of sets and an equivalence relation ρ on E and an element $y \in E$, associate first the two following predicates:

2°) $Q_{E,\rho,y}(x) \equiv$ "x is equivalent to y mod ρ"

(the defined class being called an equivalence subclass of E mod ρ, more precisely, the y-equivalence subclass of E mod ρ) and second,

3°) $Q_{E,\rho}(x) \equiv$ "x is an equivalence subclass of E mod ρ"

defining a quotient class E/ρ.

We also would like that the predicate

4°) "$x = x$"

be collectivizing which is equivalent to the existence, in our system, of a universal class U such that, for every class x, $x \in U$ holds, and in particular $U \in U$. For this universal class must hold the equality

$U = \mathfrak{P}U$.

We would also like to be able to speak of the power α of a class A, in particular, of the power υ of the universal class U. Then, the power class $\mathfrak{P}A$ would have power 2^{α} and we have indicated in [6], p. 134:

(a) that for those hypertransfinite numbers, the theorem of Cantor can fail, so that there exist necessarily such powers α for which $\alpha = 2^{\alpha}$ and (b) how still keep the system free of contradiction. In particular, υ must be the maximal hypertransfinite number, so that $\upsilon = 2^{\upsilon}$ must hold. The way of avoiding the usual contradiction in this case is to admit that the Russell predicate

$$R(x) \equiv \text{"}x \notin x\text{"}$$

applied to all classes is not collectivizing (and similarly any predicate associated by the diagonal method to any automorphism of $\{U\}$ considered as a bijection between U and $\mathcal{P}U$). However, it seems possible to admit that the predicate

$$S(x) \equiv \text{"}x \in x\text{"}$$

is collectivizing and defines a class $C_S = V$ (possibly the singleton $\{U\}$). It is therefore clear that, in this system, a subclass B of a class A cannot always have a complement. In other words, the predicate $P_{A,B}$ (one assumes $B \subset A$):

$$P_{A,B} \equiv \text{"}x \in A \quad \text{and} \quad x \notin B\text{"}$$

cannot always be collectivizing. As a matter of fact, if $A = U$, $B = V$, then $P_{U,V}$ is just the Russell predicate. This also shows that it was essential, in 2°) and 3°) above, to restrict ourselves to classes E containing only sets. Indeed, if we take $E = U$, and consider the equivalence relations σ such that

$$\text{"}x \sim_\sigma y\text{" if and only if}$$

$$\text{"simultaneously} \quad x \in x \quad \text{and} \quad y \in y \quad \text{or} \quad x \notin x \quad \text{and} \quad y \notin y\text{"} ,$$

then $Q_{U,\sigma,\emptyset}$ is the Russell predicate, while $Q_{U,\sigma,U}$ is the predicate S. In this case, U/σ cannot exist as one of its elements would be the "impossible" Russell class.

A LOGICAL MODEL WITH UNIVERSAL CLASS

We shall now derive from ZF a model D (as da Costa) with a universal class U containing itself and without Russell class (i.e., in D the Russell predicate is not collectivizing). This is however a very elementary (or rudimentary) model not adapted to category theory. However this model should give confidence in the real flexibility of the methods of symbolic logic, and therefore also confidence in statements like the one quoted above above of Ch. Ehresmann. It is therefore reasonable to expect that further exploration of axiomatic possibilities like in Church's proposed program should relatively easily lead to proper logical system fully adapted to the needs of category theory.

Let U be a Grothendieck-Sonner universe as a symbol of ZF. To define D, we limit ourselves to variables varying on all sets $x \in U$ in the sense of ZF plus the set U. We keep unchanged all logical symbols $\forall$, $\exists$, $\neg$, etc..., except the membership symbol $\in$ to be replaced by a new one ε which we define as follows.

1°) If x,y are elements of D such that $x \in y$ holds in ZF (which implies that at least x differs from U but y can be U) then we write in D

$$x \; \varepsilon \; y \; ;$$

2°) We also write in D:

$$U \; \varepsilon \; U \; .$$

We note that, of course, in ZF the relation $U \notin U$ holds.

Let us observe that, instead of a Grothendieck-Sonner universe we can limit D to the elements

$$0,1,2,\ldots,\omega \; ,$$

write $m \; \varepsilon \; n$ if $m < n$ and $\omega \; \varepsilon \; \omega$. This produces a model of the same type, with $U = \omega$, more easy to handle.

In D, we use the abbreviation

$$"\neg x \; \varepsilon \; y" = "x \not\varepsilon \; y" \; .$$

Therefore, for every $x \neq U$, the following holds true:

$$x \not\varepsilon \; x \tag{5}$$

and these x's are called "sets", while any x cr U is called a "class" (so that U is the unique, proper class). We also see that, in D, there exists no "set or class of all sets"; in other words the relation "$x \neq U$" equivalent to (5) is not collectivizing; still in other words there exists no class I in D (I as impossible) such that

$$I = \{x | x \neq U\} = \{x | x \not\varepsilon \; x\}$$

or such that

$$"x \varepsilon I" = "x \notin x".$$

However, in D, U is the "class of all classes" in the sense that

$$"x \text{ is a class}" \equiv "x \varepsilon U".$$

Also, if one defines $y \subset z$ as

$$"y \subset z" \equiv "\forall x(x \varepsilon y \Rightarrow x \varepsilon z)" ,$$

then for every class x there exists a power class $\mathfrak{P}x$ and

$$\mathfrak{P}U = U ,$$

in the sense that

$$"x \subset U" \equiv "x \varepsilon U".$$

However, there is no singleton $\{U\}$ in D. Indeed that singleton should be a class (i.e. a set or U) such that

$"x \varepsilon \{U\}"$ would be equivalent to $"x = U"$.

Neither, if x, y, z,... are sets, but not all of them, there is, in D, no symbol for

$$\{x, y, z, \dots U\}$$

(a class containing exactly x, y, z, ... and U). However, for a predicate P and a set z, the following comprehension principle holds in D: there exists a set $y = y_{P,z}$ such that

$$"x \varepsilon y" \equiv "P(x) \wedge x \varepsilon z" .$$

This model D is fit to develop the theory of U-small categories A which satisfy the relation $A \in U$, in the sense of ZF, and thus also "$A \varepsilon U$ and $A \neq U$", in the sense of D. It is however not fit to construct U-large categories, the set of objects of which is contained in U without being a member of U (in the sense of ZF). Indeed no symbol exists in D to represent such objects, so that, in order to be able to consider them, the system D should be enlarged (and thus completely modified) by adding more "proper classes" than the single U and by enacting appropriate rules for the (also modified) ε-relation. Then D has to be transformed in another system D'. It should not be difficult to make D' possess a singleton $\{U\}$ or classes $\{x, y, z, \dots, U\}$ as above.

But D is already treacherous as it tries to imitate naive set intuition and at the same time definitely diverges from it. It allows both mathematical (internal) and metamathematical (outer) interpretations. And the problem is not to invent abstract systems for fun, but systems which are useful in category theory practice. We must realize that these so far apparently do not exist, so that this note has more of a plead and of a survey or a dream than it has mathematical or logical substance, as was already the case of [6]. This possibly sufficiently justifies the "it does not really matter" attitude of pragmaticians quoted above. Or Bertrand Russell's to the effect of which, "mathematics is a science in which you do not know whether what you do is correct nor even what you are doing". This, however, should not prevent further and deeper efforts towards the clarification of this fundamental problem.

AN AXIOMATIZATION OF D

We conclude by an axiomatization of the system D just described, as it has been communicated to me by Newton da Costa. I thank him for his permission to include this paragraph.

The System D

The primitive symbols of D are the following: 1) connectives: $\vee$ (or) and $\neg$ (not); $\Rightarrow$ (implication), $\wedge$ (and), and $\Longleftrightarrow$ (equivalence) are introduced as usual; 2) The existential quantifiers: $\exists$ (we define $\forall$, the universal quantifier, as usual); 3) A denumerably infinite collection of individual variables; 4) An individual constant U; 5) Two binary predicate symbols = (equality) and $\in$ (membership) ; 6) Parentheses.

The common syntactical concepts (of formula, of proof, etc.) are defined as it is standard; of course, the underlying logic of D is the (classical) first-order predicate calculus with equality.

Specific axioms of D:

A_1) Axiom of the universal class:

$\forall x(x \in U)$

A_2) Axiom of extensionality:

$\forall x(x \in y \Leftrightarrow x \in z) \Rightarrow y = z$

A_3) Axiom of pairing:

$$(u \neq U \wedge v \neq U) \Rightarrow \exists y \forall x (x \in y \Leftrightarrow (x = u \vee x = v))$$

A_4) Axiom for union:

$$x \neq U \Rightarrow \exists y \forall t \ (t \in y \Leftrightarrow \exists v \ (t \in v \wedge v \in x))$$

A_5) Axiom of power-set:

$$x \neq U \Rightarrow \exists y \ \forall t \ (t \in y \Leftrightarrow \forall v \ (v \in t \Rightarrow v \in x))$$

A_6) Axiom for separation:

$$z \neq U \Rightarrow \exists y \ \forall x \quad (x \in y \Leftrightarrow (F(x) \wedge x \in z)) \ ,$$

with obvious restrictions. In A_5, we write $y = \mathfrak{P}x$.

If $x \neq U$ and $y \neq U$, we define $x \cup y$, $x \cap y$, $x - y$, $\{x\}$, $\{x,y\}$, $\cup x$, etc., as in ZF. By definition, we put $\mathfrak{P}U = U$, i.e., the power-class of U is U. The empty set is denoted by ϕ.

A_7) Axiom of infinity:

$$\exists y(\phi \in y \wedge \forall x \ (x \in y \Rightarrow x \cup \{x\} \in y) \wedge y \neq U).$$

The axioms of substitution (A_8), of choice (A_9), and of regularity (A_{10}) are formulated as in ZF, with clear adaptations. For instance, we have:

A_{10}) Axiom of regularity:

$$(x \neq U \wedge x \neq \phi) \Rightarrow \exists y \ (y \in x \wedge x \cap y = \phi).$$

In D we have: $U \in U$ and $x \ (x \neq U \Longleftrightarrow x \in x)$. On the other hand, D satisfies all the intuitive conditions of the section "A Logical Model with Universal Class" above.

Remarks: 1) It is easy to modify D in order that $\{x_1, x_2, \ldots, x_n, U\}$ be a set for any $x_1, x_2, \ldots, x_n$, $n \geq 0$; we have only to change the axiom of pairing. 2) By the indroduction of new axioms, we can obtain a system reflecting some of the informal ideas described in pages 77, 78, and 79 (similar systems are studied in Church [2], da Costa [3] and [4], da Costa and Dias [5], and Hatcher [11]). 3) D and ZF are equiconsistent.

NOTE

1. "Wir unterscheiden zwischen Mengen und Klassen; eine Menge ist auch eine Klasse. Die Klasse aller Mengen ist keine Menge. Wir lassen für

Klassen dieselben Operationen zu wie für Mengen. Wir vermeiden also nicht die Bildung von gewissen Klassen von Teilklassen einer Klasse. Sollten sich durch diese Begriffsbildungen Widersprüche ergeben, so wäre es immer möglich, Beschränkungen einzuführen, um im Rahmen der Mengenlehre zu bleiben; dadurch würde die Theorie aber umständlicher werden".

REFERENCES

1. Diana Brignole and Newton C. A. da Costa, On Supernormal Ehresmann-Dedecker Universes, Math. Zeitschr., Bamd. 122 (1971), pp. 342-350.

2. Alonzo Church, Set Theory with a Universal Set, Proceedings of the Tarski Symposium (U. California Berkeley, June 23-30. 1971), Amer. Math. Soc., Proc. of Coll. in Pare Math. XXV, 1974.

3. Newton C.A. da Costa, On Two Systems of Set Theory, Proc. Kon. Nederl. Acad. Wetensch., s.A, vol. 68 (1965), pp. 95-98.

4. Newton C.A. da Costa, Un nouveau système formel suggéré par Dedecker, Comptes rendus Ac. Sc. Paris, s.A, t. 265 (1967), pp. 85-88.

5. Newton C.A. da Costa et M.F. Dias, Sur le système D* de la théorie des ensembles, Comptes rendus Ac. Sc. Paris, s.A, 282 (1976), pp. 5-7.

6. Paul Dedecker, Introduction aux structurales locales. Appendice sur les fondements, Colloques du Centre belge de Recherches Mathématiques, Geométrie différentielle, Université catholique de Louvain, 1958, pp. 129-135.

7. Charles Ehresmann, Gattungen von lokalen Strukturen, Jahresbericht der Deutschen Math. Vereinigung, Bd. 60 (1957), pp. 49-77.

8. Solomon Feferman and G. Kreisel, Set Theoretical Foundations of Category Theory, Reports of the Midwest Category Seminar III, Springer Lecture Notes in Math., # 106, Berlin, Heidelberg, New York, (1969), pp. 201-247.

9. Robert Goldblatt, Topoi the Categorial Analysis of Logic, North Holland Publ. Cy., Amsterdam, New York, Oxford, 1979.

10. William S. Hatcher, Foundations of Mathematics, W.B. Saunders Company, Philadelphia, London, Toronto, 1968.

11. William S. Hatcher, Sur un système de da Costa, Comptes rendus, Acad. Sc. Paris, s.A, 260 (1969), pp. 1443-1446.

12. William S. Hatcher, The Logical Foundation of Mathematics, Pergamon Press, Oxford, New York, Toronto, 1982.

13. P.T. Johnstone, Topos Theory, Academic Press, London, New York, San Francisco, 1977.

14. F.W. Lawvere, The Category of Categories as a Foundation of Mathematics, Proc. of the Conf. on Categorical Algebra, Springer Verlag, Berlin-Heidelberg 1966, pp. 1-20.

15. Saunders Maclane, Categories for the working mathematician, Graduate Texts in Mathematics #5, Springer Verlag, Berlin, Heidelberg, New York, 1971.

16. Johann Sonner, On the formal definition of categories, Math. Zeitschr. Bd. 80 (1962), pp. 163-176.

SOME ASPECTS OF THE THEORY OF LARGE CARDINALS

Carlos A. Di Prisco/Departamento de Matemáticas, Instituto Venezolano de Investigaciones Científicas, Caracas, Venezuela

Wiktor Marek*/Departamento de Matemáticas, Instituto Venezolano de Investigaciones Científicas, Caracas, Venezuela

CONTENTS

*Current Affiliation: Department of Computer Science, University of Kentucky, Lexington, Kentucky

INTRODUCTION

These notes treat various aspects of the theory of large cardinals including combinatorial properties of some sets (or spaces) naturally related to them. We will be mainly concerned with three sorts of large cardinals: measurable, supercompact and huge. The sets associated are respectively, a regular cardinal κ, the space $P_\kappa(\lambda)$ and $[\lambda]^\kappa$ (or $(\lambda)^\kappa$).

Many investigations on these large cardinal properties suggested the study of the associated spaces, and many ideas and methods used under the large cardinal assumptions were adapted to study the spaces on their own without any additional set theoretical hypothesis. In this way a number of interesting combinatorial results on these spaces and some collections of their subsets were obtained. For example, using the techniques developed to study measurable cardinals, many important properties of closed unbounded subsets and stationary subsets of κ were proved. The remarkable paper by Baumgartner, Taylor and Wagon presents a survey of these techniques and results for a regular cardinal κ. For the spaces $P_\kappa(\lambda)$ and $[\lambda]^\kappa$ the study is not so developed. After Jech introduced the "absolute" objects corresponding to normal measures on $P_\kappa(\lambda)$, a great number of results on the structure of the space followed, but until today no comprehensive study analogous to the Baumgartner, Taylor and Wagon paper exists.

For the space $[\lambda]^\kappa$ it was only recently that and "absolute" notion corresponding to ultrafilters on $[\lambda]^\kappa$ was exhibited.[1]

These notes are organized in the following way. Chapter 1 is devoted to the space κ (κ an uncountable regular cardinal). We had necessarily to choose some topics from the immense amount of material available, we present several facts concerning ultrafilters and closed unbounded and stationary sets.

In Chapter 2, we treat the space $P_\kappa(\lambda)$. We consider normal measures (supercompactness) and the closed unbounded and stationary subsets.

Chapter 3 includes some material dealing with the so called huge cardinals and several related concepts. We study the combinatorial structure of the space $[\lambda]^\kappa$.

Throughout the notes we bear in mind the analogy expressed in the following table

Normal measure on κ (κ measurable)	Normal measure on $P_\kappa(\lambda)$ (κ λ-supercompact)	Normal measure on $[\lambda]^\kappa$ κ huge
Closed unbounded sets in κ	Jech's filter on $P_\kappa(\lambda)$	Filter $F_{\kappa,\lambda}^{(1)}$

Most of the preliminaries necessary to read these notes can be found in the paper by Kanamori and Magidor titled "The Evolution of Large Cardinal Axioms in Set Theory." ([16])

These notes were prepared for the Seminar of the VI Latin American Symposium on Mathematical Logic which was held in Caracas during July and August, 1983[2].

1.1. MEASURABLE CARDINALS

As uncountable cardinal κ is said to be measurable if and only if there exists a uniform κ-complete ultrafilter on κ. Let $\mathcal{U}$ be such an ultrafilter. Note that $\mathcal{U}$ is necessarily nonprincipal.

The ultrapower $V^\kappa/\mathcal{U}$ is well founded and so it is isomorphic to a transitive (proper) class M. If $i : V \to V^\kappa/\mathcal{U}$ is the canonical injection of V into the ultrapower and $\pi : V^\kappa/\mathcal{U} \to M$ is the isomorphism of the ultrapower onto M, we call $j = \pi \circ i : V \to M$ the resulting elementary embedding of V into M.

Since $\mathcal{U}$ is κ-complete, $j(\alpha) = \alpha$ for all $\alpha < \kappa$. For κ we have that $j(\kappa) \neq \kappa$. Indeed, given $f \in {}^\kappa\kappa$, let $[f] \in V^\kappa/\mathcal{U}$ be the equivalence class of f with respect to $\mathcal{U}$. By Łos' fundamental theorem of ultraproducts, $\pi([f]) \in j(\kappa)$. If $f \in {}^\kappa\kappa$ satisfies the property $(\alpha)_\kappa(\{\beta | f(\beta) > \alpha\} \in \mathcal{U})$ (which holds if f is the identity function) then $(\alpha)_\kappa$ $(\alpha < \pi([f]) < j(\kappa))$ and so $\kappa < j(\kappa)$. We have proved:

Theorem: If a measurable cardinal exists then there is an elementary embedding $j : V \to M$, $j \neq id$, of the universe into a transitive proper class M. □

The converse holds as well. Given $j : V \to M$ an alementary embedding different from the idendity of V into a transitive class M, since j is not the indendity there must be an ordinal moved by j. If κ is the least such ordinal (this is in fact the least rank of an element moved by j) we can define $\mathcal{U}_j = \{X \subseteq \kappa \mid \kappa \in j(X)\}$. It is easy to verify that $\mathcal{U}_j$ is a κ-complete uniform ultrafilter. This ultrafilter has an interesting additional property:

If $f \in {}^{\kappa}\kappa$ and $A \subseteq \kappa$, we say that f is regressive on A if $(\xi)_A(f(\xi) < \xi)$. If $A \in \mathcal{U}_j$ and f is regressive on A then there is $B \subseteq A$, $B \in \mathcal{U}_j$ such that f is constant on B. (In other words there is $\xi < \kappa$ such that $f^{-1}\{\xi\} \in \mathcal{U}_j$).

To prove this note that $\{\xi | f(\xi) < \xi\} \in \mathcal{U}_j$ i.e. $\kappa \in j(\{\xi \mid f(\xi) < \xi\})$ which in turn means that $j(f)(\kappa) < \kappa$. Set $\xi = j(f)(\kappa)$. This is the required ordinal.

Ultrafilters which satisfy this property are called normal. We showed that if κ possesses a κ-complete ultrafilter then it also possesses a normal one.

EXERCISE: Given a κ-complete ultrafilter on κ, A function $f \in {}^{\kappa}\kappa$ is called incompressible if for any $g \in {}^{\kappa}\kappa$ such that $\{\alpha | g(\alpha) < f(\alpha)\} \in \mathcal{U}$ there is an ordinal ξ such that $g^{-1}\{\xi\} \in \mathcal{U}$.

a) Show that if $\mathcal{U}$ is an ω-complete ultrafilter on κ there is an incompressible function.

b) Show that if f is an incompressible function with respect to a κ-complete ultrafilter $\mathcal{U}$ on κ the set $\{X \subseteq \kappa \; f^{-1} * X \in \mathcal{U}\}$ is a normal κ-complete ultrafilter on κ.

Let now $\mathcal{U}$ be a normal, κ-complete ultrafilter on κ. We will investigate closure properties of $M \approx V^{\kappa}/\mathcal{U}$.

<u>Theorem</u>: M is closed under sequences of length κ, in symbols, ${}^{\kappa}M \subseteq M$. To prove this theorem let $\langle x_\xi \mid \xi < \kappa\rangle$ be a sequence of elements of M. For each $\xi < \kappa$ let $f_\xi : \kappa \to V$ such that $x_\xi = \pi([f_\xi])$. Build now a function $g : \kappa \to V$ putting $g(\xi) = \langle f_\eta(\xi) | \eta < \xi\rangle$ for each $\xi < \kappa$. Since $g(\xi)$ is a sequence of length ξ, $[g]$ represents a sequence of length $[id]$, thus $\pi[g]$ is a sequence of length $\pi[id] = \kappa$. (See next theorem).

Let us now show that the value of this sequence at η is $\pi([f_\eta])$. The η term of $\pi[g]$ is represented by the class of the function which takes the same values as f_η almost everywhere. □

This theorem implies immediately that the class M contains all of $V_{\kappa+1}$ and so $V^M_{\kappa+1} = V_{\kappa+1}$. This in turn implies the Π^2_1 - indescidability of κ : if $\phi(\kappa,\overline{X})$ is a true Π^2_1 - statement about κ and subsets $\overline{X}$ of V_κ, then there is some $\xi < \kappa$ such that $\phi(\xi,\overline{X} \cap V_\xi)$ is also true.

We will now give several characterizations of normal ultrafilters. We define the diagonal intersection of a sequence $\{X_\xi | \xi < \kappa\}$ of subsets of κ as the set $\Delta_{\xi<\kappa} X_\xi = \{\xi < \kappa | \xi \in X_\eta$ for all $\eta < \xi\}$. We say that a sequence $\{X_\xi | \xi < \kappa\}$ is continuous if and only if for all λ limit, $X_\lambda = \bigcap_{\xi<\lambda} X_\xi$. We state without proof the following result.

Theorem: If $\mathcal{U}$ is a κ-complete uniform ultrafilter on κ then the following are equivalent:

i) $\mathcal{U}$ is normal.

ii) $\mathcal{U}$ is closed under diagonal intersection (i.e. if $X_\xi \in \mathcal{U}$ for all $\xi < \kappa$ then $\Delta_{\xi<\kappa} X_\xi \in \mathcal{U}$).

iii) If $\{X_\xi | \xi < \kappa\}$ is a continuous descending family of elements of $\mathcal{U}$ then $\{\xi | \xi \in X_\xi\} \in \mathcal{U}$.

iv) The function id $\in {}^\kappa\kappa$ is such that $\pi[\text{id}] = \kappa$.

v) $\mathcal{U} = \mathcal{U}_j$ where j is the embedding from V into $M \approx V^\kappa/\mathcal{U}$.

The class M is - obviously - included in V. Also since j is elementary $M \models$ "$j(\mathcal{U})$ is an $j(\kappa)$-complete uniform ultrafilter in $j(\kappa)$". Thus we can form, in M, the ultrapower $V^{j(\kappa)}/_{j}(\mathcal{U})$. This class is wellfounded (in M, but this notion is clearly absolute) and so there is a class M_2 and an elementary imbedding $j_{1,2}$ of M (which we will call now M_1) into M_2. Composing j (which we call now j_{01}) with $j_{1,2}$ we get an elementary imbedding j_{02} of V into M_2. In this fashion we get the first taste of iterated ultrapowers (this notion is due to H. Gaifman). We can iterate this construction by taking in the limit step the direct limit of the models obtained previously. A result, due to Gaifman, is that the family of models $\langle M_\xi \rangle_{\xi \in \text{ORD}}$ consists of well-founded models and is strictly descending. We also have a family $\langle j_{\alpha,\beta} \rangle_{\alpha < \beta}$ of elementary imbeddings $j_{\alpha,\beta} : M_\alpha \to M_\beta$. Looking back we see that $j_{02} : V \to M_2$ is the identity of V_κ. In fact it is quite evident that $j_{\alpha\beta}$ is the identity on $V^{M_\alpha}_{j_{0\alpha}(\kappa)}$. Thus setting: $\kappa_\alpha = j_{0\alpha}(\kappa)$ and $T_\alpha = V^{M_\alpha}_{\kappa_\alpha}$ we have: For all $\beta \geq \alpha$, $T_\alpha \subseteq M_\beta$. Let $K_\mathcal{U} = \bigcap_{\xi \in \text{ORD}} M_\xi$. Then, since $T_\alpha = V^{M_\beta}_{\kappa_\alpha}$ for $\beta > \alpha$ and since the family $\langle M_\xi \rangle_{\xi \in \text{ORD}}$ is decreasing we have:

Theorem: $\bigcup_{\xi \in ORD} T_\alpha = \bigcap_{\xi \in ORD} M_\xi = K_{\mathcal{U}}$.

Also, it is quite easy to show that $\langle \kappa_\xi \rangle_{\xi \in ORD}$ is a sequence of elements order-indiscernible for $K_{\mathcal{U}}$. Finally, since $j_{\alpha\beta}(\kappa_\alpha) = \kappa_\beta$, $T_\alpha \prec T_\beta$. From this and a reflection principle, plus the fact that $\{\kappa_\xi\}_{\xi \in ORD}$ is a continuous sequence, it follows that for every α, $T_\alpha \prec K_{\mathcal{U}}$. In particular $V_\kappa \prec K_{\mathcal{U}}$. Thus if κ is the least measurable cardinal $K_{\mathcal{U}} \models$ "There is no measurable cardinal" but if κ is large and $\{\xi \in \kappa \mid \xi$ is measurable$\} \in \mathcal{U}$ then $K_{\mathcal{U}} \models$". There are arbitrarity large measurable cardinals".

1.2. ULTRAFILTERS AND ELEMENTARY EMBEDDINGS

We know that if $j : V \to M$ is an elementary embedding then $\mathcal{U}_j = \{X \subset \kappa \mid \kappa \in j(X)\}$ is a normal ultrafilter in κ. It is equally simple to show that if $\xi \in [\kappa, j(\kappa))$ then $\mathcal{U}_\xi = \{X \mid \xi \in j(X)\}$ is also a κ-complete ultrafilter although it does not have to be normal; in fact $\mathcal{U}_{\kappa+1}$ is never normal since $\{\xi \mid (E\eta)(\eta \in Lim \ \& \ \xi = \eta + 1) \in \mathcal{U}_{\kappa+1}$. If $\mathcal{U}$ is a normal ultrafilter we can easily show that all the ultrafilters $\mathcal{U}_\xi (\xi \in (\kappa, j\kappa))$ generated by j (determined by $\mathcal{U}$) are different. In fact they are exactly of the form $\{f^{-1} * X : X \in \mathcal{U}\}$ for some permutation f of κ. For this assume $\xi_1, \xi_2 \in [\kappa, j\kappa)$ (j is composed of $j^1 : V \to V^\kappa/\mathcal{U}$ and $\pi : V^\kappa/\mathcal{U} \to M$). Now let $\xi_1 = [f_1]$, $\xi_2 = [f_2]$. One immediately sees that: $X \in \mathcal{U}_{\xi_i} \Longleftrightarrow f_i^{-1} * X \in \mathcal{U}$ ($\mathcal{U}$ precedes $\mathcal{U}_{\xi_i}$ in the Rudin-Keisler order, $\mathcal{U} \lneq_{RK} \mathcal{U}_\xi$). Consider now the partition Π_i of κ canonically related to f_i. Then, since $\mathcal{U}_{\xi_i}$ is non-principal, the sets of the partition Π_i are not in $\mathcal{U}$. Thus by normality of $\mathcal{U}$, the selector of least elements, call it Z_i, belongs to $\mathcal{U}$. By construction $h_i = f_i \mid Z_i$ is an injection. Let $T_i = Rg\ f_i$, so $h_i : Z_i \to T_i$. Then, for $Y \subseteq T_i$, $Y \in \mathcal{U}_{\xi_i} \Longleftrightarrow f_i^{-1} * Y \in \mathcal{U}$. Now, if $\mathcal{U}_{\xi_1} = \mathcal{U}_{\xi_2}$ then both $T_1, T_2 \in \mathcal{U}_{\xi_1}$. Thus $T_1 \cap T_2 = T \in \mathcal{U}_{\xi_1}$. Thus $T_1' = h_1^{-1} * T \in \mathcal{U}$, $T_2' = h_2^{-1} * T \in \mathcal{U}$, $h_2^{-1} \circ h_1 : T' \xrightarrow[\text{onto}]{1-1} T_2'$. But since $\mathcal{U}$ is normal, $T_1', T_2' \in \mathcal{U}$ therefore there is $S \subseteq T_1'$, $S \in \mathcal{U}$ such that $h_2^{-1} \circ h_1 | S = id$. Therefore $h_2 | S = h_1 | S$ and so $f_1 | S = f_2 | S$ i.e. $f_1 \approx_{\mathcal{U}} f_2$ thus $\xi_1 = \xi_2$.

Our reasoning (using essentially the argument of the Schröder-Bernstein theorem) may be used to construct in each case a permutation ϕ_i of κ such that

$$X \in \mathcal{U}_{\xi_i} \iff \phi_i^{-1} * X \in \mathcal{U}$$

This fact is described as saying that $V^\kappa/\mathcal{U}$ possesses just one constellation, or in other words that all $\mathcal{U}_\xi$'s are $R\kappa$-equivalent. (See Kanamori, 1977) Since, as noted before, $P(\kappa)^M = P(\kappa)$, $j(\kappa) < 2^\kappa$. Thus we proved that if κ is measurable there are always at least 2^κ κ-complete ultrafilters in $P(\kappa)$. This result rises some questions; for instance how many κ-complete ultrafilters on $\mathcal{U}$ are there. Also, for j different from our $j_\mathcal{U}$ can a given $\mathcal{U}$ appear more than once - and if so then how often? It is quite easy to construct an embedding j in which an ultrafilter $\mathcal{U}$ appears more than once. In fact this happens if we consider imbeddings constructed by means of iterated ultrapower. Also, Kunen proved that under appropriate set theoretical assumptions ($V = L[\mathcal{U}]$) there are exactly 2^κ, κ-complete ultrafilters, and they all appear as $\mathcal{U}_\xi$'s for $j = j_{0\omega}$ in the process of the iteration of the ultrapower. In section 2.1 we will show that under other assumptions there are 2^{2^κ} normal ultrafilters on κ.

Below we show that certain elementary embeddings have the property that every $\mathcal{U}$, κ-complete ultrafilter on κ appears as $\mathcal{U}_\xi$. But first we need some preliminary fact. If $j : V \to M$ is an elementary imbedding generated by ultrapower with respect to an ultrafilter $\mathcal{U}$, then $\mathcal{U}$ is of the form $\mathcal{U}_\xi$ for $\xi = [id]$. But what about other ultrafilters? Assume that $J : V \to M$ is an elementary embedding and that M is closed under sequences of the length 2^κ. The result we will prove now is that every κ-complete ultrafilter in κ is of the form $\mathcal{U}_\xi$ for some ξ. In fact given $\mathcal{U}$ κ-complete ultrafilter in κ, $\{\xi | \mathcal{U}_\xi = \mathcal{U}\}$ is cofinal in $j(\kappa)$. First of all notice that since M is closed under sequences of the length 2^κ, $P(\kappa) \subseteq M$ so $P(\kappa) \in M$ and that also $P(P(\kappa)) \subseteq M$, $P(P(\kappa)) \in M$ and thus $P(P(\kappa))^M = P(P(\kappa))$. Since measurable cardinals are inaccessible we know that $j(\kappa)$ is inaccessible in sense of M and so in M, $2^{2^\kappa} < j(\kappa)$ since $\kappa < j(\kappa)$. But then $2^{2^\kappa} < j(\kappa)$ is true. Now let $\mathcal{U}$ be given. Our task is to find ξ such that $\mathcal{U} = \mathcal{U}_\xi$. For this consider $j(\mathcal{U})$. In M, $j(\mathcal{U})$ is a $j(\kappa)$-complete uniform ultrafilter on $j(\kappa)$. We can thus construct a corresponding ultrapower of M with respect to $j(\mathcal{U})$ (within M). Let N be transitive contraction of $M^{j(\kappa)}/j(\mathcal{U})$ (remember that only the functions in M are used) and let j' be the corresponding elementary imbedding $j': M \to N$. Set $j'' = j' \circ j$, $j'' : V \to N$. We claim first that N is also closed under sequences of length 2^κ. Indeed let $S \in {}^{2^\kappa}N$. Then $S \in {}^{2^\kappa}M$ and so $S \in M$. But $2^\kappa < j(\kappa)$ (as noted above) and $M \models {}^{j(\kappa)}N \subset N$ so $M \cap ({}^{j(\kappa)}N) \subseteq N$ in particular $M \cap ({}^{2^\kappa}N) \subseteq N$. But ${}^{2^\kappa}N \subseteq {}^{2^\kappa}M \subseteq M$ so ${}^{2^\kappa}N \subseteq N$. Now look, $\mathcal{U} \subseteq P(\kappa)$ so $|\mathcal{U}| \leq 2^\kappa$. Thus $\mathcal{U} \in N$. Let R be $V^\kappa/\mathcal{U}$ or to be precise its transitive contraction. Our next claim is that: $j \mid R$ is an elementary imbedding of R into N and in fact the following diagram is commutative:

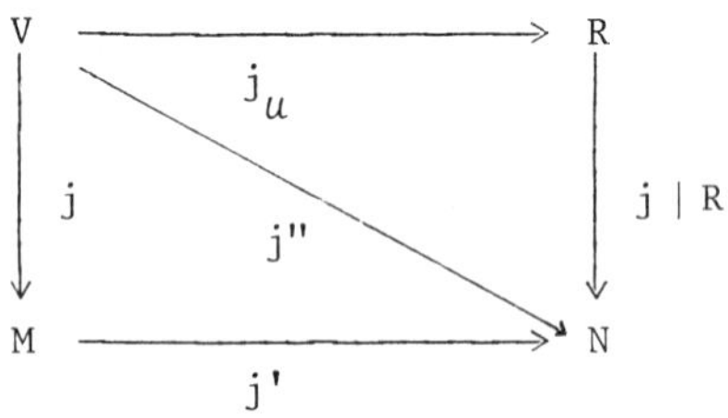

This needs checking, but it is just manipulation of the fact that $j' = j_{j(\mathcal{U})}$. Now, let $\gamma = [Id_\kappa]_{\mathcal{U}}$.

By definition $\kappa \leq \gamma \leq j_{\mathcal{U}}(\kappa)$. Applying j to this inequality we get:

$$j(\kappa) \leq j(\gamma) < j(j_{\mathcal{U}}(\kappa)) = j \circ j_{\mathcal{U}}(\kappa) \overset{*}{=} j_{j_{\mathcal{U}}} \circ j(\kappa) = j''(\kappa)$$

(The step * uses commutativity of the diagrams). Now we will see how $\mathcal{U}$ is characterized by the properties of N.

First of all notice that by Łos' theorem, as mentioned above:

$$X \in \mathcal{U} \iff \{\alpha : \alpha \in X\} \in \mathcal{U} \iff \{\alpha : Id_\kappa(\alpha) \in X\} \in \mathcal{U} \iff [Id]_{\mathcal{U}} \in j_{\mathcal{U}}(X)$$

Since $\gamma = [Id]_{\mathcal{U}}$ we have:

$$X \in \mathcal{U} \iff \gamma \in j_{\mathcal{U}}(X)$$

Since j is elementary, applying j to the left hand side we get:

$$X \in \mathcal{U} \iff j(\gamma) \in j(j_{\mathcal{U}}(X)) \text{ , and}$$

By commutativity of the diagram:

$(\dagger)\ X \in \mathcal{U} \Longleftrightarrow j(\gamma) \in j''(X)$

Now, let $j_1 = j \mid P(\kappa)$, $j_1'' = j'' \mid P(\kappa)$. Then both j_1 and j_1'' are sequences of type 2^κ (or can be represented as such) and by the closure properties of M and N respectively:

$$j_1 \in M \text{ and } j_1'' \in N$$

We claim now that $j_1'' = j'(j_1)$. To see this notice that $j'(j_1)$ is a function with domains $j'(\text{Dom } j_1)$. But $\text{Dom } j_2 = P(\kappa)$ and clearly $j'(P(\kappa)) = P(\kappa)$. So what is left to be checked is that $j_1''(X) = j'(j_1)(X)$. But $j_1''(X) = j''(X)\ \ j''(j(X)) = j'(j_1(X)) = j'(j_1)(j'(X)) = j'(j_1)(X)$. The last passage follows from the fact that j' is the identity on $j(\kappa)$ and thus on κ and $P(\kappa)$ as well. Thus our claim is proved. By the property $(\dagger)$ and the fact that for $X \in P(\kappa)\ j''(X) = j_1''(X)$ we have:

$$(X)_{P(\kappa)}(X \in \mathcal{U} \Longleftrightarrow j(\gamma) \in j_1''(X))$$

Since $P(\kappa) \in N$ and $\mathcal{U} \in N$ we have

$$N \models (X)_{P(\kappa)}(X \in \mathcal{U} \Longleftrightarrow j(\gamma) \in j_1''(X))$$

Now we are ready to prove our result: Pick any $\alpha < j(\kappa)$ then since $j(\kappa) \leq j(\gamma)\ \ \alpha < j(\gamma)$ therefore:

$$N \models (E\xi)(\alpha < \xi < j''(\kappa)\ \&\ (X)_{P(\kappa)}(X \in \mathcal{U} \Longleftrightarrow \xi \in j_1''(X))$$

(Indeed $j(\gamma)$ is our ξ).

Now we modify our formula taking into account

$$j''(\kappa) = j'(j(\kappa)),\ j'(\kappa) = \kappa\ ,\ j'(\mathcal{U}) = \mathcal{U}$$

$$j_1'' = j'(j_1) \text{ and } j''(\alpha) = \alpha$$

so we get

$$N \models (E\xi)(j'(\alpha) < \xi < j'(j(\kappa))\&(X)_{j'(P(\kappa))}(X \in j'(\mathcal{U}) \Longleftrightarrow \xi \in j'(j_1(X)))$$

Now we come back to M;

$$M \models (E_\xi)(\alpha < \xi < j(\kappa)\ \&\ (X)_{P(\kappa)}(X \in \mathcal{U} \Longleftrightarrow \xi \in j_1(X)))$$

So

$$(E_\xi)(\alpha < \xi < j(\kappa)\ \&\ (X)_{P(\kappa)}(X \in \mathcal{U} \Longleftrightarrow \xi \in j_2(X)))$$

but $j_1(X) = j(X)$ so:

$$(E\xi)(\alpha < \xi < j(\kappa)\ \&\ (X)_{P(\kappa)}(X \in \mathcal{U} \Longleftrightarrow \xi \in j(X))$$

Thus, there are arbitrarily large ξ's below $j(\kappa)$ such that $U = U_\xi$. In particular, as $U \in M$, $\{\xi : U = U_\xi\} \in M$ since it is definable from j_1 which belongs to M as well. So $\overline{\{\xi : U = U_\xi\}} = j(\kappa)$ since in M $j(\kappa)$ is regular and M is a model of ZFC. $\square$

A. Kanamori pointed out that the result 1.2 may be proved in much simpler way.

Assume $j: V \to M$ is an elementary imbedding, M closed under 2^κ-sequences, $j(\kappa) > 2^\kappa$. Let U be any κ-complete ultrafilter. Then $j(U)$ is a $j(\kappa)$-complete ultrafilter in M. Consider now $T = \{j(X) : X \in U\}$. Then $|T| = 2^\kappa$, $T \subseteq M$. Consequently $T \in M$, $T \subseteq j(U)$ and as $2^\kappa < j(\kappa)$, $\cap T \in j(U)$. Consequently $\cap T$ is confinal in $j(\kappa)$. For every $\xi \in \cap T$, $U_\xi = U$.

(See Ketonen (1972) and Sureson (1981) for related results.)

1.3. THE CLOSED UNBOUNDED FILTER AND STATIONARY SETS

In this section we will study a combinatorial structure which is closely related to measurable cardinals.

Since the existence of a κ-complete ultrafilter on a cardinal κ cannot be proved in set theory, if we want a set theoretic structure on cardinals which corresponds to κ-complete normal measures, but which exists independently of measurable cardinals, we must be willing to drop some of the properties of κ-complete normal measures. We already know that dropping normality is not enough. With the property of κ-completeness something similar happens. We would like to maintain ω-completeness but if there is an ω-complete ultrafilter on a cardinal then there is a κ-complete ultrafilter on some cardinal κ. So we drop the property of ultrafilter and look for a natural κ-complete normal filter on an uncountable regular cardinal κ.

A good candidate is the filter we now define. Let κ be a regular cardinal. We say that $X \subseteq \kappa$ is unbounded in iff $\cup X = \kappa$. X is closed iff for every $\alpha < \kappa$, $\cup (X \cap \alpha) \in X$. The last clause simply means that the increasing enumeration function for X is continuous in the natural topology of κ. Let $\xi < \kappa$ and $\langle X_\eta \rangle_{\eta < \xi}$ be a family of closed unbounded subsets of κ. We leave as an exercise to show that $\bigcap_{\eta < \xi} X_\eta$

is again closed and unbounded. From this it follows immediately that the $\langle X_\eta \rangle_{\eta<\kappa}$ is a sequence of closed unbounded sets then $\Delta_{\eta<\kappa} X_\eta =$ $\{\xi \mid (\eta)_\xi (\xi \in X_\eta)\}$ is again closed and unbounded. To see this, consider any $\alpha_o < \kappa$. Since $\bigcap_{\beta<\alpha_o} X_\beta$ is unbounded pick α_1 in this set, $\alpha_o < \alpha_1$. Repeat the construction ω-many times; $\alpha_\omega = \bigcup_{\eta\in\omega} \alpha_\eta$ is an element of $\Delta_{\eta<\kappa} X_\eta$ above α_o. Closedness $\Delta_{\eta<\kappa} X_\eta$ follows easily: if $\langle \alpha_\xi \rangle_{\xi<\rho}$ is an increasing sequence of elements of $\Delta_{\eta<\kappa} X_\eta$, $\alpha = \bigcup_{\xi<\rho} \alpha_\xi$ then, for any $\nu < \alpha$ some final segment of $\langle \alpha_\xi \rangle_{\xi<\rho}$ is contained in X_ν. Thus $\alpha \in X_\nu$ i.e. $\alpha \in \Delta_{\eta<\kappa} X_\eta$. So we have that the collection of closed unbounded subsets of κ generates a normal κ-complete uniform filter on κ. Call $X \subseteq \kappa$ stationary iff for every Y closed and unbounded, $X \cap Y \neq \phi$. Thus X is stationary iff there exists an ultrafilter U such containing X and all the closed unbounded subsets of κ.

Theorem (Fodor): if $f : \kappa \to \kappa$ is regressive on a stationary set $X \subseteq \kappa$ then there exists $Y \subseteq X$, Y stationary, such that f is constant on Y. In particular $f^{-1} * \{\xi\}$ is stationary for some ξ. (Proof: otherwise for every $\xi \in \kappa$ there must be a closed and unbounded C_ξ such that

$$C_\xi \cap (X \cap f^{-1} * \{\xi\}) = 0$$

Consider $C = \Delta_{\eta<\kappa} C_\eta$. It is closed and unbounded. But if $\xi \in C \cap X$ then $(\eta)_\xi (\xi \in C_\eta)$ i.e. since $\xi \in X, f(\xi) \neq \eta$. Thus $f(\xi) \geq \xi$ contradicting the fact that f is regressive on X).

The final segments in κ are sets of form $\{\xi \mid \xi \geq \alpha\}$. The closed unbounded filter i.e. the filter generated by the closed unbounded sets is the least normal filter containing final segments. To see this let C be closed and unbounded and let $f : \kappa \to \kappa$ be the increasing enumeration; given $\alpha \in \kappa$ let $X_\alpha = \{\beta \mid \beta > f(\alpha)\}$. $\Delta_{\alpha\in\kappa} X_\alpha$ is a closed an unbounded set. Now if $\beta \in \Delta_{\alpha\in\kappa} X_\alpha$ then $(\gamma)_\beta (\beta \in X_\gamma)$ i.e. $(\gamma)_\beta (\beta > f(\gamma))$. We can safely assume that β is limit so $\beta \geq f(\beta)$. But $\beta \leq f(\beta)$ always (as f is increasing) thus $\Delta_{\alpha\in\kappa} X_\alpha \cap \mathrm{Lim} \subseteq \{\alpha \mid \alpha = f(\alpha)\} \subseteq C$.

Now if F is any normal filter containing final segments then $\mathrm{Lim} = \Delta_{\alpha\in\kappa} \{\beta|\beta > \alpha\}$ is in F, also $\Delta_{\alpha\in\kappa} X_\alpha \in F$ and so $C \in F$. Thus all closed and unbounded sets are in F. Actually our proof shows also that for C closed and unbounded $\{\alpha|\alpha = f(\alpha)\}$ is again closed and unbounded. We also proved that the family of sets of form $\Delta_{\xi<\kappa} X_\xi$ where each X_ξ is a final segment of κ forms a basis for the filter generated by the closed unbounded sets. We will show now that stationary sets have a very strong property, namely, that every stationary set is the disjount union of κ-many stationary subsets. This result, due to Solovay has been an important point in the development in set theory. To prove this fact and exhibit the main ideas of the argument we will examine different cases.

Let $\kappa = \mu^+$ for a cardinal μ and $X \subseteq \kappa$ a stationary subset of κ. We can assume that $X \subseteq [\mu,\kappa)$. Given $\xi \in X$ fix $f_\xi : \mu \to \xi$ a bijection.
Claim: There is an $\eta < \mu$ such that for all $\gamma < \kappa$ the set $\{\xi \in X|f_\xi(\eta) > \gamma\}$ is stationary. Suppose the contrary, so for all $\eta < \mu$ there is an ordinal γ_η and a closed unbounded set $C_\eta \subseteq \kappa$ such that if $\xi \in C_\eta \cap X$ then $f_\xi(\eta) \leq \gamma_\eta$. Since κ is regular $\gamma = \bigcup_{\eta<\mu} \gamma_\eta < \kappa$. Put $C = \bigcap_{\eta<\mu} C_\eta$. Since C is closed and unbounded $C \cap X$ is stationary and therefore unbounded. But if $\xi \in C \cap X$ then $f_\xi(\eta) \leq \gamma$ for all $\eta < \mu$. This says that $\xi < \gamma$. So $X \subseteq \gamma + 1$, a contradiction. This establishes the claim.

Fix η as in the claim, and consider the regressive function $f : X \to \kappa$ defined by $f(\xi) = f_\xi(\eta)$. Since X is stationary, there is $X_1 \subseteq X$ stationary and $\gamma_1 \subset \kappa$ such that $f(\xi) = \gamma_1$ for all $\xi \in X_1$. But the claim says that $\{\xi|f_\xi(\eta) > \gamma_1\}$ is stationary so we can repeat the construction starting from this set to obtain X_2 stationary and $\gamma_2(\gamma_2 > \gamma_1)$ such that $f(\xi) = \gamma_2$ for all $\xi \in X_2$. Doing this κ-many times we obtain the desired κ-many pairwise disjoint subsets of X.

We have thus finished with the case in which κ is a successor cardinal. Assume now that κ is a limit regular cardinal and that $X \subseteq \kappa$ is a stationary subset such that there is a $\zeta < \kappa$ for which

$X \subseteq \{\alpha < \kappa | \text{cf } \alpha = \zeta\}$. To partition X we fix for each $\xi \in X$, a cofinal sequence $\{\alpha_\eta^\xi \mid \eta < \zeta\}$.

As before, there is $\eta < \zeta$ such that for all $\gamma < \kappa$ the set $\{\xi \in X | \alpha_\eta^\xi > \gamma\}$ is stationary. So the function $f : X \to \kappa$ given by $f(\xi) = \alpha_\eta^\xi$ allows us to construct the desired stationary subsets of X.

Now assume κ is a regular cardinal such that $\{\alpha \in \kappa | \alpha \text{ is regular}\}$ is not stationary in κ. Then for every stationary $X \subseteq \kappa$ there is a $\zeta < \kappa$ such that $X \cap \{\alpha | \text{cf}(\alpha) = \zeta\}$ is stationary. Indeed, since $\text{Reg} \cap \kappa$ is not stationary then there is a closed unbounded $C \subseteq \kappa$ such that $(\alpha)_C(\text{cf}(\alpha) < \alpha)$. Take $X \cap C$ and apply regressive function $f : \alpha \to \text{cf}(\alpha)$.

The cardinals κ which have the property that $\{\alpha \in \kappa | \alpha \text{ is a regular}\}$ is stationary are called Mahlo cardinals. All what was done above amounts to say that if κ is not Mahlo then we know how to partition any stationary set. And if κ is Mahlo then we also know how do to this providing $X - \{\alpha | \alpha \text{ is regular}\}$ is stationary. Thus we are left with the case: κ is Mahlo, $X \subseteq \{\alpha | \alpha \text{ is regular}\}$. Now let us notice the following: If C is closed unbounded in κ than Cantor's derivative of C, $C' = \{\alpha : \cup(C \cap \alpha) = \alpha\}$ is again closed and unbounded. Also C' consists of limit ordinals only. Solovay's trick to prove the result is the following: if x is stationary and $X \subseteq \{\alpha \in \kappa : \alpha \text{ is a regular}\}$ (it is enough to assume that $X \subseteq \{\alpha < \kappa | \text{cof}(\kappa) > \omega\}$). Then $S = \{\xi \in X : X \cap \xi \text{ is nonstationary in } \xi\}$ is stationary. To see this take any closed and unbounded C, its derivative C' and let $\xi = \cap (X \cap C')$. We show $\xi \in S$: to this end we need to find a closed and unbounded set in ξ omitting $X \cap \xi$. Look, $C \cap \xi$ is unbounded in ξ and it is closed. Also $C' \cap \xi = (C \cap \xi)'$. Thus $C' \cap \xi$ is closed and unbounded in ξ and by construction, as ξ is the least in $C' \cap X$, $(X \cap \xi) \cap (C' \cap \xi) = \phi$. Thus $X \cap \xi$ is not stationary in ξ i.e. $\xi \in S$ and $\xi \in C$ so $S \cap C \neq \phi$.

Now we are armed to complete the proof of Solovay's theorem. We proceed similarly to previous cases. Let $X \subseteq \kappa$ be given we can assume that X consists of regular cardinals and that for all $\xi \in X$, $X \cap \xi$ is nonstationary in ξ. Thus for every $\xi \in X$ there is a continuous sequence $\langle a_\alpha^\xi \rangle_{\alpha<\xi}$ outside of X converging to ξ, $\bigcup_{\alpha<\xi} a_\alpha^\xi = \xi$. We claim that there is ν such that for all $\eta < \kappa$ $\{\xi \in X | a_\nu^\xi \geq \eta\}$ is stationary. Indeed, if not then for every $\nu < \kappa$ there is a closed unbounded C_ν and η_ν such that:

$(\xi)_{X \cap C_\nu}(a^\xi_\nu < \eta_\nu)$. Set $C = \Delta_{\nu<\kappa} C_\nu$. If $f \in {}^\kappa\kappa$ then $\{\xi | (\alpha)_\xi (f(\alpha) < \xi\}$ is closed and unbounded (we leave the proof to the reader) so $F = \{\xi | (\alpha)_\xi (\eta_\alpha < \xi)\}$ is closed and unbounded. Consider now the first two numbers in $C \cap F \cap X$, say $\alpha < \beta$. If $\xi < \alpha$ then $a^\beta_\xi < \eta_\xi < \alpha$. By continuity of $\langle a^\beta_\eta \rangle_{\eta<\beta}$, $a^\beta_\alpha = \alpha$. But, then as $\alpha \in X$, $a^\beta_\alpha \in X$ which contradicts the fact that $a^\xi_\eta \in X$ for every ξ and η. Our claim granted we proceed very much a before; the function $f : \xi \to a^\xi_\nu$ is regressive and with its help we construct the required collection of disjoint stationary subsets.

1.4. THE ALGEBRA $P(\kappa)_{NS_\kappa}$ AND VARIOUS OTHER FACTS CONCERNING STATIONARY SETS.

The nonstationary sets in κ form a κ-complete ideal, called NS_κ. This is the ideal dual to the club filter. As usual, we will denote by NS^+_κ the set of subsets of κ not in NS_κ. The theorem of Solovay proved in 3 shows that NS_κ is nowhere κ-saturated; every set in NS^+_κ can be partitioned into κ-many objects in NS^+_κ. Our principal interest in this section is the study of the algebra $P(\kappa)/NS_\kappa$. This Boolean algebra is atomless and κ-complete (which is immediate from the fact that NS_κ is κ-complete). The elements of $P(\kappa)/NS_\kappa$ are equivalence classes of the equivalence relation: $X \sim Y \iff X \dot{-} Y \in NS_\kappa$. This relation is - in turn - the canonical equivalence related with the preordering: $X \leq Y \iff X - Y \in NS_\kappa$. We will denote by [X] the equivalence class of $X \subseteq \kappa$. We will think of these equivalence classes as degrees of stationariness. The degrees form a κ-complete Boolean algebra. Its zero is the degree of the empty set (or equivalently of a nonstationary set). Its unity is the degree of closed and unbounded sets. Our purpose now is to see that the algebra of degrees is in fact κ^+-complete and that its Boolean meet corresponds to the diagonal intersection. This result is due to Baumgartner-Hajnal and Maté. Some further facts, explaining the relationships between Boolean sum and set theoretical sum in $P(\kappa)/NS_\kappa$ are due to ourselves. First we show the following lemma (due to Baumgartner, Taylor and Wagon). Suppose that $\langle X_\xi \rangle_{\xi<\kappa}$, $\langle Y_\xi \rangle_{\xi<\kappa}$ are sequences of subsets of κ such that all $\xi < \kappa$, $X_\xi \sim Y_\xi$. Then: $\Delta_{\xi<\kappa} X_\xi \sim \Delta_{\xi<\kappa} Y_\xi$.

Proof:

Let $A = \Delta_{\xi<\kappa} X_\xi$ and $B = \Delta_{\xi<\kappa} Y_\xi$. If $A \dot{-} B$ is stationary then, without loss of generality we can suppose $A - B$ is stationary. For $\eta \in A - B$ set $f(\eta)$ = the least α such that $\eta \notin Y_\alpha$. Since $B = \Delta_{\xi<\kappa} y_\xi$, f is regressive. Therefore there is $\alpha_o \in \kappa$ and $C \subseteq A - B$, C stationary, such that f is constant on C with value α_o. Thus $C \cap Y_{\alpha_o} = \phi$. Let $C' = C \cap \{\beta : \beta > \alpha_o\}$. Then, as $C' \subseteq \Delta_{\xi<\kappa} X_\xi$, $C' \subseteq X_{\alpha_o}$. But C' is stationary and $C' \subseteq X_{\alpha_o} - Y_{\alpha_o}$. This is absurd. □

This fact tells us that Δ is in fact defined on degrees! Also, Δ embodies $\cap$ as we see in the following lemma: Let $\langle X_\alpha \rangle_{\alpha<\eta}$, $\eta < \kappa$ be a family of sets and $\langle Y_\xi \rangle_{\xi<\kappa}$ be another sequence of sets of length κ such that: $(\alpha)_\eta (E\xi)(X_\alpha \sim Y_\xi)$ and $(\xi)_\kappa (E\alpha)_\eta (X_\xi \sim Y_\alpha)$. Then $\Delta_{\xi<\kappa} Y_\xi \sim \cap_{\xi<\eta} X_\xi$. To prove this, set as above $A = \Delta_{\xi<\kappa} Y_\xi$, $B = \cap_{\alpha<\eta} X_\alpha$. We show that neither $A - B$ nor $B - A$ is stationary. Indeed, suppose $A - B$ is stationary. For $\xi \in A - B$ set $f(\xi)$ = least $\alpha, \xi \notin X_\alpha$. Then as $Rg\ f \subseteq \eta < \kappa$ there is a stationary $C \subseteq A - B$ on which f is constant with value - say - α_o. Pick the least ρ such that $Y_\rho \sim X_{\alpha_o}$. Then $C - \rho \subseteq Y_\rho$, $C \cap X_{\alpha_o} = \phi$. As $C - \rho$ is stationary this contradicts $Y_\rho \sim X_{\alpha_o}$. If $B - A$ is stationary, we proceed as in the proof of the result of Baumgartner, Taylor and Wagon.

These two facts allow us to introduce the operation Π in $P(\kappa)/NS_\kappa$ as follows:

Given $\langle d_\xi \rangle_{\xi<\eta}$ $(\eta \leq \kappa)$ define: $\Pi_{\xi<\eta} d_\xi = [\Delta_{\xi<\kappa} X_\xi]$ where $\langle X_\xi \rangle_{\xi<\eta}$ is any system of representatives for $\langle d_\xi \rangle_{\xi<\eta}$ such that if $\eta < \kappa$ then X_ξ, for $\eta \leq \xi \leq \kappa$ is X_o. The lemmas above say that our operation Π is well-defined. It remains to be shown that Π is the meet in our algebra $P(\kappa)/NS_\kappa$. For this we need to prove:

a) $\Pi_{\xi<\kappa} d_\xi \leq d_\xi$ for all $\xi < \kappa$

b) If $e \leq d_\xi$ for all $\xi < \kappa$ then $e \leq \Pi_{\xi<\kappa} d_\xi$

a) is in fact obvious. Suppose that $[X] = \prod_{\xi<\kappa} d_\xi$. We can assume that

$X = \Delta_{\xi<\kappa} X_\xi$, $[X_\xi] = d_\xi$ for all $\xi < \kappa$. Then $X - \xi \subseteq X_\xi$ and as

$(X - \xi) \sim X$ we are done.

b) Assume $[X] = e$, $[X_\xi] = d_\xi$ (for every $\xi < \kappa$). By the lemma of Baumgartner, Taylor and Wagon it is enough to show $X - \Delta_{\xi<\kappa} X_\xi \in NS_\kappa$. For this, repeat the argument of the lemma of Baumgartner, Taylor and Wagon.

We have thus proved that $P(\kappa)/NS_\kappa$ is κ^+-complete and that for sequences of length κ, Δ plays the role of Boolean meet. Nevertheless, for families of the length κ, Π is not related to intersection. As it is shown by the following trivial example: $X = \kappa - \xi$. Then $\bigcap_{\xi<\kappa} X_\xi = \emptyset$ $\prod_{\xi<\kappa} [X_\xi] = \prod_{\xi<\kappa} 1 = 1$, $[\bigcap_{\xi<\kappa} X_\xi] = 0$. To help understand the relationship among the Boolean operations in $P(\kappa)/NS$ and set theoretical operations we will consider the dual operation $\Sigma(= - \Pi -)$. First of all notice that since $P(\kappa)/NS_\kappa$ is κ^+ complete, the supremum of a family of degrees of length κ is always equal to a supremum of a pairwise disjoint family of such length. Furthermore notice that if $\langle d_\xi \rangle_{\xi<\kappa}$ is a family of pairwise disjoint degrees then there is a system of representatives $\langle X_\xi \rangle_{\xi<\kappa}$ which is again pairwise disjoint. Our purpose now is to determine a connection between $d_1 = \sum_{\xi<\kappa} [X_\xi]$ and $d_2 = [\bigcup_{\xi<\kappa} X_\xi]$. Clearly $d_1 \leq d_2$. However as we will see soon, they may be different. To see the exact connection between d_1 and d_2 we introduce the notion of canonical selector for $\langle X_\xi \rangle_{\xi<\kappa}$. If $X = \langle X_\xi \rangle_{\xi<\kappa}$ is a family of pairwise disjoint stationary sets, let S_X be the set consisting of the least elements of the X_ξ's. Therefore $[S_X] \wedge [X_\xi] = \emptyset$ and so $[S_X] \wedge \sum_{\xi<\kappa} [X_\xi] = \emptyset$. We have the following property: If $X = \langle X_\xi \rangle_{\xi<\kappa}$ is a family of pairwise disjoint nonempty sets then:

$$[\bigcup_{\xi<\kappa} X_\xi] = \sum_{\xi<\kappa} [X_\xi] \vee [S_X]$$

Clearly $\geq$ holds, since $S_X \leq \bigcup_{\xi<\kappa} X_\xi$ and so $[S_X] \leq [\bigcup_{\xi<\kappa} X_\xi]$. The other inequality is a bit more complicated; we just need to show that the difference $(\bigcup_{\xi<\kappa} X_\xi) \setminus S_X$ is of degree $\sum_{\xi<\kappa} [X_\xi]$. Clearly this amounts to showing $[(\bigcup_{\xi<\kappa} X_\xi) - S_X] \leq \sum_{\xi<\kappa} [X_\xi]$ (the other inequality is obvious), this is the same as:

$$\prod_{\xi<\kappa} [-X_\xi] \leq -[(\bigcup_{\xi<\kappa} X_\xi) - S_X] \text{ , or}$$

$$[\underset{\xi<\kappa}{\Delta} (\kappa - X_\xi)] \leq [\bigcap_{\xi<\kappa} (\kappa - X_\xi) \cup S_X] \text{ , the proof of}$$

this last thing we leave to the reader.

Thus, given $\langle d_\xi \rangle_{\xi<\kappa}$, a family of nonzero degrees, using a simple construction it is possible to show that the degree of the canonical selectors for families $\{X_\xi | \xi < \kappa\}$ such that $[X_\xi] = d_\xi$ (for all $\xi < \kappa$) give all degrees between $\mathbb{0}$ and $-\sum_{\xi<\kappa} d_\xi$.

Consider now a pairwise disjoint family $\{d_\xi | \xi < \kappa\}$ of degrees. We will indicate how to obtain a set of degree $\sum_{\xi<\kappa} d_\xi$. Pick a pairwise disjoint family X representatives. (This is always possible since for any family of representatives $\{X_\xi | \xi < \kappa\}$, putting $X'_\xi = X_\xi - \bigcup_{\eta<\zeta} X_\eta$ we get a pairwise disjoint family). The set $\bigcup_{\xi<\kappa} X_\xi - S_X$ is of degree exactly $\sum_{\xi<\kappa} d_\xi$.

We conclude this section proving a result due the Shelah and Pelletier on functions on stationary sets. First we show a lemma due to Weglorz: Assume I is an ideal on κ, $A \notin I$, $0 \notin A$, $f : A \to \kappa$ is regressive on A. Then there exists $B \subseteq A$, $B \notin I$ such that $f \upharpoonright B : B \to \kappa - B$. (For I normal this lemma is trivial - in view of Fodor theorem - but we will use it for other ideals).

<u>Proof</u>:

As f is regressive there must be, for every $\xi \in A$, a natural number n_ξ such that $f^{n_\xi}(\xi) \notin A$. Then $A = \{\xi \in A | n_\xi \text{ is even}\} \cup \{\xi \in A | n_\xi \text{ is odd}\}$. As $A \notin I$ at least one of those sets is not in I, and f transforms each part into its complement! Using this lemma we shall prove the following fact due to Pelletier and Shelah: If f,g are two functions defined on a stationary $X \subseteq \kappa$ and such that $(\xi)_X(f(\xi) \neq g(\xi))$ then there is $Y \subseteq X$, Y stationary such that $(f * Y) \cap (g * Y) \neq \phi$.

Proof:

Without loss of generality we can assume that $f,g \in {}^{\kappa}\kappa$. Since $(\xi)_X(f(\xi) \neq g(\xi))$, $X = \{\xi | f(\xi) < g(\xi)\} \cup \{\xi | g(\xi) < f(\xi)\}$. Again, we can assume that $\{\xi | f(\xi) < g(\xi)\}$ is stationary. And in fact we shall deal only with this set. There are two cases. Case one is when for a certain ξ, $g^{-1} * \{\xi\}$ is stationary. In this case take $Y = g^{-1} * \{\xi\}$. For $\eta \in Y$, $f(\eta) < g(\eta) = \xi$, thus $(f * Y) \cap (g * Y) = (f * Y) \cap \{\xi\} = \phi$. Now we consider case two: for all ξ, $g^{-1} * \{\xi\} \in NS_\kappa$. Let D be a selector of least elements for $\langle g^{-1} * \{\xi\}\rangle_{\xi \in Rgg}$. Then D is stationary, in fact $[D] = [X]$. Now, g restricted do D is one-to-one. Define

$$J = \{X \leq \kappa | (g|D)^{-1} * X \in NS_\kappa\}$$

(while computing $(g|D)^{-1}$ we <u>do not</u> take the whole g^{-1} but only those things which are in D!)

Let $A = g * D$ and define $h : A \to \kappa$ as follows: $h(\xi) = f((g|D)^{-1}(\xi))$. Clearly $A \in J$. Indeed, if $A \in J$ then $(g|D)^{-1} * (g*D) \in NS_\kappa$ i.e. $D \in NS_\kappa$ (as $g|D$ is one-to-one on its domain) which is not true. Now, in order to use Weglorz' lemma we show that h is regressive on A. Indeed $h(\xi) = f((g|D)^{-1}(\xi))$. But $A = g * D$ so $\xi = g(\zeta)$ so $h(\xi)$ is f of this <u>unique</u> ζ as $f(\xi) < g(\zeta) = \xi$ and $h(\xi) = f(\zeta)$ we are done. Now we apply the lemma and get $A' \subseteq A$, $A' \notin J$ such that $h|A' : A' \to \kappa - A'$. We claim $Y = (g|D)^{-1} * A'$ is our desired stationary set. First of all Y is stationary since if $Y \in NS_\kappa$ then $A' \in J$ which is not true. If $\xi \in Y$ then $g(\xi) \in A'$ by definition. But $f(\xi) = f((g^{-1}|D)(g(\xi))) = h(g(\xi)) \notin A'$ since $g(\xi) \in A'$ and h transforms A' into its complement. □

2.1. SUPERCOMPACT CARDINALS

Let κ be a regular cardinal and $\lambda \geq \kappa$. We will denote by $P_\kappa(\lambda)$ the set $\{p \subseteq \lambda | \ |p| < \kappa\}$. Given $p \in P_\kappa(\lambda)$ let $\hat{p}$, the cone over p, be the set $\{q \in P_\kappa(\lambda) | p \subseteq q\}$. Given a collection of subsets of $P_\kappa(\lambda)$, $\{X_\xi | \xi < \lambda\}$, its diagonal intersection, denoted by $\Delta_{\xi<\lambda} X_\xi$ is the set $\{p \in P_\kappa(\lambda) | p \in X_\xi$ for all $\xi \in p\}$.

We say that κ is λ-supercompact iff there is a non-principal ultrafilter on $P_\kappa(\lambda)$ which is κ-complete, normal (i.e. closed under diagonal intersections), and fine (i.e. contains the cones $\hat{p}$ for $p \in P_\kappa(\lambda)$. There is a notion of regressiveness related to normality (we shall introduce other different notions later): $f : p_\kappa(\lambda) \to \lambda$ is regressive on X for $X \subseteq P_\kappa(\lambda)$ iff for all $p \in X$, $f(p) \in p$. A filter $\mathcal{U}$ on $P_\kappa(\lambda)$ is normal iff for every function f regressive on a set $X \in \mathcal{U}$ there exists $Y \subseteq X$, $Y \in \mathcal{U}$ such that f is constant on Y. Note that κ is κ-supercompact iff κ is measurable. However, in oposition to the case of measurability, the existence of a fine κ-complete filter on $P_\kappa(\lambda)$ does not, in general, imply λ-supercompactness of κ. Normality of a filter $\mathcal{U}$ implies a property apparently stronger. Let $f : P_\kappa(\lambda) \to {}_\kappa(\lambda)$ be such that for all $p \in X$, $|f(p)| < |p \cap \kappa|$ where $X \in \mathcal{U}$. Then there exists $Y \subseteq X$, $Y \in \mathcal{U}$ such that f is constant on Y. This property generalizes normality because for every normal measure on $P_\kappa(\lambda)$ there exists a measure one set Z such that $p \cap \kappa$ is a cardinal for all p in Z. To prove our result notice that $\{p \in X | p \cap \kappa \in \mathrm{Ord}\}$ is of measure one (just use normality) and so the auxiliary function $f'(p) = |f(p)|$ is regressive let $X' \subseteq X$ on which f' is constant with value α and then order $f(p)$ is type α for every $p \in X'$. Now step-by-step fix elements of $f(p)$.

A basic result due to Reinhardt and Solovay is that κ is λ-supercompact iff there exists an inner model M and an elementary imbedding $j : V \to M$ such that κ is critical for j, $j(\kappa) > \lambda$ and ${}^\lambda M \subseteq M$. In fact given such an elementary imbedding, $\mathcal{U}$ is defined "generically" in a manner similar to the case of measurability; $X \in \mathcal{U} \iff j * \lambda \in j(X)$. A cardinal κ is supercompact iff it is λ-supercompact for all λ. This is a very strong large cardinal property. As shown by Vopenka and Hrbaček, a weaker property (strong compactness) already implies that for any set a, $V \neq L[a]$. In our case the proof is particularly simple; assume a is such that $V = L[a]$ and $a \subseteq \nu$ for some ordinal ν (we can assume this since $L[a] \models AC$). Since κ is ν-supercompact, there is an elementary embedding $j : V \to M$ with κ critical point and ${}^\nu M \subseteq M$. This last property implies that $a \in M$, and since $M \subseteq V = L[a]$ we have that $M = L[a]$. From this we can find a contradiction in several ways. First, this contradicts Kunen's result that there is no elementary embedding from V into V. Also, a contradiction follows from the fact that the ultrafilter on $P_\kappa(\nu)$ obtained from j is not in M (as in Scott's proof of $MC \to V \neq L$).

Another way is along Silver's argument that $V = L[a] \Rightarrow a^{\#}$ does not exist.

Our next task is to investigate the number of normal ultrafilters on $P_\kappa(\lambda)$. We will show that if κ is $2^{\lambda^{<\kappa}}$-supercompact then there are as many normal ultrafilters on $P_\kappa(\lambda)$ as there can be, namely, $2^{2^{\lambda^{<\kappa}}}$ normal ultrafilters. The proof, due to Solovay, is outlined below.

Notice that if κ is ν-supercompact and $\kappa \leq \lambda \leq \nu$ then κ is λ-supercompact as well. In fact j witnessing to ν-supercompactness witnesses to λ-supercompactness. We know that $U = \{X \subseteq P_\kappa(\nu) \mid j * \nu \in j(X)\}$ is a suitable ultrafilter in $P_\kappa(\lambda)$. Also $U_\lambda = \{X \subseteq P_\kappa(\lambda) \mid j_*{}^{\lambda} \in j(X)\}$ is a suitable ultrafilter in $P_\kappa(\lambda)$. The connection between these ultrafilters is as follows:

$$X \in U_\lambda \Longleftrightarrow \{p \in P_\kappa(\nu) \mid p \cap \lambda \in X\} \in U.$$

Then we show that the ultrapowers $V^{P_\kappa(\lambda)}/U_\lambda$ and $V^{P_\kappa(\nu)}/U$ are closely connected; the function $k : [f] \to [g_f]$ defined by $g_f(p) = f(p \cap \lambda)$ is an elementary embedding from $V^{P_\kappa(\lambda)}/U_\lambda$ into $V^{P_\kappa(\nu)}/U$ which makes the following diagram commute:

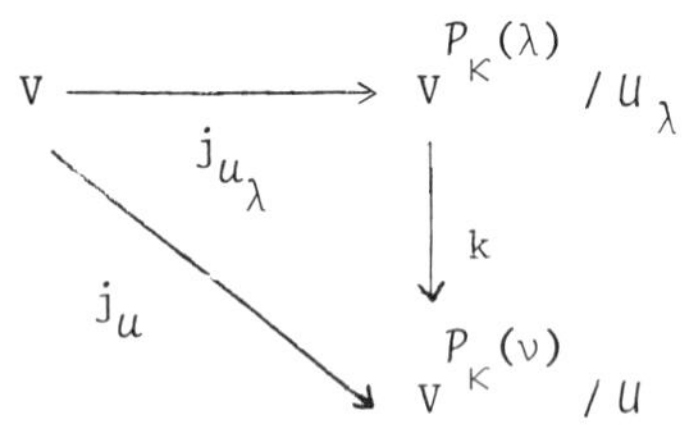

There are four facts which we need in the course of the argument and leave to the reader check:

1º For $\alpha \leqslant \lambda$, $k(\alpha) = \alpha$

2º For $p \in P_\kappa(\lambda)$, $k(p) = p$

3º $k(P_\kappa(\lambda)) = P_\kappa(\lambda)$

4º If $X \subseteq P(P_\kappa(\lambda))$ and $X \in V^{P_\kappa(\lambda)}/U_\lambda$ then $k(X) = X$

With these facts we are able to prove a crucial lemma: If κ is $2^{\lambda^{<\kappa}}$-supercompact and $X \in PPP_\kappa(\lambda)$ then there exists a normal ultrafilter

V in $P_\kappa(\lambda)$ such that $X \in M_V (\approx V^{P_\kappa(\lambda)}/V)$.

To see this consider U a normal ultrafilter on $P(2^{\lambda^{<\kappa}})$ and the diagram

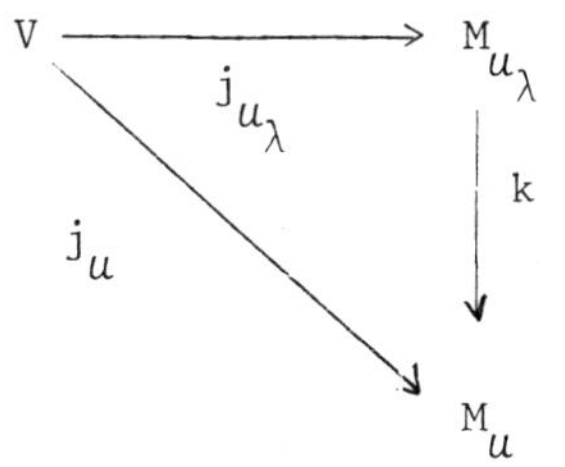

where M_{U_λ}, M_U are transitive and isomorphic to $V^{P_\kappa(\lambda)}/U_\lambda$ and $V^{P_\kappa(2^{\lambda^{<\kappa}})}/U$ respectively.

If the lemma is not true then there exists $X \in PPP_\kappa(\lambda)$ such that for every normal V in $P_\kappa(\lambda)$ $X \notin PPP_\kappa(\lambda) \cap M_V$. Abbreviate the above as: $(EX)\, \phi(X,\kappa,\lambda)$. But if this holds it also holds in M_U. Since all normal ultrafilters on $P_\kappa(\lambda)$ belong to M_U (by the closure properties of M_U). Thus, since $k(\kappa) = \kappa$ and $k(\lambda) = \lambda$, $M_{U_\lambda} \models (EX)\, \phi(X,\kappa,\lambda)$ pick $X \in M_{U_\lambda}$ making (in M_{U_λ}) $\phi(X,\kappa,\lambda)$ true. Passing to M_U and back to V we find that $\phi(X,\kappa,\lambda)$ is true which is absurd since U_λ shows that $\neg\phi(X,\kappa,\lambda)$ is true. From the lemma, the theorem follows by a cardinality argument: If V is normal on $P_\kappa(\lambda)$ then $|j_V(\kappa)| = 2^{\lambda^{<\kappa}}$ (just calculate the number of functions with values smaller than κ) so: $j_V(\kappa) < (2^{\lambda^{<\kappa}})^+$. Now $M_V \models$ "$j_V(\kappa)$ is inaccessible" so $(2^{2^{\lambda^{<\kappa}}})^{M_V} < j_V(k)$. Thus $(2^{2^{\lambda^{<\kappa}}})^{M_V} < (2^{\lambda^{<\kappa}})^+$.

But by the above lemma: $PP(P_\kappa(\lambda)) = \bigcup_V PPP_\kappa(\lambda) \cap M_V$ as $|PP(P_\kappa(\lambda))| = 2^{2^{\lambda^{<\kappa}}}$ and each of terms of the above union has at most $2^{\lambda^{<\kappa}}$ elements therefore the index-set must be of cardinality $2^{2^{\lambda^{<\kappa}}}$. □

If $\lambda = \kappa$ then since κ is inaccessible then $\kappa^{<\kappa} = \kappa$. Thus, if κ is 2^κ-supercompact then there are 2^{2^κ} normal ultrafilters on κ.

This result was further strengthened by Menas (for $\lambda > \kappa$) (see section 2.4).

The result, proved in 1.2 tells us that all ultrafilters (in particular those normal ones constructed above) appear, and with many repetitions, in the collection $\{\mathcal{U}_\xi \mid \xi < j(\kappa)\}$ where j witnesses 2^κ-supercompactness.

This result generalizes to the context of $P_\kappa(\lambda)$. Given $j : V \to M$ an elementary embedding, with critical point κ, $\lambda \geq \kappa$ and $A \subseteq j(\lambda)$, $A \in M$, $|A|^M < j(\kappa)$. Define $\mathcal{U}_{j,A} = \{X \subseteq P_\kappa(\lambda) : A \in j(X)\}$. The ultrafilters $\mathcal{U}_{j,A}$ are κ-complete. Besides $\mathcal{U}_{j,A}$ is fine iff $j * \lambda \subseteq A$.

Added in Proof:
The argument of Kanamori provided in Section 1.2 shows, in addition, that $\{A:\mathcal{U}=\mathcal{U}_{j,A}\}$ is unbounded in $[P_{j(\kappa)}(j(\lambda))]^M$.

If κ is $2^{\lambda^{\underset{\smile}{\kappa}}}$-supercompact and j witnesses to this then every κ-complete ultrafilter in $P_\kappa(\lambda)$ is of the form $\mathcal{U}_{j,A}$. This is proved just as in 1.2

To end this section, let us show the reflection property of supercompact cardinals. We assume that the reader is familiar with Levy's set theoretical hierarchy. First of all notice that any supercompact κ is inaccessible, so $H_\kappa = V_\kappa$ and thus Σ_1-formulas are absolute for V_κ, thus Π_2 formulas reflect downward to V_κ, in symbols $V_\kappa \prec_1 V$. We show now that Σ_2 formulas are also preserved downwards and so: Σ_2-formulas are absolute with respect to V_κ. Assume ϕ : Ex ψ where ψ is a Π_1-formula. Let $y \in V_\kappa$ and pick x making $\psi(\cdot,y)$ true where $x \in V_\nu$. By supercompactness we have $\lambda > \nu$ and $j : V \to M_\nu$ such that $x \in M_\nu$. ψ being Π_1, $M_\nu \models \psi(x,y)$ and so as $j(\kappa) > \nu$ $V^M_{j(\kappa)} \models \psi(x,y)$. Thus $V^M_{j(\kappa)} \models (Ex)\,\psi\,(x,y)$ and as $V_\kappa \prec V^M_{j(\kappa)}$, $V_\kappa \models (Ex)\,\psi\,(x,y)$ which was to be proved. As Σ_3 formulas reflect upwards and Π_3 formulas reflect downwards, therefore formulas Δ^{ZF}_3 (ZF-provably both Σ_3 and Π_3) are absolute with respect to V_κ.

2.2. CLOSED UNBOUNDED SETS IN $P_\kappa(\lambda)$

The notion of closed and unbounded subset of $P_\kappa(\lambda)$ was introduced by Jech. A subset X of $P_\kappa(\lambda)$ is called unbounded iff for every $p \in P_\kappa(\lambda)$ there is $q \supseteq p$, $q \in X$. In other words an unbounded set is one which meets all cones. A subset X of $P_\kappa(\lambda)$ is closed iff for every $\subseteq$- chain $\langle p_\xi \rangle_{\xi<\eta}$ $(\eta < \kappa)$ of elements of X, $\bigcup_{\xi<\kappa} p_\xi \in X$. Actually the original definition of Jech required closure of X under sums of directed systems and not only of chains. Magidor and Solovay proved that this is just the same notion. The proof of this fact is done by induction on the cardinality of the directed set D and is based on the following two general facts:

1º Assume $\langle p_\xi \rangle_{\xi<\kappa}$ (where κ is a regular uncountable cardinal) is a directed system. Then there are arbitrarily large $\eta<\kappa$ such that $\langle p_\xi \rangle_{\xi<\eta}$ is directed.

2º Assume D is an infinite directed family and $E \subseteq D$ any subfamily. Then there exists $E \subseteq F \subseteq D$, $|F| = |E| + \aleph_0$, F directed.

From these facts the result follows by induction: For finite D there is nothing to prove; if $|D| = \aleph_0$ we order D in type $\aleph_0$ and find inductively a subchain of D with same union. For $|D| = \kappa$, κ uncountable regular we use 1º and the inductive assumption and for κ uncountable singular we order D in type κ and then, using 2º, change the ordering obtaining one with the property that every initial segment whose type is a cardinal is directed.

The family of closed and unbounded subsets of $P_\kappa(\lambda)$ is closed under intersections of length less than κ. To prove this we use a hand-over-hand argument i.e. a choice of elements from consecutive subsets of $P_\kappa(\lambda)$. An interesting fact is that closed and unbounded sets are also closed under diagonal intersections. i.e. if all X_ξ's $(\xi < \lambda)$ are closed and unbounded then so is $\Delta_{\xi<\lambda} X_\xi$. Indeed given $p \in P_\kappa(\lambda)$ choose $p_0 \in \bigcap_{\xi \in p} X_\xi$, $p \subseteq p_0$ then $p_1 \in \bigcap_{\xi \in p_0} X_\xi$, $p_0 \subseteq p_1$, etc. $q = \bigcup_{n\in\omega} p_n$ belongs to $\Delta_{\xi<\lambda} X_\xi$ and covers p This shows unboundedness. If $p_0 \subseteq p_1 \subseteq \ldots \subseteq p_\xi \subseteq \ldots\ \xi < \eta < \kappa$, then $p = \bigcup_{\xi<\eta} p_\xi = \bigcup_{\alpha<\xi<\eta} p_\xi$ for every $\alpha < \eta$. But then p_ξ for $\xi < \alpha$ belongs to every X_ν for ν in p_α. This $p \in X_\alpha$ for every α in p. This shows closedness. The filter $\mathrm{CLUB}_{\kappa,\lambda}$ generated by the closed and unbounded sets is, thus, κ-complete and normal. Also every cone $\hat{p}$ is both closed and unbounded thus our filter is fine. In the next section we will prove that $\mathrm{CLUB}_{\kappa,\lambda}$ is the least κ-complete, fine, normal filter in $P_\kappa(\lambda)$. For a moment let us deal with the stationary sets in $P_\kappa(\lambda)$ and their splitting problem. A subset $X \subseteq P_\kappa(\lambda)$ is called stationary iff it meets every $\text{club}_{\kappa,\lambda}$ set. Equivalently iff there exists an ultrafilter containing $\{X\} \cup \text{CLUB}_{\kappa,\lambda}$. Normality of $\text{CLUB}_{\kappa,\lambda}$ implies a version of the Fodor theorem for stationary subsets of $P_\kappa(\lambda)$. If f is regressive on a stationary set $X \subseteq P_\kappa(\lambda)$ then there exists $Y \subseteq X$, Y stationary and $f|Y$ is constant. Below, we prove Jech's splitting theorem for the case when κ is a successor cardinal (without additional assumptions) and Baumgartner's splitting theorem (every stationary set can be split if $0^{\#}$ does not exist). Jech proved that if $\kappa = \mu^+$ and λ is regular then every stationary set in $P_\kappa(\lambda)$ can be split into λ many pairwise disjoint stationary. To see this let $X \subseteq P_{\mu^+}(\lambda)$ be given. For every $p \in X$ let $\langle a^p_\xi \rangle_{\xi<\mu}$ be an enumeration of p (we can assume this as the set $\{p : |p| = \mu\}$ is closed and unbounded in $P_{\mu^+}(\lambda)$). We claim now that there exists $\xi < \mu$ such that for every $\nu < \lambda$ $\{p \in X : a^p_\xi > \nu\}$ is stationary. Otherwise for every $\xi < \mu$ there exists a $\nu_\xi < \lambda$ and C_ξ closed and unbounded such that for all $p \in X \cap C_\xi$ $a^p_\xi \leq \nu_\xi$. Set $C = \bigcap_{\xi<\mu} C_\xi$, $\nu = \bigcup_{\xi<\mu} \nu_\xi$. Then by regularity of λ, $\nu < \lambda$ and for $p \in C \cap X$, for all $\xi < \mu$, $a^p_\xi < \nu$. Thus if $p \in C \cap X$ then $p \subseteq \nu$. This shows that $C \cap X$ is bounded, contradiction. Once ξ is given, using the regressive function $p \to a^p_\xi$ we consecutively construct λ-many pairwise disjoint stationary subsets of X.

Baumgartner shows that under assumptions limiting the width of universe (say $\neg 0^{\#}$) and if κ and λ are regular $\kappa < \lambda$ then every set can be split. To see this we show a following lemma: Let $B = \{p \in P_\kappa(\lambda) \mid |p \cap \kappa| = |p \setminus \kappa|\}$. Then B is stationary. This is shown constructing by induction of length ω a sequence $\langle p_n \rangle_{n\in\omega}$ such that $p_{2n} \in B$, p_{n+1} belongs to a given closed unbounded set C. $p = \bigcup_{n\in\omega} p_n$ belongs to both B and C. Further we show that if $X \subseteq B$, X stationary then X can be split. Indeed given $p \in X$ we fix $f_p : p\setminus\kappa \xrightarrow[\text{onto}]{1-1} p \cap \kappa$. For $\alpha \in \lambda \setminus \kappa$ we set $f_\alpha(p) = f_p(\alpha)$. The fact that f_α is not defined on the whole X does not make harm, the domain of f_α being $X \cap \{\hat{\alpha}\}$ is stationary. f_α is regressive on its domain thus constant with value

$\nu_\alpha < \kappa$ on $X_\alpha \subseteq X$, X_α stationary. Now, since there are λ many α's and only κ many possible ν_α and since $cf(\lambda) > \kappa$ therefore there is $\nu < \kappa$ being ν_α for λ many α's. We claim that $\{X_\alpha \mid \nu_\alpha = \nu\}$ is a desired family of λ many pairwise disjoint stationary subsets of X. Indeed, if $p \in X_{\alpha_1} \cap X_{\alpha_2}$, $\alpha_1 \neq \alpha_2$ then $f_{\alpha_1}(p) = f_{\alpha_2}(p) = \nu$ i.e. $f_p(\alpha_1) = f_p(\alpha_2)$. But f_p was a bijection, contradiction.

Now, under assumption like $\neg 0^{\#}$ we show that B belongs to $CLUB_{\kappa,\lambda}$ filter. To see this enumerate L_λ with λ in such a way that L_κ is enumerated by κ. In this fashion an element of $P_\kappa(\lambda)$ is (a code of) a substructure of $P_\kappa(L_\lambda)$. Now the collection C of codes for elementary substructures of L_λ is clearly closed and unbounded in $P_\kappa(\lambda)$. We show that if $C \not\subset B$ then $0^{\#}$ exists. As is well known $0^{\#}$ exists iff there exists an L-ultrafilter. If $C \not\subset B$ we construct one. Indeed assume $p \in C - B$, this means that $|p \cap L_\kappa| < |p \setminus L_\kappa|$. As $p \prec L_\lambda$ there is a unique $\nu < \kappa$ such that p is isomorphic to L_ν. The converse to the Mostowski's isomorphism, π^{-1} embeds elementarily L_ν into L_λ. Besides, as $|p \cap L_\kappa| < \kappa$, π (and π^{-1}) are not identity functions. Let ρ be the first ordinal moved by π^{-1}. If $\rho^+ \leq \nu$ then $P^L(\rho) \subseteq L_\nu$ and so $\{X : \rho \in \pi^{-1}(X)\}$ is a desired L-ultrafilter. Thus we just need to show that $\rho^+ \leq \nu$. But ρ is either image of an element ζ of $p \cap L_\kappa$ or, in the worst case, image of the first element of $p \setminus L_\kappa$. In both cases there are more ordinals in $p \setminus \zeta$ then in $p \cap \zeta$. This completes the proof.

Finally, for κ inaccessible, λ regular bigger than κ, Di Prisco provided a method of splitting some stationary subsets of $P_\kappa(\lambda)$. We say that $X \subseteq P_\kappa\lambda$ is weakly closed iff it is closed under sums of $\subseteq$-chains $\langle p_\xi \rangle_{\xi<\nu}$ $(\nu < \kappa)$ such that $\xi_1 < \xi_2 \implies |p_{\xi_1}| < |p_{\xi_2}|$. Clearly every closed set is weakly closed but as we show below the converse need not be true. We leave to the reader checking that weakly closed sets (in our situation; κ inaccessible) generate a κ-complete normal, fine filter. The filer $WCLUB_{\kappa,\lambda}$ being larger than $CLUB_{\kappa,\lambda}$ there are less sets being stationary with respect to $WCLUB_{\kappa,\lambda}$. Those are called superstationary.

An important tool in our further task is the following: Weakly closed sets are closed under a different "diagonal intersection" Δ^c. If $\langle X_\xi \rangle_{\xi<\kappa}$ (notice that the length of the family is short now)

$\Delta^c_{\xi<\kappa} X_\xi = \{p|(\eta)_{|p|}(p \in X_\eta)\}$. The $WCLUB_{\kappa,\lambda}$ filter is closed under $\Delta^c_{\xi<\kappa} X_\xi$. We leave this fact to the reader and show that every superstationary set X can be split into λ many superstationary sets. To this end order each $p \in X$ into its $\langle a^p_\alpha \rangle_{\alpha<|p|}$. Now we show that there is $\xi < \kappa$ such that for all $\nu < \lambda$ $\{p : a^p_\xi > \nu\}$ is superstationary. To see this, assuming that no such ξ exists we have, for every $\xi < \kappa$ ordinal ν_ξ and a weakly closed set C_ξ such that for all $p \in X \cap C_\xi$, $a^p_\xi < \nu_\xi$. Setting $\nu = \bigcup_{\xi<\kappa} \nu_\xi$ and $C = \Delta^c_{\xi\in\kappa} C_\xi$ we have that C is weakly closed and unbounded, thus $X \cap C$ is superstationary, but for $p \in X \cap C$, $p \subseteq \nu$. This is a contradiction.

Once ξ as above is obtained we proceed with the construction of pairwise disjoint family superstationary subsets of X as in the Jech's proof.

This result can be obtained directly from Baumgartner's argument since the set B considered in it is weakly closed and unbounded but not closed and unbounded. We leave the proof to the reader.*

If κ is λ-supercompact and μ is a normal measure in $P_\kappa(\lambda)$ then all closed unbounded sets $P_\kappa(\lambda)$ are in μ. To see this assume C is a closed unbounded set outside of μ. Then $P_\kappa(\lambda) - C \in \mu$. We claim that if $p \notin C$ then there exists $q \in P_{|p\cap\kappa|}(p)$ such that whenever $r \in C \cap P_{|p\cap\kappa|}(p)$ then $q \not\subseteq r$. Indeed, if not then for some $p \notin C$ all sets q in $P_{|p\cap\kappa|}(p)$ have the property that there is r in $C \cap P_{|p\cap\kappa|}(p)$ such that $q \subseteq r$. But then the collection $C \cap P_{|p\cap\kappa|}(p)$ is a directed system with sum p and so $p \in C$ which is a contradiction. So, for every $p \in P_\kappa(\lambda) - C$ we set $f(p) = q$ where q is a above. As $q \subseteq p$ and $|q| < |p \cap \kappa|$ we have f constant on a measure-one set $D \subseteq (P_\kappa(\lambda) - C)$ (see section 2.1). So there is s such that on all $p \in D$, $f(p) = s$. Now, C being unbounded, there is $t \in C$ such that $s \subseteq t$. Finally we choose u in D, such that $t \cup |t|^+ \subseteq u$. Then $f(u) = s$ as $u \in D$ and so: for all $r \in C \cap P_{|u\cap\kappa|}(u)$, s _ r. But $|u\cap\kappa| \geq |t|^+ > |t|$ thus $t \in C \cap P_{|u\cap\kappa|}(u)$. Therefore $s \subseteq t$ which is in contradiction with our construction.

*M. Gitik has recently showed in an unpublished manuscript that the consistency of the existence of a super compact cardinal implies that it is consistent for some regular cardinal κ there is a stationary $S \subseteq P_\kappa(\kappa^+)$ which cannot be split into κ^+ disjoint stationary subsets.

2.3. MINIMALITY OF THE CLUB FILTER IN $P_\kappa(\lambda)$. RELATED PROPERTIES

We will show that the $CLUB_{\kappa,\lambda}$ filter is the least normal, fine κ-complete filter in $P_\kappa(\lambda)$. The argument due to Carr (based on results of Menas), has set-theoretical flavor. If $f : \lambda^n \to \lambda$ then $C(f) = \{p \mid (\overline{x})_p (f(\overline{x}) \in p\}$ is a closed unbounded set in $P_\kappa(\lambda)$. This set - however - possesses necessarily small elements; for there exist always denumerable sets in $C(f)$ and in fact every infinite cardinality below κ is attained by $C(f)$. If, however $f : \lambda^n \to P_\kappa(\lambda)$ then the set $C(f) = \{p \mid (\overline{x})_p (f(\overline{x}) \subseteq p)\}$ is also closed and unbounded and has none of the drawbacks mentioned above. Our strategy is now as follows. First we exhibit, for every closed and unbounded X an operation $f : \lambda^{<\omega} \to P_\kappa(\lambda)$ such that $C(f) \subseteq X$. Then we show that every such operator can be "refined" in order to get $f' : \lambda^2 \to P_\kappa(\lambda)$ such that $C(f') \subseteq X$. This is turn implies easily that $CLUB_{\kappa,\lambda}$ is the least normal, κ-complete, fine filter in $P_\kappa(\lambda)$. Thus first we exhibit $f : \lambda^{<\omega} \to P_\kappa(\lambda)$ such that $C(f) \subseteq X$. The operation is defined by induction on the length of a tuple as follows: $f(\phi)$ is any infinite element of X. If f is defined for all tuples of the length $\leq n$, $f(x_o,\ldots,x_n)$ is defined as follows: take the union of the values of f on all n-subsets of $\{x_o,\ldots,x_n\}$, add $\{x_o,\ldots,x_n\}$ and find, in X, an infinite set containing it. Pick one such set as a value. This function has the following properties: (i) $f(x_o,\ldots,x_{n-1}) \in X$, (ii) $\{x_o,\ldots,x_{n-1}\} \subseteq f(x_o,\ldots,x_{n-1})$ and
(iii) $\{x_o,\ldots,x_n\} \subseteq \{y_o,\ldots,y_m\} \Rightarrow f(x_o,\ldots,x_n) \subseteq f(y_o,\ldots,y_m)$.
Now we claim that $C(f) \subseteq X$. Indeed, given $p \in C(f)$, p is infinite and thus p is a direct limit of all its finite subsets. But as f is monotone and p is closed under f, p is a direct limit of sets of form $f(x_o,\ldots,x_n)$ for $\{x_o,\ldots,x_n\} \in p^{<\omega}$. Now, as every $f(x_o,\ldots,x_n) \in X$, $p \in X$ as well. Every function $f : \lambda^{<\omega} \to P_\kappa(\lambda)$ can be viewed as a sequence of functions $\langle f_n \rangle_{n\in\omega}$, $f_n : \lambda^n \to P_\kappa(\lambda)$. For every such function, $n \geq 2$ we construct $f'_n : \lambda^2 \to P_\kappa(\lambda)$ such that $C(f'_n) \subseteq C(f)$. To this end we use induction to reduce f_n to f'_n by finding intermediary f''_n's of less arguments; specifically we show that:
Given $g : \lambda^{n+1} \to P_\kappa(\lambda)$, $n \geq 2$ there is an $g' : \lambda^n \to P_\kappa(\lambda)$ $C(g') \subseteq C(g)$.
Thus let $g : \lambda^{n+1} \to P_\kappa(\lambda)$. Let j be a pairing function for λ^2 and let K,L be its converses: i.e. $j : \lambda^2 \xrightarrow[\text{onto}]{1-1} \lambda$, $j(K(\alpha), L(\alpha)) = \alpha$.

The set $C' = C(g) \cap C(j)$ is closed and unbounded and included in $C(g)$. Define g' as follows: given $\alpha_o,\ldots,\alpha_{n-1}$, set $g'(\alpha_o,\ldots,\alpha_{n-1})$ as any element of C' containing $g(K(\alpha_o), L(\alpha_o),\alpha_1,\ldots,\alpha_{n-1}) \cup \{\alpha_o,\ldots,\alpha_{n-1}\}$. The elements of $C(g')$ are closed under j. Indeed if $p \in C(g')$, $\alpha_o,\alpha_1 \in p$ then complete α_o,α_1 any fashion to $\alpha_o,\ldots,\alpha_{n-1}$ and since $g'(\alpha_o,\ldots,\alpha_{n-1}) \subseteq p$, $g'(\alpha_o,\ldots,\alpha_{n-1}) \in C(j)$ and $\alpha_o,\alpha_1 \in g'(\alpha_o,\ldots,\alpha_{n-1})$ we are done. Now pick any $p \in C(g')$, $\alpha_o,\alpha_1,\ldots,\alpha_n \in p$. Then $j(\alpha_o,\alpha_1) \in p$, $g'(j(\alpha_o,\alpha_1),\alpha_2,\ldots,\alpha_n) \subseteq p$ but $g(\alpha_o,\alpha_1,\ldots,\alpha_n) \subseteq g'(j(\alpha_o,\alpha_1),\alpha_2,\ldots,\alpha_n)$ so $g(\alpha_o,\ldots,\alpha_n) \subseteq p$. Thus $p \in C(g)$. Having provided for each $n \geq 2$, $f'_n : \lambda^2 \to P_\kappa(\lambda)$ such that $C(f'_n) \subseteq C(f_n)$ we show finally that there is one $f' : \lambda^2 \to P_\kappa(\lambda)$ such that $C(f) \subseteq \bigcap_{n\in\omega} C(f'_n)$. This would complete our work; well, $f(\alpha,\beta)$ such that simply $\bigcup_{n\in\omega} f_n(\alpha,\beta)$. Clearly if p is closed under f then p is closed under each f_n as well. Thus we proved: Given $C \in \mathrm{Club}_{\kappa,\lambda}$ there exists $f : \lambda^2 \to P_\kappa(\lambda)$ such that $C(f) \subseteq C$. We will call this a basis result, the sets of the form $C(f)$ for $f : \lambda^2 \to P_\kappa(\lambda)$ constitute a basis for the filter $\mathrm{CLUB}_{\kappa,\lambda}$.

From this we show that $\mathrm{CLUB}_{\kappa,\lambda}$ is the least fine, normal, κ-complete filter in $P_\kappa(\lambda)$. Indeed assume it is not the least one. Then there is a κ-complete, fine, normal filter $F \not\subseteq \mathrm{CLUB}_{\kappa,\lambda}$. But then for some $f : \lambda^2 \to P_\kappa(\lambda)$ there must be $C(f) \notin F$ (for otherwise $\mathrm{CLUB}_{\kappa,\lambda} \subseteq F$). So pick such an f. If $C(f) \notin F$ then $A = P_\kappa(\lambda) - C(f)$ must intersect every element of F (i.e. has to be F-stationary). For if $A \cap D = \phi$ for some $D \in F$ then $D \subseteq P_\kappa(\lambda) - A = C(f)$ and so $C(f) \in F$. Now, as F is normal filter, Fodor's theorem holds for F-stationary sets. One important feature of A is that for every $p \in A$, p is not closed under f so there are $\alpha,\beta \in p$ such that $f(\alpha,\beta) \not\subseteq p$. By applying normality of F twice we fix α,β in λ and get an F-stationary $A' \subseteq C$ such that $A' \subseteq \widehat{\{\alpha,\beta\}}$ and $f(\alpha,\beta) \not\subseteq p$ for all $p \in A'$. But look, F is fine and so A' is unbounded. Thus there is $p \in A'$ such that $f(\alpha,\beta) \subseteq p$. This is a desired contradiction.

Let us exhibit now another basis for $\mathrm{CLUB}_{\kappa,\lambda}$. This is a result due to Baumgartner. As noticed above, for $f : \lambda^n \to \lambda$, the sets $C(f)$ do not form a basis for the $\mathrm{Club}_{\kappa,\lambda}$ filter. Yet they "almost form" it in the following sense: There exists a closed and unbounded set C_o such that for every closed unbounded set C there exists a unary function $t : \lambda \to \lambda$ such that $C(t) \cap C_o \subseteq C$. To see this we need to exhibit C_o and then, given C, construct t. $C_o = C(j) \cap C(K) \cap C(L) \cap \{p \mid p \cap \kappa \in ORD\}$. The fact that $\{p \mid p \cap \kappa \in ORD\}$ is closed and unbounded is left as an exercise. Now we construct the desired t as follows. By Menas' basis result there exists $f : \lambda^2 \to P_\kappa(\lambda)$ such that $C(f) \subseteq C$. Now, fix for every $p \in P_\kappa(\lambda)$, $n_p : |p| \underset{onto}{\overset{1-1}{\to}} p$

$$\text{Set: } g(\alpha) = \begin{cases} n_{f(K(L(\alpha)),\ L(L(\alpha)))}(K(\alpha)) & \text{if } K(\alpha) \in |f(K(L(\alpha)),L(L(\alpha)))| \\ 0 & \text{otherwise} \end{cases}$$

$$h(\alpha) = |f(K(\alpha),\ L(\alpha))|$$

Finally $t(\alpha) = j(g(\alpha),\ h(\alpha))$.

We claim t is the required function. Thus let $p \in C(t) \cap C_o$ we show $p \in C(f)$. Let $\alpha,\beta \in p$, we prove $f(\alpha,\beta) \subseteq p$ so let $\xi \in f(\alpha,\beta)$. Since $\alpha,\beta \in p$, $j(\alpha,\beta) \in p$. Let $\gamma = j(\alpha,\beta)$. Then $t(\gamma) \in p$ and so $L(t(\gamma)) = h(\gamma) \in p$ this simply means $|f(K(\gamma),\ L(\gamma))| = |f(\alpha,\beta)| \in p$. But $|f(\alpha,\beta)| \in \kappa$ and $p \cap \kappa$ is an ordinal so $|f(\alpha,\beta)| \subseteq p$ (since $p \in C_o$!). Thus there is η such that $n_{f(\alpha,\beta)}(\eta) = \xi$ and this $\eta \in p$ (as $\eta < |f(\alpha,\beta)| \subseteq p$). Let $\delta = j(\eta, j(\alpha,\beta))$. Then $\delta \in p$ and thus $t(\delta) \in p$ and so $K(L(\delta)) \in p$ therefore $g(\delta) \in p$. But then

$n_{f(K(L(\delta)),\ L(L(\delta)))}(K(\delta)) \in p$ i.e.

$n_{f(\alpha,\beta)}(\eta) \in p$ i.e. $\xi \in p$. Thus $f(\alpha,\beta) \subseteq p$

Menas' basis result has the following interesting collorary. Let $R_o = \{\hat{p} \mid p \in p_\kappa(\lambda)\}$ be the collection of cones. Define $R_\xi = \bigcup\{R_\eta \mid \eta < \xi\}$ if ξ is a limit ordinal and $R_{\xi+1} = \Delta R_{\xi+1}$ is the family of diagonal intersections of elements of R_ξ). Clearly for some $\xi \leq \lambda$ we must have $R_{\xi+1} = R_\xi$, and R forms a basis for $CLUB_{\kappa,\lambda}$. In the case of the closed unbounded filter of subsets of a regular cardinal κ the stabilization point of the analogous hierarchy is 1(i.e. $\Delta\{\kappa-\alpha \mid \alpha \in \kappa\} = CLUB_\kappa$). D. Carr has show that for $CLUB_{\kappa,\lambda}$ the stabilization occurs after two steps so $\Delta\Delta\{\hat{p} \mid p \in p_\kappa(\lambda)\}$, the filter generated by R_2, is just $CLUB_{\kappa,\lambda}$.

Indeed, for $f:\lambda^2 \to P_\kappa(\lambda)$, $C(f) = \Delta_{\beta<\lambda}(\Delta_{\alpha<\lambda} \widehat{f(\alpha,\beta)})$. To show that $\Delta\{\hat{p} \mid p \in P_\kappa(\lambda)\} \subsetneq \mathrm{CLUB}_{\kappa,\lambda}$ we consider some other notions.
Obviously the cones generate a κ-complete filter $P_\kappa(\lambda)$. It is not normal as there are closed unbounded sets not containing cones. Another filter which we consider here is the filter generated by $\cup$-closed and unbounded sets. A set $D \subseteq P_\kappa(\lambda)$ is $\cup$-closed iff for every $E \subseteq D$, $|E| < \kappa \Rightarrow E \in D$. The $\cup$-closed and unbounded sets generate a κ-complete filter. The sets of the form $\Delta_{\xi<\lambda} \hat{p_\xi}$ form a basis for this filter. To see this first show that set of the form $\Delta_{\xi<\lambda} \hat{p_\xi}$ are $\cup$-closed. Indeed take any $q_0, \ldots q_\xi, \ldots,\ \xi < \eta < \kappa$ a collection of elements in $\Delta_{\xi<\lambda} \hat{p_\xi}$, let $q = \bigcup_{\xi<\eta} q_\xi$. We want to show $q \in \Delta_{\xi<\lambda} \hat{p_\xi}$ i.e. that $\alpha \in q \Rightarrow p_\alpha \subseteq q$. But if $\alpha \in q$ then for some $\xi < \eta$, $\alpha \in q_\xi$. Hence $p_\alpha \subseteq q_\xi$ and so $p_\alpha \subseteq q$. On the other hand, if C is $\cup$-closed and unbounded then, for every $\alpha < \lambda$ pick $p_\alpha \in C$, $\alpha \in p_\alpha$. We show $\Delta_{\xi<\lambda} \hat{p_\xi} \subseteq C$. Indeed, if $q \in \Delta_{\xi<\lambda} \hat{p_\xi}$ then $\xi \in q$ implies $p_\xi \subseteq q$. As $\xi \in p_\xi$ this is turn implies $q = \bigcup_{\xi \in q} p_\xi$. As every $p_\xi \in C$ and C is $\cup$-closed, $q \in C$.

Thus the $\cup$-closed and unbounded sets generate the same filter as the sets of the form $C(f)$ where f is a <u>unary</u> function mapping λ in $P_\kappa(\lambda)$ (Since $C(f) = \Delta_{\xi<\eta} f(\xi)$). Menas result tells us that $C(f)$ for <u>binary</u> functions form a basis for the $\mathrm{CLUB}_{\kappa,\lambda}$. Thus we have to ask, could Menas' result be improved? In other words, do $C(f)$ for f unary form a basis for $\mathrm{CLUB}_{\kappa,\lambda}$ or equivalently: is $\cup\text{-CLUB}_{\kappa,\lambda} = \mathrm{CLUB}_{\kappa,\lambda}$?
If $|\lambda| = \kappa$ then this happens; indeed if f is a bijection of κ and λ then $\{p \in P_\kappa(\lambda) \mid f^{-1} * p$ is an ordinal$\}$ is a closed and unbounded set in $P_\kappa(\lambda)$ and so essentially the space $\mathrm{CLUB}_{\kappa,\lambda}$ is nothing else but κ and there $\cup$-CLUB is exactly CLUB. However if $\kappa < |\lambda|$ then $\cup\text{-CLUB}_{\kappa,\lambda} \not\subseteq \mathrm{CLUB}_{\kappa,\lambda}$. It is enough to exhibit one set in $\mathrm{CLUB}_{\kappa,\lambda} - \cup\text{-CLUB}_{\kappa,\lambda}$. One such set is $C_0 = C(j) \cap C(K) \cap C(L)$. The elements in C_0 are characterized as those for which $j \mid p : p \xrightarrow[\text{onto}]{1-1} p$. Indeed if $p \in C_0$ then if $\alpha \in p$ then $\alpha = j(K(\alpha), L(\alpha))$ so $j|p$ is a surjection and $j|p$ (if well-defined) is always an

injection. Now C_o is completely non-$\cup$-closed: If $p,q \in C_o$, $p \setminus q \neq \phi \neq q \setminus p$ then $p \cup q \notin C_o$. Indeed if $\alpha \in p \setminus q$, $\beta \in q \setminus p$ then $j(\alpha,\beta) \notin p \cup q$. On the other hand any unbounded set in $P_\kappa(\lambda)$, if $\kappa < |\lambda|$, contains incomparable sets. To see this let D be a an unbounded set in $P_\kappa(\lambda)$. Pick a chain E in D of the length exactly κ. $|\cup E| = \kappa$ so $\lambda \setminus \cup E \neq \phi$. Take $\delta \in \lambda \setminus \cup E$ and pick $p \in D$, $\delta \in p$. Then p must be incomparable with (almost) all elements of E. Thus if $\langle p_\xi \rangle_{\xi<\lambda}$ is any collection of elements of $P_\kappa(\lambda)$, $\underset{\xi<\lambda}{\Delta} p_\xi$ contains incomparable sets. This implies that $\underset{\xi\in\lambda}{\Delta} \hat{p_\xi}$ is not contained in C_o for if $\underset{\xi\in\lambda}{\Delta} \hat{p_\xi} \subseteq C_o$ then for incomparable p,q in $\underset{\xi\in\lambda}{\Delta} \hat{p_\xi}$, $p \cup q \in \underset{\xi\in\lambda}{\Delta} \hat{p_\xi} \setminus C_o$, contradiction.

2.4. PARTITION PROPERTIES IN κ AND $P_\kappa(\lambda)$

We say that an ultrafilter U in κ possesses the partition property iff for every $n \leq \omega$ and $f : [\kappa]^n \to 2$ there exists $X \in U$ such that $|f * [X]^n| = 1$. If U is an ω_1-complete ultrafilter then the above mentioned partition property implies that if $f : [\kappa]^{<\omega} \to \alpha$ then there exists $X \in U$ such that for every n, $|f * [X]^n| = 1$. Normal ultrafilters on possess the partition property. The proof is done by induction on n; for n = 1 it is obvious. If $f : [\kappa]^{n+1} \to 2$ then f can be viewed as a sequence $\langle f_\alpha \rangle_{\alpha\in\kappa}$ of functions $f_\alpha : [\kappa-(\alpha+1)]^n \to 2$, namely $f_\alpha(\beta_o,\ldots,\beta_{n-1}) = f(\alpha,\beta_o,\ldots,\beta_{n-1})$. Now for every $\alpha < \kappa$ there is $X \subseteq \kappa - (\alpha + 1)$ and $\varepsilon_\alpha \in 2$ such that f_α is constant on $[X_\alpha]^n$ with value ε_α. Let $X = \underset{\xi<\kappa}{\Delta} X_\xi$. Then $X \in U$. Moreover one of the sets: $\{\alpha \in X | \varepsilon_\alpha = 0\}, \{\alpha \in X | \varepsilon = 1\}$ is in U. This is the desired homogeneous set. This result is due to Rowbottom. Notice that the above serves for partitions not only in 2 parts but also in any $< \kappa$ number of parts. The partition property of ultrafilters is not equivalent to normality; there are non-normal ultrafilters with the partition property. In particular, whenever $\pi \in \kappa!$ then $\pi_* U = \{\pi_*^{-1} X | X \in U\}$ is an ultrafilter with a partition property.

In $P_\kappa(\lambda)$ the partition property is formulated analogously but not entirely identically. The point is that in κ all elements (viewed as subsets) are comparable, here, in $P_\kappa(\lambda)$ this is no longer true, for we know that if $\kappa < |\lambda|$ then in any unbounded set there is a plenty p and q's such that $p \not\subseteq q$ and $q \not\subseteq p$. For $X \in P_\kappa(\lambda)$ define $[X]^2 = \{\langle p,q\rangle \in X^2 \mid p \subsetneq q\}$. We say that a measure U on $P_\kappa(\lambda)$ possesses the partition property iff for every $f : [P_\kappa(\lambda)]^2 \to 2$ there exists $X \in U$ such that $|f * [X]^2| = 1$. Rowbottom's result says that every normal measure in $P_\kappa(\kappa)$ has the partition property (Since $P_\kappa(\kappa)$ is essentially κ). We will see that the same holds in $P_\kappa(\lambda)$ for some ordinals $\lambda > \kappa$ but not necessarily for all of them. For example, if $\lambda = \kappa^+$ then every normal measure on $P_\kappa(\lambda)$ has the partition property.

Before we prove these and other related facts let us introduce some characterizations of the partition property. Let μ be a normal κ-complete, fine measure on $P_\kappa(\lambda)$. The following are equivalent: (a) μ possesses the partition property. (b) There exists $A \in P_\kappa(\lambda)$ such that $A \in \mu$ and for all $p,q \in A$, $p \subsetneq q \implies |p| < |q \cap \kappa|$, (c) For all $h : P_\kappa(\lambda) \to P_\kappa(\lambda)$ such that $\{p | p \subseteq h(p)\} \in \mu$ there is a set $X \in \mu$ such that if $p,q \in X$, $p \neq q$ then $h(p) \subseteq q$. Let us assume (a) i.e. that μ possesses the partition property. Define $f : [P_\kappa(\lambda)]^2 \to 2$ by: $f(p,q) = 0 \iff |p| < |q \cap \kappa|$. There is a measure 1 homogeneous set for f. It is easy to show that if $p,q \in A$, $p \subseteq q$ then $f(p,q) = 0$. For this it is enough to notice that in any measure 1 set we can find p,q such that $p \neq q$ and $|p| < |q \cap \kappa|$(it just follows from unboundedness of sets in the ultrafilter). This shows b). The implication a) $\implies$ c) is obtained the same way. Define $f : [P_\kappa(\lambda)]^2 \to 2$ by: $f(p,q) = 0 \iff h(p) \subseteq q$ and reason the same way as above. The implications c) $\implies$ b) is quite obvious. Let $h : P_\kappa(\lambda) \to P_\kappa(\lambda)$ be as required above and in addition $|h(p) \cap \kappa| > |p|$ for all p. If $A \subseteq P_\kappa(\lambda)$ is such that $A \in \mu$ and $p,q \in A$, $p \subsetneq q \implies h(p) \subseteq q$ then also $|p| < |q \cap \kappa|$. The last implication to be proved is (b) $\implies$ (a). Let $f : [P_\kappa(\lambda)]^2 \to 2$ be given. If A is as in (b) define, for each $p \in A$, $f_p : \hat{p} \to \alpha$ by: $f_p(q) = f(p,q)$. f_p is unary and defined on a measure 1 set into 2 so it is constant on a measure 1 set A_p with the value ε_p. Let B be a subset of A such that for $p,q \in A$, $\varepsilon_p = \varepsilon_q$. Set now $C = \{q \in B | q \in A_p$ for all $p \in B$, $p \neq q\}$. Then C is the desired homogeneous set. To show that C has measure 1 one uses strong from of normality introduced in 2.1.

It will be useful for the sequel to have another property related to the partition property. We first show a result due to Solovay. If X is a set then we say that f is an ω-Jonsson function on X if $f : [X]^{\omega} \to X$ and for every $Y \subseteq X$, if $|Y| = |X|$ then $f * [X]^{\omega} = X$. Erdös and Hajnal proved that if λ is any infinite cardinal then there is an ω-Jonsson function on λ. (For a proof see the paper of Solovay, Reinhardt and Kanamori). We have the following interesting property of normal measures on $P_{\kappa}\lambda$. If μ is normal and if is an ω-Jonsson on λ then $\{p | f \upharpoonright [p]^{\omega}$ is ω-Jonsson on $p\} \in \mu$. To prove this notice first that $\{p | f \upharpoonright [p]^{\omega}$ takes values in $p\}$ has measure 1 (just use normality ω-many times). Suppose now that $A = \{p | f \upharpoonright [p]^{\omega}$ is not ω-Johnsson on p but takes values in $p\}$ has measure 1. Then for every $p \in A$ there is a subset $q_p \subseteq p$ $|q_p| = |p|$ and an ordinal $\xi_p \in p$ such that $\xi_p \notin f * [q_p]^{\omega}$. By normality we can find a unique ξ which equals to ξ_p for p in a measure 1 set B. Now define $Y \subseteq \lambda$ as follows: $\alpha \in Y$ iff for almost all p, $\alpha \in q_p$. If $s \in [Y]^{\omega}$ then $f(s) \neq \xi$ because if $f(s) = \xi$ then by κ-completeness (in fact ω_1-completeness) we can find $p \in B$ such that $s \subseteq q_p$ which is a contradiction since ξ is left our by $f * [q_p]^{\omega}$ for all $p \in B$. We have to show now that $|Y| = \lambda$. For each $p \in B$ pick

$h_p : q_p \xrightarrow[\text{onto}]{1-1} p$ and define $\phi(\alpha) = \beta$ if $\beta = h_p(\alpha)$ for almost every p. h is well defined and h is "onto" since for $\beta \in \lambda$, the function $j : \{\hat{\beta}\} \to \lambda$ defined by $j(p) = h_p^{-1}(\beta)$ is regressive and so almost constant (with some value, which is the desired (ordinal) mapped by h to β). This conclude the proof.

An immediate corollary is that if μ is a normal measure in $P_{\kappa}(\lambda)$ then there is a set $A \in \mu$ such that $p,q \in A$, $p \not\subseteq q \Rightarrow |p| < |q|$. This is just the set constructed above, for if $p,q \in A$, $p \subsetneq q$ and $|p| = |q|$ then $f * [q]^{\omega}$ is p as f is ω-Jonsson on p and $q \subsetneq p$, $|q| = |p|$ and $f * [q]^{\omega} = q$ as f is ω-Jonsson on q. Thus $p = q$, contradiction.

Let us notice, finally, that there is a direct, metamathematical argument (using an elementary embedding witnessing to λ-supercompactness of κ) to prove this result. We gave the above proof because of its combinatorial flavor.

We introduce now a certain sufficient condition for the partition property. Let $f : \kappa \to \kappa$, we say that $X(f,\mu)$ holds iff $\{p | f(|p \cap \kappa|) = |p|\} \in \mu$ and we say that $X(\mu)$ iff $(Ef)(X(f,\mu))$. We prove now a result due to Menas: $X(\mu) \Rightarrow \mu$ possesses the partition property. Let us prove this fact by showing that if a function making $X(f,\mu)$

exists then there is a measure one set A such that for $p,q \subseteq A$, $p \subsetneq q \Rightarrow |p| < |q \cap \kappa|$. Notice first that $\{p | f * |p \cap \kappa| \subseteq |p \cap \kappa|\}$ has measure one (in fact it contains a closed unbounded set) Let $B = \{p | f(|p \cap \kappa|) = |p|$ and $f * |p \cap \kappa| \subseteq |p \cap \kappa|\}$. Then, let C be a measure 1 set such that for $p,q \in C$, $p \subsetneq q \Rightarrow |p| < |q|$ (such C exists by Solovay's result proved above). Then let $A = B \cap C$. Then if $p,q \in A$ then $|p \cap \kappa| < |q \cap \kappa|$ for otherwise $|p| = f(|p \cap \kappa|) = f(|q \cap \kappa|) = |q|$ as $q \in B$, $f * |q \cap \kappa| \subseteq |q \cap \kappa|$ so, $|p \cap \kappa| < |q \cap \kappa|$, $f(|p \cap \kappa|) < |q \cap \kappa|$ and $|p| = f(|p \cap \kappa|)$, $|p| < |q \cap \kappa|$. □

Pelletier proved that the property X is strictly stronger than the partition property.

We can show now that if κ is supercompact then for some $\lambda > \kappa$ all normal measures in $P_\kappa(\lambda)$ have the partition property. Suppose for the moment that $\lambda = \kappa^+$ and that μ is any normal in $P_\kappa(\lambda)$. We show that $\{p | (p \cap \kappa)^+ = p\}$ belongs to μ. Clearly $|p \cap \kappa| \leq |p|$. We show first that for arbitrary $\lambda \geq \kappa^+$ and any normal measure on $P_\kappa(\lambda)$ that $\{p | \ |p \cap \kappa| < |p|\}$ has measure 1. Suppose $C = \{p | \ |p \cap \kappa| = |p|\}$ has measure 1. We will show that $|\lambda| = \kappa$ i.e. construct a surjection of κ onto λ. For every $p \in C$ pick a bijection $f_p : p \cap \kappa \to p$. For $\alpha \in \kappa$ let $f_\alpha : \{\hat{\alpha}\} \cap C \to \lambda$ be defined by $f_\alpha(p) = f_p(\alpha)$. Obviously, each f_α is regressive, thus constant on $Z_\alpha \subseteq C\{\hat{\alpha}\}$ with a value ξ_α. Now define $f : \kappa \xrightarrow{\text{onto}} \lambda$ putting $f(\alpha) = \xi_\alpha$. We need to show that every element of λ is a value. If $\xi \in \lambda$ then the function $h : \{\hat{\xi}\} \to \kappa$ defined by $h(p)_p^{-1}(\xi)$ is regressive thus constant almost everywhere such value α. Clearly $f(\alpha) = \xi$.

Now, in case $\lambda = \kappa^+$, fix for every $\kappa \leq \xi < \lambda$ a bijection $f_\xi : \kappa \to \xi$. Now, if $\kappa < \xi < \lambda$, $A = \{p | \xi \in p$ & $f_\xi \upharpoonright p \cap \kappa$ is a bijection onto $p \cap \xi\}$ has measure 1. (Clearly, $p | f_\xi \upharpoonright p \cap \kappa$ maps $p \cap \kappa$ into $p \cap \xi\}$ is closed unbounded and by a normality argument we can trim it down as desired) Now the set $\underset{\xi<\kappa}{\Delta} A_\xi$ is a measure 1 set. If $p \in \underset{\xi<\lambda}{\Delta} A_\xi$ then $|p \cap \kappa| = |p \cap \xi|$ for all $\xi \in p \setminus \kappa$. The intersection of $\underset{\xi<\lambda}{\Delta} A_\xi$ with $\{p | (p \cap \kappa) < |p|$ has measure 1 and for each of its elements, $(p \cap \kappa)^+ = |p|$.

This says that the function $f : \kappa \to \kappa$ defined by $f(\alpha) = \alpha^+$ satisfies $X(f,\mu)$ for any normal measure μ. Therefore μ has the partition property, as μ was arbitrary, we are done.

Let us remark that there is also a metamathematical argument to obtain this result; using it one proves the following: assume $\phi(\cdot,\cdot)$ is a formula such that ZFC $\vdash (x)(E!y)\phi(x,y)$, $\phi(\kappa,\lambda)$ holds and for every M, if ${}^{\lambda}M \subseteq M$ then $M \models \phi(\kappa,\lambda)$ then: Every normal measure in the space $P_\kappa(\lambda)$ possesses the partition property; in fact every much measure has the property χ.

The next result we shall prove is that, given κ supercompact then for every $\lambda \geq \kappa$ there are normal measures in $P_\kappa(\lambda)$ with the partition property. To this end we try to employ the same argument which we used showing that there are $2^{2^{\lambda^{\kappa}}}$-many normal ultrafilters in $P_\kappa(\lambda)$. In this case we need an additional twist using the property $X(\mu)$.

Let us write $\Psi(X,\kappa,\lambda)$ for: $X \in P(P(P_\kappa(\lambda))$ and for every ultrafilter μ on $P_\kappa(\lambda)$ such that $\chi(\mu)$, $X \notin (P(P_\kappa(\lambda))) \cap M_\mu$.

We will show that for every λ, if $X \in P(P(P_\kappa(\lambda)))$ then there is an ultrafilter μ such that $X(\mu)$ and $X \in M_\mu$. From this our result follows exactly as in the section 2.

Let $\phi(\alpha,\beta)$ stands for: β is the least cardinal such that α is β-supercompact and there is X such that $\Psi(X,\alpha,\beta)$. Assume $\phi(\kappa,\lambda)$ and search for contradiction.

Let $\mathcal{U}$ be a normal ultrafilter in $P_\kappa(2^{\lambda^{\kappa}})$. Consider the diagram

$$\begin{array}{ccl} V & \longrightarrow & M_0 \approx V^{P_\kappa(\lambda)}/\mathcal{U} \restriction \lambda \\ & \searrow & \downarrow k \\ & & M_{\mathcal{U}} = V^{P_\kappa(2^{\lambda^{\kappa}})}/\mathcal{U} \end{array}$$

We already know that $k(\alpha) = \alpha$ for $\alpha \leq \lambda$, $k(p) = p$ for $p \in P_\kappa(\lambda)$, $k(X) = X$ for $X \in P(P_\kappa(\lambda))$ and also $k(X) = X$ for $X \in P(P(P_\kappa(\lambda)))$. Now, as $\phi(\kappa,\lambda)$ holds, it holds in $M_{\mathcal{U}}$ as well and so also $M_0 \models \phi(\kappa,\lambda)$. Let $X \in M_0$ make $\psi(X,\kappa,\lambda)$ true in M_0.

Since $k(X) = X$, $M_{\mathcal{U}} \models \psi(X,\kappa,\lambda)$. Thus we will get a contradiction if we show $\chi(\mathcal{U} \upharpoonright \lambda)$. To this end define

$$f(\alpha) = \begin{cases} \text{The least } \beta \geq \alpha \text{ such that } \phi(\alpha,\beta) \text{ (if there exists one)} \\ 0 \quad \text{otherwise} \end{cases}$$

Since $M_0 \models \phi(\kappa,\lambda)$, $\{p \mid \phi(|p \cap \kappa|, |p|)\} \in \mathcal{U} \upharpoonright \lambda$. Thus $\{p \in P_\kappa(\lambda) \mid f(|p \cap \kappa|) = |p|\} \in \mathcal{U} \upharpoonright \lambda$ thus $\chi(f,\mathcal{U} \upharpoonright \lambda)$.

Finally we show that under suitable assumptions there are normal measures in $P_\kappa(\lambda)$ without partition property. Suppose κ is λ-supercompact and λ is measurable. We construct a normal measure in $P_\kappa(\lambda)$ without partition property. To this end let μ' be any normal measure on $P_\kappa(\lambda)$ and η a normal measure in λ. μ' may or may not possess the partition property but we construct out of μ' and η a normal measure without partition property. For each $\kappa < \alpha < \lambda$ let μ_α be $\mu' \upharpoonright P_\kappa(\alpha)$ i.e.: $\mu_\alpha(A) = \mu'(\{p \in P_\kappa(\lambda) \mid p \cap \alpha \in A\})$ for $A \subseteq P_\kappa(\lambda)$. The desired measure without partition property is the "glue together measure" defined by:

$$A \in \mu \Longleftrightarrow \{\alpha \mid A \cap P_\kappa(\alpha) \in \mu_\alpha\} \in \eta$$

To show this we will prove that if $A \in \mu$ then there are $r,s \in A$, $r \subsetneq s$ and $|r| \geq |s \cap \kappa|$. But look, if $A \in \mu$ and $C = \{\alpha < \lambda \mid (A \cap P_\kappa(\alpha) \in \mu_\alpha\}$, define for each $\alpha \in C$ the set $D_\alpha = \{p \in P_\kappa(\lambda) \mid p \cap \alpha \in A\}$. Now: $C \in \eta$ (by the definition of μ) and $\mu'(D_\alpha) = 1$ for each $\alpha \in C$ by the definition of C and μ_α. Now pick $\alpha,\beta \in C$ and above κ, $p \in D_\alpha \cap D_\beta$ such that $p \cap (\beta-\alpha) \neq \phi$. This is possible as $D_\alpha \cap D_\beta$ is unbounded. Put $r = p \cap \alpha$, $s = p \cap \beta$. Both r and s are in A and:

$$|r| = |p \cap \alpha| \geq |p \cap \alpha \cap \kappa| = |p \cap \beta \cap \kappa| = |s \cap \kappa| \qquad \square$$

2.5. PARTITION PROPERTIES IN $P_\kappa(\lambda)$ CONTINUED. FLIPPING PROPERTIES

In the previous section we introduced partition properties of the ultrafilters. Very much as in case of κ we can introduce "weaker" partition properties: We say that Part (κ,λ) holds iff for all $F : [P_\kappa(\lambda)]^2 \to 2$ there is an unbounded subset A of $P_\kappa(\lambda)$ which is homogeneous for F (i.e. $|F * [A]^2| = 1$). The result of Menas proved in section 2.4, implies that if κ is supercompact then for every

$\lambda \le \kappa$ Part (κ,λ) holds since there is a normal, fine measure in $P_\kappa(\lambda)$ with the partition property and since the sets in the normal fine measure are unbounded. In fact every set in U is stationary and so if κ is supercompact then a stronger partition actually holds namely, "having a stationary homogeneous set" for every partition. This last property we call Part $*(\kappa,\lambda)$. One can generalize the partition properties considering Part $^3(\kappa,\lambda)$, the partition property for triples, and similarly Part $^{*,3}(\kappa,\lambda)$ etc.

We shall prove that Part $^3(\kappa,2^{\lambda^\kappa})$ implies that κ is λ-compact i.e. that there is a λ-complete fine filter in $P_\kappa(\lambda)$. The exact strength of Part $^2(\kappa,\lambda)$ is not known. Consider now $*$-partition properties. The statements (λ) Part $*(\kappa,\lambda)$, (λ) Part $^{*,3}(\kappa,\lambda)$ and even (λ) Part $^{*,<\omega}(\kappa,\lambda)$ are equivalent and in fact they are equivalent to supercompactness. We prove below that (λ) Part $*(\kappa,\lambda)$ implies that κ is supercompact. Let us show first that if is supercompact then (λ) Part $^{*,\omega}(\kappa,\lambda)$ holds. The argument here is similar to one used in 2.4 to show that normal ultrafilters possess the partition property, only that now we can assume that U already possesses the partition property (for pairs). Thus there exists a set $M \in U$ such that $(p,q)_M(p \subsetneq q \Rightarrow |p| < |q \cap \kappa|)$. While looking for a homogeneous set we can safely assume that se set X whose $[X]^n$ is partitioned is included in M. Then we proceed very much as in the proof of partition theorem for κ. We can assume $n \ge 2$. Let $f : [X]^{n+1} \to 2$. Our inductive assumption is that if we partition $[Y]^n$ for $Y \in U$ into two parts then there exists an homogeneous set in U. Now given $p \in M \cap X$ define $f_p : [\hat{p} \cap X]^n \to 2$ by: $f_p(q_0,\ldots,q_{n-1}) = f(p,q_0,\ldots,q_{n-1})$. Let Z_p be homogeneous for f_p. Remember that $Z_p \subseteq M \cap \hat{p}$ and so if $q \in Z_p$ and $p \subsetneq q$ then $|p| < |q \cap \kappa|$. Then, by a result from 2.1 $Z' = \Delta_{p \in P_\kappa \lambda} Z_p \in U$. Now we split Z' according to the value of constant value of f_p on Z_p and the "large" part is the desired stationary homogeneous set for f.

Our line of attack for showing that (λ) Part $*(\kappa,\lambda)$ implies κ is supercompact is to introduce yet another property of $P_\kappa(\lambda)$ and then to show that this property is implied by (λ) Part $*(\kappa,\lambda)$ and that it implies the supercompactness of κ. The property we have in mind is the so called flipping property. Generally, if $f : X \to P(Z)$ then

$g : X \to P(Z)$ is called a flip of f (in symbols $f \sim g$) iff for all $x \in X$, $g(x) = f(x)$ or $g(x) = Z \setminus f(x)$. In our situation $X = \lambda$ and $Z = P_\kappa(\lambda)$. We say that κ has the λ-flipping property iff for all $t : \lambda \to P(P_\kappa(\lambda))$ there exists t' such that $t \sim t'$ and $\Delta_{\alpha<\lambda} t'(\alpha)$ is stationary. Now we will show that if Part $*(\kappa,\lambda)$ then κ has the λ-flipping property and then that if κ has the λ flipping property for all $\lambda \geq \kappa$ then κ is supercompact. More precisely we show that if κ has 2^{λ^κ} flipping property then κ is λ-supercompact.

The first argument is due to Magidor. Let $t : \lambda \to P(P_\kappa(\lambda))$. Define $F : [P_\kappa(\lambda)]^2 \to 2$ by: $F(p,q) = 0 \iff [\{\alpha \in p | p \in t(\alpha)\} \neq \{\alpha \in p | q \in t(\alpha)\}$ & the least ordinal β in $q \dot{-} p$ has the property: $q \in t(\beta)]$. Let $A \subseteq P_\kappa(\beta)$ be stationary and homogeneous for F. There are two possible cases. Cases one is when $F * [A]^2 = \{0\}$. Define $B \subseteq \lambda$ by induction as follows: $\alpha \in B$ if and only if there is a set p in A such that $p \in t(\alpha)$ and $\{\beta \in p \cap | p \in t(\beta)\} = B \cap p \cap \alpha$. If such $p \in A$ exists pick one and call it p_α. Otherwise $\alpha \notin \beta$. Now define a flip t' as follows: $t'(\alpha) = t(\alpha)$ if $\alpha \in \beta$, $t'(\alpha) = P_\kappa(\lambda) - t(\alpha)$ if $\alpha \notin B$. We claim that $\Delta_{\alpha<\lambda} t'(\alpha)$ is stationary. Otherwise there is a closed and unbounded C such that: if $p \in C$ then $p \notin t'(\alpha)$ for some $\alpha \in p$. As the set $C \cap A$ is stationary, defining $h : C \cap A \to \lambda$ by $h(p) =$ first ξ such that $p \notin t'(\xi)$ we have a stationary $D \subseteq C \cap A$ and such that h is constant with value α on D. Also, for $p \in D$, $\{\beta \in p \cap \alpha | p \in t(\beta)\} = p \cap \alpha \cap B$. Now: assume $\alpha \notin B$. Then $t'(\alpha) = P_\kappa(\lambda) - t(\alpha)$ and if $p \in D$, $p \in t(\alpha)$ contradicting $\alpha \notin B$. So $\alpha \in B$ and then $t'(\alpha) = t(\alpha)$ so $p \notin t(\alpha)$ for all $\alpha \in D$. Pick $q \in D$ such that $p_\alpha \not\subseteq q$, thus $F(p_\alpha, q) = 0$. By the definition of p_α, $\{\beta \in p_\alpha \cap \alpha | p_\alpha \in t(\beta)\} = B \cap p_\alpha \cap \alpha$ and $p_\alpha \in t_\alpha$. Since $q \in D$, $\{\beta \in q \cap \alpha | q \in t(\beta)\} = B \cap q \cap \alpha$. Also, since $p_\alpha \subseteq q$, $\{\beta \in p_\alpha \cap \alpha | q \in t(\beta)\} = \{\beta \in p_\alpha \cap \alpha | q \in t(\beta) |\}$ and $q \notin t(\alpha)$. This contradicts $F(p_\alpha, q) = 0$ and thus proves the claim. This leaves us with the second possibility that $F * [A]^2 = \{1\}$. Then define B, also by induction, as follows: $\alpha \notin B$ if and only if there is a $p \in A$ such that $p \in t(\alpha)$ and $\{\beta \in p \cap \alpha | p \in t(\beta)\} = B \cap p \cap \alpha$. In this case, pick one such p and call it p_α. Otherwise $\alpha \in B$. We define now $t'(\alpha) = t(\alpha)$ if $\alpha \in B$ and $t'(\alpha) = P_\kappa(\alpha) - t(\alpha)$ otherwise. Again we claim that $\Delta_{\alpha<\lambda} t'(\alpha)$ is stationary. Otherwise there is a closed and unbounded $C \subseteq P_\kappa(\lambda)$ such that for all $p \in C$, for some in p, $p \notin t'(\alpha)$. As above define $h : C \cap A \to \lambda$ as $h(p)$ being the least $\xi \in p$ such that $p \notin t'(\xi)$. Again we apply Fodor's property (as $C \cap A$ is stationary) getting $D \subseteq C \cap A$ and $\alpha \in \lambda$ such that h is constant on D with the value α. Now, for $p \in D$, $\{\beta \in p \cap \alpha | p \in t(\beta)\} = B \cap p \cap \alpha$. Suppose $\alpha \in B$. Then if $p \in D$, $p \in t(\alpha)$. But $t'(\alpha) = t(\alpha)$, a contradiction. So $\alpha \notin B$ and we can pick $q \in D$ such that $p_\alpha \subsetneq q$ (So $F(p_\alpha, q) = 1$). But since $q \in D$, $q \notin t'(\alpha) = P_\kappa(\alpha) - t(\alpha)$. So $q \in t(\alpha)$ and: $\{\beta \in p_\alpha \cap \alpha | p_\alpha \in t(\alpha)\} = \{\beta \in p_\alpha \cap \alpha | q \in t(\alpha)\}$. But then $F(p_\alpha, q) = 0$ and not 1, again a contradiction. This concludes the argument.

Now we show that if κ has the flipping property for $\gamma = 2^{\gamma^\kappa}$ then κ is λ-supercompact. What we need is a normal, κ-complete, fine measure in $P_\kappa(\lambda)$. So let $\gamma = 2^{\gamma^\kappa}$ and let us well-order $P(P_\kappa(\lambda))$ in type γ. Let A_α be the $\alpha^{\underline{th}}$ subset of $P_\kappa(\lambda)$. Given $A \subseteq P_\kappa(\lambda)$ let $F(A) \subseteq P_\kappa(\gamma)$ be $\{q \in P_\kappa(\gamma) | q \cap \lambda \in A\}$. Now set $t : \gamma \to P(P_\kappa(\lambda))$ by: $t(\alpha) = F(A_\alpha)$ (for $\alpha < \gamma$). By the γ-flipping property there is $t' \sim t$ such that $\underset{\alpha<\gamma}{\Delta} t'(\alpha)$ is stationary in $P_\kappa(\gamma)$.

We define now a family $U \subseteq P(P_\kappa(\lambda)$ as follows: Let $A \subseteq P_\kappa(\lambda)$, $A = A_\alpha$. We put $A \in U \iff t'(\alpha) = t(\alpha)$. We show that U is a κ-complete, fine, normal ultrafilter in $P_\kappa(\lambda)$. There are four things to check: First, we show U is fine. Pick $p \in P_\kappa(\lambda)$, $\hat{p} = A_\alpha$. Now $\hat{p}$ in $P_\kappa(\gamma)$ is $F(\hat{p})$ (where $\hat{p}$ was taken in $P_\kappa(\lambda)$) . If $t'(\alpha) = P_\kappa(\gamma) - t(\alpha)$ then $\underset{\xi<\lambda}{\Delta} t'(\xi)$ is bounded (as it is disjoint with $\hat{p}$) so it is no stationary. To prove κ completeness, suppose that for all $\alpha \in p$, $A_\alpha \in U$. If $\bigcap_{\alpha \in p} A_\alpha \notin U$ then for β such that $\bigcap_{\alpha \in p} A_\alpha = A_\beta$,

$t'(\beta) = P_\kappa(\gamma) - F(A_\beta) = F(P_\kappa(\lambda) - A_\beta)$. Pick $q \in P_\kappa(\gamma)$ such that $p \cup \{\beta\} \subseteq q$. If we show that $q \notin \underset{\xi<\gamma}{\Delta} t'(\xi)$ then we have that

$\underset{\xi<\gamma}{\Delta} t''(\xi) \cap \widehat{p \cup \{\beta\}} = \phi$ so $\underset{\xi<\gamma}{\Delta} t'(\xi)$ is bounded, a contradiction. So

suppose $q \in \underset{\xi<\gamma}{\Delta} t'(\xi)$. Then $q \in F(A_\xi)$ for $\xi \in p$ and

$q \in F(P_\kappa(\lambda) - A_\beta)$. Thus $q \cap \lambda \in A_\alpha$ for all $\alpha \in p$ and $q \cap \lambda \in P_\kappa(\lambda) - A_\beta = P_\kappa(\lambda) - \bigcap_{\xi \in p} A_\xi$. This is absurd, so U is κ-complete.

Before we prove normality we have to show that U is an ultrafilter. To this end pick $A_\alpha \notin \mathcal{U}$. Then $t'(\alpha) = P_\kappa(\gamma) - F(A_\alpha)$. Now, if is such that $A_\beta = P_\kappa(\lambda) - A_\alpha$ and $A_\beta \notin \mathcal{U}$ then $t'(\beta) = {}_\kappa(\gamma) - F(A_\beta) = F(A_\alpha)$. But then $\underset{\xi<\gamma}{\Delta}\ t'(\alpha) \cap \widehat{\{\alpha,\beta\}} = \phi$. Hence $A_\beta \in \mathcal{U}$. The same argument shows that if $A_\alpha \in \mathcal{U}$ then $A_\beta \notin \mathcal{U}$, hence $\mathcal{U} \neq P(P_\kappa(\lambda))$.

Finally we show that $\mathcal{U}$ is normal. For this pick $\langle B_\xi \rangle_{\xi<\lambda}$ a collection of sets from $\mathcal{U}$ and let $f : \lambda \to \gamma$ be defined by $f(\xi)$ is such that $B_\xi = A_{f(\xi)}$. As usual it is immediate that $C = \{p \in P_\kappa(\gamma) \mid (\xi)_{p\cap\lambda}(f(\xi) \in p)\}$ is closed and unbounded in $P_\kappa(\gamma)$. Now, if $\underset{\xi<\lambda}{\Delta} B_\xi$ is not in $\mathcal{U}$ then $A_\beta = P_\kappa(\lambda) - (\underset{\xi<\lambda}{\Delta} B_\xi) \in \mathcal{U}$ so $t'(\beta) = F(A_\beta) = \{q \in P_\kappa(\gamma) \mid (E\alpha)_{q\cap\lambda}(q\cap\lambda \notin B_\alpha)\}$. Let $p \in \underset{\xi<\gamma}{\Delta}\ t'(\alpha) \cap \{\hat\beta\}$. Then there is $\alpha \in p \cap \lambda$ such that $p \cap \lambda \notin B_\alpha$, hence $p \notin C$. Thus if $\underset{\xi<\lambda}{\Delta} B_\xi \notin \mathcal{U}$ the $\underset{\xi<\gamma}{\Delta}\ t'(\xi) \cap \{\hat\beta\}$ is nonstationary! This is the desired contradiction. This completes the proof. Thus the flipping property (for $\gamma = 2^{\lambda^\kappa}$) implies that κ is λ-supercompact (with in turn implies the λ-flipping property, just flip according to the ultrafilter).

We conclude this section showing that Part $^3(\kappa, 2^{\lambda^\kappa})$ implies that κ is λ-compact i.e. that there exists a κ-complete fine ultrafilter in $P_\kappa(\lambda)$. To see this let us introduce another flipping property in $P_\kappa(\lambda)$ which we call the mild flipping property: κ possesses the mild λ-flipping property iff for every $t : \lambda \to P(P_\kappa(\lambda))$ there is flip $t' \sim t$ such that for all $p \in P_\kappa(\lambda)$, $\underset{\xi\in p}{\cap} t'(\xi)$ is unbounded. Clearly if κ is λ-strongly compact then κ possesses the mild λ-flipping property (just flip according to ultrafilter, the intersection is again in the ultrafilter and, as the latter is fine, it is unbounded). Our argument splits into 2 parts: The first one is to show that Part $^3(\kappa,\gamma)$ implies that κ has the mild γ-flipping property. The second part is to show that if κ has the mild 2^{λ^κ}-flipping property, the κ is λ-compact. For this second part we use the argument given above, just notice that the only place we used the full 2^{λ^κ}-flipping property was to show normality, everything else follows from the mild 2^{λ^κ}-flipping property. Thus we need just to show that Part $^3(\kappa,\lambda)$ implies that possesses the mild λ-flipping property. To this end let: $\lambda \to P(P_\kappa(\lambda))$. Given $P \in P_\kappa(\lambda)$

p belongs to some of the $t(\xi)$ so let Z_p be the characteristic function of the set of those ξ that $p \in t(\xi)$. Now let $F[P_\kappa(\lambda)]^3 \to 2$ be defined as follows: $F(p,q,r) = 0 \iff Z_q \restriction p = Z_r \restriction p$. Thus $F(p,q,r) = 0$ says q and r behave same way on the positions from p. Pick A unbounded and homogeneous for F. We claim that $F * [A]^3 = \{0\}$. Suppose that $F * [A]^3 = \{1\}$. Fix $p \in A$ then if $q \subseteq r$, both in $\hat{p} \cap A$, the set of positions η in p for which $q \in t(\eta)$ is different from the analogous set corresponding to r. Thus we may construct an injection of κ into $P(p)$. But this is a absurd since $\text{Part}^3(\kappa,\lambda)$ implies $\text{Part}(\kappa,\lambda)$ which in turn implies that κ is inaccessible. Once we know that $F * [A]^3 = \{0\}$ we can set $t' : \lambda \to P(P_\kappa(\lambda))$ by $t'(\alpha) = t(\alpha)$ iff $A \cap t(\alpha)$ is unbounded; otherwise set $t'(\alpha) = P_\kappa(\lambda) - t(\alpha)$. First of all notice that $A \cap t'(\alpha)$ is always unbounded, for if $A \cap t(\alpha)$ is bounded then there is a cone $\hat{p}$ omitting $A \cap t(\alpha)$ but then $\hat{p} \cap A \subseteq A \cap (P_\kappa(\lambda) - t(\alpha))$ and as $\hat{p} \cap A$ is unbounded so is $A \cap t'(\alpha)$. Let us show $\bigcap_{\alpha \in p} t'(\alpha)$ is unbounded for each $p \in P_\kappa(\lambda)$. Notice that since A is unbounded, it is enough to show that $\bigcap_{\alpha \in p} t'(\alpha)$ is unbounded whenever $p \in A$.

Now, given $\alpha \in p$, there is $q_\alpha \in A \cap t'(\alpha)$ such that $p \subsetneq q_\alpha$ and for for every $q \in A$ such that $\bigcup_{\alpha \in p} q_\alpha \subsetneq q$ we have $q \in \bigcap_{\alpha \in p} t'(\alpha)$. This happens since $q_\alpha \in t'_\alpha$, and for each $\alpha \in p$, $F(p,q_\alpha,q) = 0$ implies that $q \in t(\alpha)$ if and only if $q_\alpha \in t(\alpha)$. Thus we have shown that if $p \in A$, $\bigcap_{\alpha \in p} t'(\alpha)$ contains $A \cap (\bigcup_{\alpha \in p} q_\alpha)^\wedge$ which is clearly unbounded and this completes the proof.

3.1. HUGE CARDINALS

In this section we will deal with cardinals whose existence implies that the universe is very wide. First let us introduce some notation: $[\lambda]^\kappa = \{P \subseteq \lambda \mid |P| = \kappa\}$, $(\lambda)^\kappa = \{P \subseteq \lambda \mid \text{ot}(P) = \kappa\}$. Clearly $(\lambda)^\kappa \subseteq [\lambda]^\kappa$. The notion of cone is introduced in analogy with $P_\kappa(\lambda)$, if $p \in P_\kappa(\lambda)$ Then $\check{p} = \{P \in [\lambda]^\kappa \mid p \subseteq P\}$. The restriction $\check{p}$ to $(\lambda)^\kappa$ is a cone in $(\lambda)^\kappa$. Diagonal intersection is also introduced in analogy with $P_\kappa(\lambda)$; if $\langle X_\xi \rangle_{\xi<\lambda}$ is a family of subsets of $[\lambda]^\kappa$ (or $(\lambda)^\kappa$) then

$\underset{\xi<\lambda}{\Delta} X_\xi = \{P \mid (\eta)_P \ (P \in X_\eta)\}$. The corresponding notion of regressive function is as in $P_\kappa(\lambda)$, f is regressive iff for every P, $f(P) \in P$. Now, let κ,λ be cardinals. We say that κ is huge with target λ, and write $\kappa \to (\lambda)$, iff there exists a κ-complete, fine, normal ultrafilter in $[\lambda]^\kappa$. Let us consider now the ultrapower $M \simeq V^{[\lambda]^\kappa}/U$. Since our ultrafilter U is normal, Fodor's theorem holds for sets in the ultrafilter and so, if $Id = \{\langle P,P\rangle \mid P \in [\lambda]^\kappa\}$ is the identity function we have the following:

$\pi[Id] = j*\lambda$ and so:

$X \in U \Longleftrightarrow j*\lambda \in j(X)$

Now, since λ is a cardinal it is a cardinal in M as well. Besides , $|j*\lambda|^M = \lambda$ as $j*\lambda$ is of order type λ.

Therefore

$$\begin{aligned} j(\kappa) &= \pi[\langle \kappa \mid P \in [\lambda]^\kappa\rangle] \\ &= \pi[\langle |p| \mid p \in [\lambda]^\kappa\rangle] \\ &= |\pi[Id]|^M \\ &= |j*\lambda|^M = \lambda \end{aligned}$$

On the other hand, $j*\lambda \in M$ and so, by the argument already used to show that if κ is measurable then ${}^\kappa M \subseteq M$, we have that ${}^\lambda M \subseteq M$. Therefore we proved that if κ is huge with target λ (κ,λ cardinals), then there is an elementary imbedding $j : V \to M$ with κ critical, $j(\kappa) = \lambda$, ${}^\lambda M \subseteq M$. The converse implication is even simpler; if $j : V \to M$ as above then define in $[\lambda]^\kappa$ a family U as follows: $X \in U \Longleftrightarrow j*\lambda \in j(X)$. Now, as ${}^\lambda M \subseteq M$, $j*\lambda \in M$ and our U is well-defined and now it is routine to check that U is a κ-complete fine, normal ultrafilter in $[\lambda]^\kappa$. Let U be as above, then we have for our j, $X \in U$ $j*\lambda \in j(X)$. But o.t. $(j*\lambda) = \lambda$ and since $j((\lambda)^\kappa) = [(j(\lambda))^{j(\kappa)}]^M = (j(\lambda))^\lambda$ and $j*\lambda \in (j(\lambda))^\lambda$, $(\lambda)^\kappa \in U$. Then the restriction of to $(\lambda)^\kappa$ is easily seen to be a κ-complete, fine, normal ultrafilter in $(\lambda)^\kappa$. Conversely if U is a κ-complete, fine, normal ultrafilter in $(\lambda)^\kappa$ then the trivial extension to $[\lambda]^\kappa$ ("measuring the trace on $(\lambda)^\kappa$") is again a κ-complete, fine, normal ultrafilter in $[\lambda]^\kappa$. Thus we could (remember λ is a cardinal) interchange safely, in the definition of hugeness, $[\lambda]^\kappa$ and $(\lambda)^\kappa$.

In the combinatorial counterpart of the theory; while studying $[\lambda]^{\kappa}$ and $(\lambda)^{\kappa}$, there will be, however, serious differences. Besides, if λ is not a cardinal then our argument does not go through. Notice, however, that if $|\xi| = |\eta|$ then any filter from $[\xi]^{\kappa}$ can be transposed to $[\eta]^{\kappa}$. All properties like κ-completeness etc. are preserved. Thus if there is a κ-complete fine, normal filter in $[\xi]^{\eta}$ then $|\xi|$ is a target for κ. If κ is huge with a target λ, then λ is measurable; for $M \models$ "$j(\kappa)$ is measurable" (as κ is measurable) and so there is a measure $\mathcal{U}'$ on $j(\kappa)$ in M. But as ${}^{\lambda}M \subseteq M$, $\mathcal{P}(\lambda) \subseteq M$ and so $\mathcal{P}(\lambda)^M = \mathcal{P}(\lambda)$. Thus $\mathcal{U}'$ is an ultrafilter. Besides, again by closure properties of M, $\mathcal{U}'$ is λ-complete. Notice that in the case of the ultrapower $V^{\kappa}/\mathcal{U}$, $j(\kappa)$ is never measurable in V, (in M it is of course measurable). If κ is huge with target λ then the associated embedding j witnesses to κ being ξ-supercompact (by equivalence of existence of embedding and existence of ultrafilter) for all $\xi < \lambda$. So, for every $\kappa \leq \xi < \lambda$ in $\mathcal{P}_{\kappa}(\xi)$ there exists a κ-complete normal fine filter. All those filters, being included in $\mathcal{P}(\mathcal{P}_{\kappa}(\xi)) \subseteq \mathcal{P}\mathcal{P}(\xi)$, belong to $V_{\xi+3}$. As λ is limit, $V_{\lambda} \models$ "For every ξ greater or equal to κ, κ is ξ-supercompact". Thus existence of a huge cardinal implies consistency of ZFC + "There exists a supercompact cardinal". The following result due to Morgenstern is interesting: If there is κ which is huge and λ which is supercompact then the least huge cardinal is strictly smaller than the least supercompact. No contradiction is involved as, for κ least huge and λ its first target (actually it is the unique target for κ), $V_{\lambda} \models$. "There is no huge cardinal". To prove the Morgenstern's result assume $\kappa \to (\tau)$ and that ν is supercompact. Lift ν above $|V_{\tau+3}|$ is such a way that ${}^{|V_{\tau+3}|}M \subseteq M$. Then $M \models \kappa$" is huge with a target τ". But then $M \models$ "There exists a huge cardinal below $j(\nu)$" by the choice of τ. Thus there is a huge cardinal below ν. □

Having considered the possibility of the existence of a cardinal and an elementary embedding $j : V \to M$ such that κ is the critical point of j and ${}^{j(\kappa)}M \subseteq M$ increases our appetite; could it happen that ${}^{j^2(\kappa)}M \subseteq M$ or that ${}^{j^{(n)}(\kappa)}M \subseteq M$? A theorem of Kunen provides a bound: if $j : V \to M$, κ critical for j and $\rho = \bigcup_{n \in \omega} j^{(n)}(\kappa)$ then ${}^{\rho}M \not\subseteq M$.

The weaker forms mentioned above does not seem to lead to

inonsistencies and if ${}^{j^{(n)}(\kappa)}M \subseteq M$ we say that κ is n-huge. There is an ultrafilter characterization of n-huge cardinals: κ is n-huge with value λ for $j^{(n)}(\kappa)$ iff there exist a sequence $\kappa = \lambda_0 < \lambda_1 < \ldots < \lambda_{n-1} < \lambda_n = \lambda$ of cardinals and a κ complete normal ultrafilter $\mathcal{U}$ in $[\lambda]^\kappa$ such that for all $i < n$, $\{x : |x \cap \lambda_{i+1}| = \lambda_i\} \in \mathcal{U}$.

The result of 1.2 generalizes as follows: If is huge and $j : V \to M$, $j(\kappa) = \lambda$ and ${}^{2^\lambda}M \subseteq M$ (this happens for instance if κ is 2-huge) then all κ-complete ultrafilters in $[\lambda]^\kappa$ are of the form $\mathcal{U}_{j,A}$ (where A is a subset of $j(\lambda)$ of cardinality λ).

3.2. MULTIHUGE CARDINALS

In the previous section 3.1 we proved that if there exists a huge cardinal and if there exists a supercompact cardinal then the least huge cardinal is smaller than the least supercompact cardinal. But what happens if there exists a huge cardinal above a supercompact? A refinement of the argument we gave ("jump above it") shows that then the set of huge cardinals is cofinal in our supercompact cardinal. A very interesting case occurs if we suppose that κ is a cardinal which is both huge and supercompact. In that case there are normal ultrafilters $\mathcal{U}_1$ and $\mathcal{U}_2$ such that $\{\xi < \kappa | \xi$ is huge$\}$ belongs to $\mathcal{U}_1$ and $\{\xi < \kappa | \xi$ is supercompact$\}$ is in $\mathcal{U}_2$. For suppose that κ is simultaneously huge and supercompact. Let $j : V \to M$ witness to hugeness of κ, $j(\kappa) = \lambda$, then in M, by its closure properties κ is still α-supercompact for every $\kappa \leq \alpha < \lambda$. Thus $V_\lambda^M \models \kappa$ is supercompact. But then, setting $X = \{\xi < \kappa | V_\kappa \models$ "ξ is supercompact"$\}$ we obtain $\kappa \in j(X)$. So $X \in \mathcal{U}_j$. Now we just need to show that if κ is supercompact then the formula "$(\cdot)$ is supercompact" is absolute with respect to V_κ. It is obviously preserved downwards. To show that it is upwards absolute, let $V_\kappa \models$ "α is supercompact" and β be arbitrary. As κ is 2^{β^α}-supercompact there is $i : V \to N$, $i(\kappa) > 2^{\beta^\alpha}$, ${}^{2^{\beta^\alpha}}N \subseteq N$. Now $V_\kappa \prec V_{i(\kappa)}^N$ and $V_{i(\kappa)}^N \models$ "α is β-supercompact" (as $\beta > i(\kappa)$). Thus $N \models$ "α is β-supercompact". By closure properties of N, α is β-supercompact. This shows that $\mathcal{U}_j$ satisfies the required property of the desired ultrafilter $\mathcal{U}_2$. On the other hand assume λ is a target for κ. As κ is supercompact we can jump above 2^{λ^κ} and so the resulting $j : V \to P$ has the property that, in P, κ is huge with a target λ. But then $\{\alpha < \kappa | \alpha \text{ is huge}\}$ belongs to $\mathcal{U}_j$. □

The formulas "κ is α-supercompact" and "κ is huge with a target λ" are Σ_2. It is quite laborius to write these formulas in the desired form. We can - though - overcome the problem writing instead $(E\delta)(Ex)(x = V_\delta \;\&\; x \models \ulcorner\theta\urcorner)$ where θ is any formula expressing "κ is α-supercompact" or "κ is huge with a target λ". The formula $x = V_\delta$ is easily seen to be Σ_2 and the formula expressing satisfaction formally is Δ_1. Both formulas: "κ is α-supercompact" and "κ is huge with a target λ" are also Π_2. To see this we just write: $(E\delta)(Ex)\ (x = V_\delta \;\&\; \kappa \in \delta \;\&\; \alpha \in \delta \;\&\; X \models \ulcorner\neg\theta\urcorner)$ where θ is as above. This is a Σ_2 formula telling that κ is not α-supercompact, so "κ is α-supercompact" is Π_2. The same argument shows "κ is huge with a target λ" is Π_2. Therefore formula "κ is supercompact" is Π_2. The statement "there exists a supercompact cardinal" is Σ_3 and not Π_3 as follows from the fact that this sentence is false in V_κ where κ is the least supercompact. We could circumvent our argument at the beginnning of this section using Levy's hierarchy but the argument we used shows the power of supercompactness.

The ultrafilters $\mathcal{U}_i$ constructed above do not coincide if κ is - say - the least cardinal which is both huge and supercompact.

However for every κ measurable there exists an ultrafilter which does not contain the set of smaller ordinals which are - say - huge. Just do this by induction; if κ is the least measurable then clearly there is such an ultrafilter. Otherwise, we have two cases;if the normal ultrafilter we get in κ by means of (any) elementary imbedding is already good i.e. $\{\xi | \xi \text{ is not huge}\} \in \mathcal{U}_j$ then we are done. Otherwise, for each $\xi < \kappa$, ξ measurable pick $\mathcal{U}_\xi$ such that $\{\eta < \xi | \eta \text{ is huge}\} \notin \mathcal{U}_\xi$. Define $\mathcal{U}'$ in κ as follows: $X \in \mathcal{U}' \iff \{\alpha | \alpha \text{ is huge} \;\&\; X \cap \alpha \in \mathcal{U}_\alpha\} \in \mathcal{U}$. Now, $\{\alpha | \alpha \text{ is huge}\} \notin \mathcal{U}'$. (It is enough to show that $\{\alpha | \alpha \text{ is not huge}\} \in \mathcal{U}'$ which is more or less obvious).. $\mathcal{U}'$ is easily seen to be normal. □

This argument is due to Solovay and it may be used to construct a normal ultrafilter κ containing $\{\alpha | \neg\phi(\alpha)\}$ if ϕ is any property implying measurability. If κ is the least huge & supercompact then as we noticed the ultrafilter U_2 does not contain $\{\xi < \kappa | \xi$ is huge$\}$. Taking into account that the property: "κ is huge $\vee$ κ is supercompact" implies "κ is measurable" we get that there is a normal (etc.) ultrafilter U' on κ which contains $\{\xi < \kappa | \xi$ is neither supercompact nor huge$\}$. Thus U' is different from U_2 (and from U_1 as well). So if κ is the least cardinal which is both huge and supercompact then there are 2 different normal ultrafilters contains $\{\xi < \kappa | \xi$ is not huge$\}$ and similarly for supercompactness.

We found that under the appropriate conditions (κ huge and supercompact) there are lots of huge cardinals below κ. Consider now the following questions: Can a given κ possess many targets? (Such κ's are called multihuge cardinals). And if so, what are the closure properties of the set of targets? First of all we have the following transitivity property: if $\kappa \to (\lambda)$ and $\lambda \to (\mu)$ then $\kappa \to (\mu)$. To see this, let $j : V \to M$ witness to $\lambda \to (\mu)$ i.e. $j(\lambda) = \mu$, ${}^{\mu}M \subseteq M$. Then as $j(\lambda) = \mu$ and $\kappa \to (\lambda)$, $j(\kappa) \to (j(\lambda))$ in M i.e. in M there is a κ-complete fine, normal measure U on $(\mu)^{\kappa}$. Now as μ is inaccessible, $\mu^{\kappa} = \mu$ and so by closure properties of M, all subsets of $[\mu]^{\kappa}$ belong to M. Therefore $P([\mu]^{\kappa})^{M} = P([\mu]^{\kappa})$ and so U is really an ultrafilter. Using the closure properties of μ again we find that U is κ-complete, normal and fine thus U witnesses to $\kappa \to (\mu)$.

The main techniques which we used until now were the following:

One way to construct a normal ultrafilter in κ containing a set $\{\xi | \xi < \kappa \ \& \ \phi(\xi)\}$ is to have an elementary imbedding $j : V \to M$ and such that $M \models \phi[\kappa]$. This implies that $\{\xi | \xi < \kappa \ \& \ \phi(\xi)\} \in U_j$. If we have a weaker property, namely that for some $\kappa < \alpha < j(\kappa)$ $M \models \phi[\alpha]$ then $\{\xi | \xi < \kappa \ \& \ \phi(\xi)\}$ is unbounded in κ.

Another technique (which is used in combination with the first one) is to obtain an embedding j such that $j : V \to M$ and M is "wide enough" to preserve some properties of κ. Combining these techniques we have a number of facts which we shall leave without proof (we do not always state the optimal result but we believe that the reader will be able to obtain the more general results).

1º Assume κ possesses 2 targets. Then there is a normal ultrafilter in κ which concentrates on cardinals which are huge.

This fact generalizes to α targets ($\alpha < \kappa$).

2º If λ is supercompact and κ is the first cardinal with ν-targets ($\nu < \lambda$) then $\kappa < \lambda$, also those targets are smaller than λ.

3º If λ is the least supercompact and κ the first huge cardinal possessing λ targets then $\lambda < \kappa$ (just notice that $\kappa = \lambda$ is impossible since λ is the least supercompact and $\kappa < \lambda$ implies that λ many targets of κ are below λ by Σ_2 absoluteness). This in particular implies that $V_\lambda \models$ "κ is supercompact" and again by Σ_2-absoluteness κ is really supercompact contradicting that λ is the first supercompact.

4º The same argument tells us that if λ is supercompact $\{\xi < \lambda \mid \xi$ is supercompact$\}$ is small (what small means should be clear from the proof) and κ is the first huge with λ targets then $\lambda < \kappa$.

For 4º, to show that $\kappa \neq \lambda$, notice that if $\kappa=\lambda$ then κ is both huge and supercompact. And for showing $\neg \kappa < \lambda$ notice that if $\kappa < \lambda, \kappa$ is again huge and supercompact so there are lots of supercompact cardinals below κ.

Let us notice that if κ is huge and if $2^\kappa = \kappa^+$ then for every target λ of κ, $2^\lambda = \lambda^+$. Indeed, let $j : V \to M$ be such that $j(\kappa) = \lambda$, $^\lambda\mu \subseteq \mu$. Then $M \models 2^\lambda = \lambda^+$ and so, since $P(\lambda)^M$ $P(\lambda)$, in M there is a $1-1$ map of $P(\lambda)$ onto $(\lambda^+)^M$. But for sure $(\lambda^+)^M < \lambda^+$ so if there is a $1-1$ map of $P(\lambda)$ onto $(\lambda^+)^M$ then by Cantor's theorem $(\lambda^+)^M \leq \lambda^+$ so $(\lambda^+)^M = \lambda^+$ and thus $2^\lambda = \lambda^+$.

Suppose now that κ possesses a proper class of targets. Let us call such κ superhuge (in analogy to supercompactness). Clearly if κ is superhuge, κ is supercompact. But then κ is both huge and supercompact and so there are many supercompact cardinals below κ. Superhuge cardinals have stronger reflection properties than supercompacts. They are Σ_3 reflecting i.e. if ϕ is Σ_3 and true then $V_\kappa \models \phi$. There are various ways to show this. As shown by It Beng Tan, superhuge cardinals have certain property called extendibility. Besides Solovary, Reinhardt and Kanamori show that extendible cardinals are Σ_3-reflecting. One can, however give a much simpler argument: Assume $(Ex)\phi(x)$ is a true Σ_3 formula. Pick x such that $\phi(x)$ and let λ be a target for κ such that rank $(x) < \lambda$. Since λ inaccessible, $H_\lambda = V_\lambda$ and by Levy absoluteness (downward), since $\phi(x)$ is Π_2 $V_\lambda \models \phi(x)$. But then $V_\lambda \models (Ex)\phi(x)$ and since $V_\kappa \prec V_\lambda$ (indeed $V_\kappa \prec V^M_{j(\kappa)}$ but $V^M_{j(\kappa)} = V_{j(\kappa)}$ by closure properties of M), $V_\kappa \models Ex\ \phi\ (x)$.

The formula "κ is superhuge" is easily seen to be Π_3 and the sentence "there exists a superhuge cardinal" is a Σ_4.

Finally let us introduce another type of large cardinal: we say that κ is doubly superhuge iff $\{\alpha|\alpha$ is a target for κ and α is superhuge$\}$ is unbounded.

We show that if κ is doubly superhuge then V_κ is Σ_4 reflecting. To see this notice that as κ is Σ_3 reflecting it is as well Π_3 reflecting and now if $(Ex)\phi$ is true then pick $j : V \to M$ such that $j(\kappa)$ is larger than rank of an x making ϕ true and $j(\kappa)$ is superhuge. Then $V_{j(\kappa)} \models \phi(x)$, $V_{j(\kappa)} \models (Ex)\phi(x)$ and so $V_\kappa \models (Ex)\phi(x)$ as $V_\kappa \prec V_{j(\kappa)}$. Now we introduce the notion of n-fold superhuge by induction; κ is 0-fold superhuge if κ is huge and α is $(n+1)$ = fold superhuge if $\{\alpha|\alpha$ is a target for κ and α is n-fold superhuge$\}$ is unbounded. We leave to the reader the proof of the following: If κ is n-fold superhuge then V_κ is Σ_{2n+1} - reflecting.

The consistency of the existence of an n-fold superhuge cardinal follows from the consistency of an n+1-huge cardinal. More precisely, we will indicate how to show that if κ is 3-huge and $j : V \to M$ such that $j(\kappa) = \lambda_1$, $j(\lambda_1)=\lambda_2$, $j(\lambda_2) = \lambda_3$ and ${}^{\lambda_3}M \subseteq M$ then $V_{\lambda_1} \models$ "κ is doubly superhuge". To see this notice first that if κ is 2-huge and $i : V \to N$ is such that $i(\kappa) = \nu_1$ $i(\nu_1) = \nu_2$ and ${}^{\nu_2}N \subseteq N$, then $N \models$"κ is huge with target ν_1" and the same holds in V. If $\mathcal{U}$ is the corresponding ultrafilter then $i(\mathcal{U})$ is an ultrafilter making ν_1 huge with target ν_2. Let k be the embedding corresponding to $i(\mathcal{U})$, $k : V \to P$, ${}^{\nu_2}P \subseteq P$ and ν_1 critical point of k with $k(\nu_1) = \nu_2$. Given $\alpha < \nu_1$, $P \models$ "There is a target $>\alpha$ for κ smaller than $j(\nu_1)$". Thus the set of targets for κ smaller than ν_1 is unbounded in ν_1.

Coming back to $j,\lambda_1,\lambda_2,\lambda_3$, notice that λ_1 is 2-huge, hence the target for λ_1 are cofinal in λ_2.

Let i witness to the fact that λ_1 is 2-huge. Then $j(\kappa) = \kappa$ and for $\alpha < \lambda_1$, $i(\alpha) = \alpha$, $i : V \to M$, $i(\lambda_1) = \lambda_2$. We know, by the closure properties of M that all the targets of λ_1 below λ_2 are still targets in M, thus $M \models$ "$(E_\xi)(\kappa \to \xi)\alpha < \xi$ & ξ possesses unboundedly many targets below $j(\lambda_1)$". So $(E\xi)(\kappa \to (\xi)$ & $\alpha < \xi$ & ξ possesses unboundedly many targets below $\lambda_2)$.

Thus $V_{\lambda_1} \models$ "κ is doubly superhuge".

3.3. CLOSED UNBOUNDED SETS IN $[\lambda]^\kappa$, COMBINATORICS OF $(\lambda)^\kappa$

The spaces κ and $P_\kappa(\lambda)$ have very nice properties; they possess, minimal κ-complete normal fine filters (the respective closed unbounded filters) and if by chance κ is measurable (or λ-supercompact) then every normal measure on κ (or $P_\kappa(\lambda)$) contains this filter. In section 3.1 we encountered the spaces $[\lambda]^\kappa$ and $(\lambda)^\kappa$. Thus the question arises; can one introduce similar notions of a closed unbounded set in - say - $[\lambda]^\kappa$ in such a way that if $\kappa \to (\lambda)$ then the closed unbounded sets have measure 1 and in addition the filter contructed in this fashion is the least κ-complete, fine, normal filter in $[\lambda]^\kappa$? (also: what should "fine" mean in this case?). We will consider below such a notion, but first let us answer the following: Since $[\lambda]^\kappa \subseteq P_{\kappa^+}(\lambda)$ and it is clearly stationary what's wrong with just restricting $CLUB_{\kappa^+,\lambda}$ to $[\lambda]^\kappa$? The point is that not all sets from $CLUB_{\kappa^+,\lambda}$ have measure 1 under normal measure $[\lambda]^\kappa$ if $\kappa \to (\lambda)$. In fact as we had shown in 3.1 every such normal measure concentrates on $(\lambda)^\kappa$. Now, if $P \subseteq \lambda$, o.t. $(P) > \kappa$, $\hat{P}$ (the cone over P in $P_{\kappa^+}(\lambda)$) is disjoint from $(\lambda)^\kappa$ thus must have measure 0. This forces us to look at some other construction.

To this end let us introduce, for $X \subseteq P_\kappa(\lambda)$, and operation A_X producing, out of X, a certain subset of $[\lambda]^\kappa$. $A_X = \{P \in [\lambda]^\kappa \mid$ there exists a directed system $D \subseteq X$ such that $\cup D = P\}$. Let us see what we obtain if we consider sets of the form A_X, for X closed and unbounded in $P_\kappa(\lambda)$. Here are some basic facts: a) If $p \in P_\kappa(\lambda)$ then $\{P \in [\lambda]^\kappa \mid p \subseteq P\}$ is of the form $A_{\hat{p}}$. b) If $\langle X\rangle_{\xi<\nu}$ $\nu < \kappa$ is a family of closed and unbounded sets in $P_\kappa(\lambda)$, then $A_{\bigcap_{\xi<\nu} X_\xi} \subseteq \bigcap_{\xi<\nu} A_{X_\xi}$.

c) If $\langle X_\xi\rangle_{\xi<\lambda}$ is a family of closed and unbounded sets in $P_\kappa(\lambda)$ then $A_{\Delta_{\xi<\lambda} X_\xi} \subseteq \Delta_{\xi<\lambda} A_{X_\xi}$. d) If $\kappa \to (\lambda)$ that μ is a normal in $[\lambda]^\kappa$ witnessing to it then for every unbounded set X, $A_X \in \mu$. All these facts are easy to prove. The filter $F_{\kappa,\lambda}$ in $[\lambda]^\kappa$ generated by sets of the form A_X for X closed and unbounded in $P_\kappa(\lambda)$ is thus a κ-complete, fine, normal filter in $P_\kappa(\lambda)$. If we show its minimality then the final criterion for a notion of closed and unbounded set will be satisfied. The argument for minimality follows the line of 2.3 (case of $P_\kappa(\lambda)$). For $f : [\lambda]^n \to P_\kappa(\lambda)$, let $C_\kappa(f) = \{P \subset [\lambda]^\kappa \mid (x)\ (x \subseteq P \Rightarrow f(x) \subseteq P)\}$. Let us show that $A_{C(f)} = C_\kappa(f)$. The inclusion $C_\kappa(f) \subseteq A_{C(f)}$ is quite obvious and uses a Skolem-Lowenheim like argument. If $P \in A_{C(f)}$, $x \subseteq P$ then since $P = \cup D$, D directed, $D \subseteq C(f)$, there is (since $\vec{x}$ is finite) $p \in D$, $\vec{x} \subseteq p$. But then $f(\vec{x}) \subseteq p \subseteq P$ thus $f(\vec{x}) \subseteq P$. This result implies the following basis property for sets in $F_{\kappa,\lambda}$: $A \in F_{\kappa,\lambda}$ iff there is $F : [\lambda]^2 \to P_\kappa(\lambda)$ such that $C_\kappa(f) \subseteq A$. The minimality of $F_{\kappa,\lambda}$ follows easily as in the case of $P_\kappa(\lambda)$. For otherwise pick $F \subsetneqq F_{\kappa,\lambda}$. Then, there must be $f : [\lambda]^2 \to P_\kappa(\lambda)$ such that $C_\kappa(f) \notin F$. Thus $R = [\lambda]^\kappa \setminus C_\kappa(f)$ is F-stationary, we fix a pair of ordinals α,β and an F-stationary R' such that each $P \in R'$, $\{\alpha,\beta\} \subseteq P$ and $f(\alpha,\beta) \not\subseteq P$. But then $R' \cap (f(\alpha,\beta))^\vee = \phi$, contradiction. We have as an exercise to check that the Magidor-Solovay result for $P_\kappa(\lambda)$ (see section 2.2) extends to this context; instead of direct limits we could take in our case sums of chains, the result is the same. We choose this definition only because it makes the proof of d) easier.

We noted above (point c) that if $\langle X_\xi \rangle_{\xi<\lambda}$ is a family of closed unbounded sets then $A_{\underset{\xi<\lambda}{\Delta} X_\xi} \subseteq \underset{\xi<\lambda}{\Delta} A_{X_\xi}$. One can in fact show the equality. This needs a bit of construction and we wish the reader good luck doing it. Once this is shown we can extend the result of Carr, section 2.3, to the context of $P_\kappa(\lambda)$. Indeed, $\check{p} = A_{\hat{p}}$ (as it was pointed out in a)) where $\check{p}$ is the cone in $[\lambda]^\kappa$ over p. Now as the sets of the form $\underset{\xi<\lambda}{\Delta}\ \underset{\eta<\lambda}{\Delta} \hat{p}_{\xi\eta}$ form a basis for the closed unbounded filter in $P_\kappa(\lambda)$, the sets form $\underset{\xi<\lambda}{\Delta}\ \underset{\eta<\lambda}{\Delta} A_{\hat{p}_{\xi,\eta}}$ form a basis for $F_{\kappa,\lambda}$ i.e. the sets of form $\underset{\xi<\eta}{\Delta}\ \underset{\xi<\lambda}{\Delta} \check{p}_{\xi,\eta}$ form a basis for $F_{\kappa,\lambda}$. □

Unfortunately $(\lambda)^\kappa$ never belongs to $F_{\kappa,\lambda}$. If $\kappa \to (\lambda)$ then $(\lambda)^\kappa$ is clearly stationary and we can, then restrict $F_{\kappa,\lambda}$ to $(\lambda)^\kappa$ (by means of intersections). However, if $\kappa \not\to (\lambda)$ in particular if our ontological assumptions make the existence of huge cardinal impossible (e.g., if

V = L) then the situation is different. Indeed, under V = L, $(\lambda)^\kappa$ is never $F_{\kappa\lambda}$-stationary. We employ a reasoning analogous to the one used in Baumgartner's partition theorem (see section 2.3). Enumerate L_λ with λ such a way that L_κ is enumerated with κ. The $\{p \in P_\kappa(\lambda) | p$ codes an elementary subsystem of $L_\lambda\}$ is closed unbounded in $P_\kappa(\lambda)$. Applying operation A to it we get $\{P \in [\lambda]^\kappa | P$ codes an elementary substructure of $L_\lambda\}$. Under V = L, it is easy to see that if $X \prec L_\lambda$, $|X| = \kappa$ then $L_\kappa \subseteq X$. Thus we easily see that $\check{\kappa} \in F_{\kappa,\lambda}$. One $\check{\kappa} \in F_{\kappa\lambda}$ then for all $P \in [\lambda]^\kappa$, $\check{P} \in F_{\kappa,\lambda}$ and this shows that $(\lambda)^\kappa$ is not stationary with respect to $F_{\kappa,\lambda}$. Note that A_X is always closed under unions of increasing chains of the length κ. This needs a construction which we leave to the reader but once proved allows us to get the following result: If V = L then $F_{\kappa,\lambda}$ is $\text{Club}_{\kappa^+,\lambda} \upharpoonright [\lambda]^\kappa$. Indeed, every set in $F_{\kappa,\lambda}$ is closed in $P_{\kappa^+,\lambda}$. Thus $F_{\kappa^+,\lambda} \subseteq \text{Club}_{\kappa^+,\lambda}$. On the other hand by Carr's result in the context of $P_{\kappa^+,\lambda}$, every closed and unbounded set in $P_{\kappa^+,\lambda}$ is of the form $\underset{\xi<\lambda}{\Delta}\ \underset{\xi<\lambda}{\Delta}\ P_{\xi,\eta}$. Under V = L, all the sets of the form $\check{P}$ are in $[\lambda]^\kappa$ and so, since $F_{\kappa,\lambda}$ is normal, $\text{Club}_{\kappa^+,\lambda} \upharpoonright [\lambda]^\kappa \subseteq F_{\kappa,\lambda}$.

This implies immediately that we cannot define (in ZFC alone) a κ-complete fine, normal, nontrivial filter G in $(\lambda)^\kappa$. For if we could, then defining G' on $[\lambda]^\kappa$ by $X \in G' \iff X \cap (\lambda)^\kappa \in G$ we obtain a filter G' extending $F_{\kappa,\lambda}$ such that $(\lambda)^\kappa \in G'$. It follows that $(\lambda)^\kappa$ is $F_{\kappa,\lambda}$-stationary, but this is impossible if V = L.

This indicates that if we want to define a κ-complete and normal filter on $(\lambda)^\kappa$ then it cannot be fine.

NOTES

1. See in Section 3.3.
2. We would like to thank A. Kanamori for many valuable comments and suggestions.

REFERENCES

1. Abramson, F.G.; L.A. Harrington; E.M. Kleinberg and W.S. Zwicker: Flipping properties: a unifying thread in the theory of large cardinals, Ann. Math. Logic 12 (1977), 25-58.
2. Barbanel, J.; C.A. Di Prisco and I.B. Tan: Many times huge and superhuge cardinals, Journal of Symbolic Logic 49 (1984), 112-122.
3. Baumgartner, J.; A. Hajnal and A. Máté: Weak saturation properties of ideals. In "Infinite and Finite Sets": Coll. Math. Soc. J. Bolyzi 10 (1973), 137-158.
4. Baumgartner, J.; A. Taylor and S. Wagon: On splitting stationary subsets of large cardinals, Journal of Symbolic Logic 42 (1977), 203-214.
5. Baumgartner, J.; A. Taylor and S. Wagon: Structural properties of ideals, Dissertationes Mathematicae 197.
6. Carr, D.J.: The minimal normal filter on $P_{\kappa}\lambda$, Proc. Amer. Math. Soc. 86 (1982), 316-320.
7. Di Prisco, C.A.: Ph.D. Thesis Massachusetts Institute of Technology, 1976.
8. Di Prisco, C.A. and W. Marek: Some properties of stationary sets, Dissertationes Mathematicae, 198 (1982).
9. Di Prisco, C.A. and W. Marek: A filter on $[\lambda]^{\kappa}$, Proc. A.M.S. 90 (1984), 591-598
10. Di Prisco, C.A. and W. Marek: On the space $(\lambda)^{\kappa}$ (to appear).
11. Di Prisco, C.A. and W. Zwicker: Flipping properties and supercompact cardinals, Fund. Math. 109 (1980), 31-36.
12. Fodor, G.: Eine Bemerkung Zur Theorie der regressiven Funktionen, Acta Sc. Math. (Szeged) 17 (1956), 139-142.
13. Jech, T.: Set Theory, Academic Press (1978).
14. Jech, T.: Some combinatorial problems concerning uncountable cardinals, Ann. Math. Logic 5 (1973), 165-198.
15. Kanamori, A.: Ultrafilters over a measurable cardinal, Ann. Math. Logic 10 (1976), 315-356.
16. Kanamori, A. and M. Magidor: The evolution of large cardinal axioms in set theory in: higher set theory. Lecture Notes in Mathematics vol. 699 (Ed. G. Muller and D. Scott) Springer-Verlag (1978), 99-275.
17. Kunen, K: Elementary embbedings and infinitary combinatoric, J. Symbolic Logic 36 (1971), 407-413.
18. Kunen, K. and D.H. Pelletier: On a combinatorial property of Menas related to the partition property for measure on supercompact cardinals, J. Symbolic Logic 48 (1983), 475-481.
19. Magidor, M.: On the role of supercompact and extendible cardinals in logic, Israel J. Math. 10 (1971), 147-157.
20. Magidor, M.: Combinatorial characterization of supercompact cardinals, Proc. Amer. Math. Soc. 42 (1974), 279-285.

21. Menas, T.K.: On strong compactness and supercompactness, Ann. Math. Logic 7 (1975), 327-359.

22. Menas, T.K.: A combinatorial property of $P_\kappa(\lambda)$, J. Symbolic Logic 41 (1976), 225-233.

23. Morgenstern, C.F.: On the ordering of certain large cardinals, J. Symbolic Logic 44 (1979), 563-565.

24. Pelletier, D.H.: A simple proof of Weglorz' characterization of normality for ideals (to appear).

25. Rowbottom, F.: Some strong axioms of infinity incompatible with the axiom of constructibility, Ann. Mat. Logic 3 (1971), 1-44.

26. Scott, D.: Measurable cardinals and constructible sets, Bull. Acad. Polon. Sci. 9 (1961), 521-524.

27. Silver, J.: Some applications of model theory in set theory. Ann. Math. Logic 3 (1971), 45-110.

28. Solovay, R.M.: Real-valued measurable cardinals in "axiomatic set theory", Proc. Symp. Pure Math. 13, I (D. Scott ed.), Amer. Math. Soc., Providence, R.I., 1971, 397-428.

29. Solovay, R.M.; W.N. Reinhardt and A. Kanomori: Strong axioms of infinity and elementary embeddings, Ann.Math. Logic 13 (1978), 73-116.

30. Sureson, C.: Thése 3$^{\underline{e}}$ Cycle, Paris VII (1981)

31. Vopénka, P. and K. Hrbáček: On strongly measurable cardinals, Bull. Acad. Polon. Sci. 14 (1966), 587-591.

32. Weglorz, B.: Some properties of filters, in Springer-Verlag, L.N.M. 619, Berlin-Heidelberg, New York 1977, 311-328.

33. Zwicker, W.: $P_\kappa(\lambda)$ combinatorics I: Stationary coding sets rationalize the club filter (to appear).

34. Ketonen, J.: Strong Compactness and other Cardinal Sins. Ann. Math. Logic 5 (1972), 47-76.

THE CONDITIONAL AND PARACONSISTENT LOGICS

Ítala Maria Loffredo D'Ottaviano and E. G. K. Lopez-Escobar*/Instituto de Matemática, Estatistica e Ciênciá da Computação, Universidade Estadual de Campinas, Campinas, São Paulo, Brazil

INTRODUCTION

A paraconsistent logic is usually defined as a logic which admits paraconsistent theories, where a theory T is paraconsistent iff (i) there is at least one sentence which is not a theorem of T and (ii) there is a sentence such that it and its negation are theorems of T.

Since paraconsistent logics grew out of classical logic (see Arruda 1980), it is not surprising to find that negation plays a very important rôle. In fact, most paraconsistent logics can be viewed as classical logic with a weakened negation. The problem is that there is no agreement on how much negation can be weakened and still merit the name of "negation".

If one considers a logic as the embodiment of its (absolute) theorems, then the above approach to paraconsistent logics is very reasonable. But there are other views of logic; one could be more interested in the inferences (deductions) than in the theorems.

Let $\vdash$ be the (extensional) representation of an inference relation, that is $\alpha_0, \ldots, \alpha_{n-1} \vdash \beta$ iff there is an inference of β from $\{\alpha_0, \ldots, \alpha_{n-1}\}$. The following rules are then valid for all inference

*Current Affiliation: Department of Mathematics, University of Maryland, College Park, Maryland.

relations (well, almost all):

(R) $\alpha \vdash \alpha$

(T) If $\alpha_o,\dots,\alpha_{n-1} \vdash \beta$ and $\beta \vdash \gamma$ then $\alpha_o,\dots,\alpha_{n-1} \vdash \gamma$.

Now a rôle for the conditional connective $\to$ (often miscalled implication) is to try to express, within the formal language, the behavior of $\vdash$. That is why we find that in many logics the conditional $\to$ satisfies the following rules:

$$\text{(DTC)} \quad \frac{\alpha_o,\dots,\alpha_{n-1}, \alpha \vdash \beta}{\alpha_o,\dots,\alpha_{n-1} \vdash (\alpha \to \beta)}$$

(Deduction Theorem and Converse).

Although negation can be problematic (specially in an operational interpretation of the connectives for then $\neg\alpha$ should attempt to express that α is not derivable) the existence of an absurdity is not so; by an absurdity we understand a propositional constant $\bot$ such that for all formulas α :

(ABD) $\bot \vdash \alpha$

In logics in which we have both (DTC) and (ABD), the negation of α, $\neg\alpha$, is often defined as $(\alpha\to\bot)$ and then we have that for all formulas β:

(CN) $\vdash \alpha \to (\neg\alpha\to\beta)$,

because

$$\cfrac{\cfrac{\cfrac{\cfrac{(\alpha\to\bot) \vdash (\alpha\to\bot)}{\alpha,(\alpha\to\bot) \vdash \bot} \qquad \bot \vdash \beta}{\alpha,\ (\alpha\to\bot) \vdash \beta}}{\alpha \vdash (\neg\alpha\to\beta)}}{\vdash \alpha\to(\neg\alpha\to\beta)}$$

Now (CN) is usually the first property of classical (and intuitionistic) negation that has to be given up in order to have paraconsistent logics. Thus if we are going to deal with such logics then we cannot satisfy both of the following conditions

(A) the negation of a formula α to be (equivalent to) $(\alpha \to \bot)$, where $\bot$ is absurdity.

(B) the conditional $\rightarrow$ to satisfy the rules (DTC).

The traditional approach to paraconsistent logics is to give up (A) - actually it is often not even considered. The aim of this paper is to show that another way is to maintain (A) but to give up (B). There are examples of logics which do not satisfy (B). The principal ones being the predicate calculus allowing empty universes (see Mostowski,1951) and the many-valued propositional calculi of Łukasiewicz. On the other hand the logics we shall consider will be predicate calculi about non-empty universes and, in a certain sense (made precise in Scott 1974) will be two-valued.

Part of the motivation for the logics developed here can be found in D'Ottaviano and Lõpez-Escobar 1982 - specially for the propositional logics.

ENTAILMENT RELATIONS

The Formal Languages

Although most of what follows only requires that a language be a non-empty set, we prefer to consider the languages to be (possible extensions of) the first-order predicate calculus and to contain at least the following:

(i) A binary predicate symbol P,

(ii) The propositional constants $\top$, $\bot$ for Truth and Absurdity respectively,

(iii) The connectives $\wedge$, $\vee$ for conjunction and disjunction respectively,

(iv) The conditional connective: $\rightarrow$,

(v) The individual variables: $x_o, x_1, \ldots, x_n, \ldots$,

(vi) The individual parameters: $a_o, a_1, \ldots, a_n, \ldots$,

(vii) A set $\mathbb{C} = \{c \mid c \in \mathbb{C}\}$ of individual constants,

(viii) The universal quantifier $\bigwedge$,

(ix) The existencial quantifier $\bigvee$.

Let L be one such language. Then $Fm\ell_L$ is the set of formulas L and Snt_L the set of sentences. If we wish to emphasize the set $\mathbb{C}$ of individual constants then instead of "L" we may write "$L(\mathbb{C})$". Given a language $L(\mathbb{C})$ and $\mathbb{C}^*$ another set of individual constants then

$L(\mathbb{C} \cup \mathbb{C}^*)$ is the language just like $L(\mathbb{C})$ except that $\mathbb{C} \cup \mathbb{C}^*$ is its set of individual constants.

We shall use lower case greek letters for formulas and uppercase greek letters for finite sets of formulas. An ordered pair $\langle \Gamma, \Sigma \rangle$ of finite sets of formulas of L is called an L-sequent and will often be denoted by "$\Gamma \Rightarrow \Sigma$".

If α is a formula then we may write "$\alpha_{d_1 \ldots d_k}$" to indicate that the individual parameters or individual constants that occur in α are included in $\{d_1, \ldots, d_k\}$. $\alpha^{t_1 \ldots t_k}_{d_1 \ldots d_k}$ is the formula that results in simultaneously substituting t_1 for $d_1, \ldots, t_k$ for d_k in α. It will be assumed, without any further ado, that all substitutions are proper.

Another familiar concept which will not be further discussed (but assumed known) is that of "alphabetic bound variants"; we shall say that α'' and α' are alphabetically equivalent, in symbols $\alpha'' \approx \alpha'$, iff they are alphabetic variants of some formula α.

A parameter substitution (or simply: a substitution) is simply a function s from the set of individual parameters into the union of the individual parameters and individual constants. We assume that a substitution s has a unique extension to a function from $Fm\ell_{L(\mathbb{C})}$ into $Fm\ell_{L(\mathbb{C})}$-which also will be denoted by "s". If A is a set of formulas and s a substitution then $sA = \{s(\alpha) \mid \alpha \in A\}$. Similarly $S = \Gamma \Rightarrow \Sigma$ then $sS = s\Gamma \Rightarrow s\Sigma$.

If $S_1 = \Gamma_1 \Rightarrow \Sigma_1$, $S_2 = \Gamma_2 \Rightarrow \Sigma_2$. Then S_1 and S_2 are alphabetic variants, in symbols: $S_1 \approx S_2$, if Γ_1, Γ_2 and Σ_1, Σ_2 consist respectively of alphabetic variants.

Abstract Entailment Relations

A set $\vdash$ of L-sequents is called an abstract entailment ralation (a.e.r.) iff it satisfies the following conditions, which we shall call "rules" (and where we write "$\Gamma \vdash \Sigma$" instead of the more cumbersome "$\Gamma \Rightarrow \Sigma \in \vdash$").

(T) $\emptyset \vdash \top$ and $\bot \vdash \emptyset$ ($\emptyset$ is the empty set),

(R) $\alpha \vdash \alpha$, for all formulas α,

(T) If $\Gamma \vdash \Sigma, \alpha$ and $\Gamma, \alpha \vdash \Sigma$ then $\Gamma \vdash \Sigma$,

(M) If $\Gamma_1 \vdash \Sigma_1$ then $\Gamma_1, \Gamma_2 \vdash \Sigma_1, \Sigma_2$,

(S) $\vdash$ is closed under substitution, that is, if $S \in \vdash$ and s is a parameter substitution $sS \in \vdash$,

(A) $\vdash$ is closed under alphabetic equivalence.

The rules (R), (T), (M) of reflexivity, transitivity and monotonicity were singled out in Scott 1974. The other rules were introduced to accomodate the propositional constants and the individuals.

An a.e.r. $\vdash$ is trivial iff every sequent belongs to it. Observe that $\vdash$ is trivial iff (i) $\emptyset \vdash \emptyset$, or (ii) There is some formula α such that $\alpha \vdash \emptyset$ and $\emptyset \vdash \alpha$, or (iii) $\emptyset \vdash \bot$.

A pre-theory is simply a pair $T = \langle A,B \rangle$ of sets of sentences of L such that $\top \in A$ and $\bot \in B$. A is the set of theorems of T and B the set of refutations of T.

$$Cn_{\vdash}(A) = \{\sigma \in Snt_L \mid \exists \Gamma_{\Gamma \subseteq A}(\Gamma \vdash \sigma)\}.$$

$$Co_{\vdash}(B) = \{\sigma \in Snt_L \mid \exists \Sigma_{\Sigma \subseteq B}(\sigma \vdash \Sigma)\}.$$

A pre-theory $T = \langle A,B \rangle$ is a $\vdash$-theory iff $Cn_{\vdash}(A) \subseteq A$ and $Co_{\vdash}(B) \subseteq B$. T is inconsistent iff $\exists \Gamma_{\Gamma \subseteq A}\ \Sigma_{\Sigma \subseteq B}(\Gamma \vdash \Sigma)$. T is trivial iff $A = Sent_L$ and T is total iff it is consistent and $A \cup B = Sent_L$.

An applied theory is a theory $T = \langle A,B \rangle$ where $B = Co_{\vdash}(\{\bot\})$.

Entailment Relations

An entailment relation (er) is an a.e.r. $\vdash$ which in addition satisfies a finite set of rules of the form:

if $S_1 \in \vdash, \ldots, S_k \in \vdash$ then $R(S_1, \ldots, S_k) \in \vdash$.

We shall say that the e.r. $\vdash$ is of type τ, or that it is a τ - e.r. iff τ encodes (the names for) the additional rules.

A τ - e.r. $\vdash$ is said to be τ-maximal (or simply: maximal) iff the only proper τ - e.r. extension is the trivial one. The following theorem is but a restatement of Lindenbaum's lemma:

THEOREM.

(1) To every non-trivial τ - e.r. $\vdash$ there corresponds at least one maximal τ - e.r. $\vdash^*$ such that $\vdash \subseteq \vdash^*$.

(2) Every τ - e.r. $\vdash$ is the intersection of all the maximal τ - e.r. which contain it.

(3) For certain τ, if $\vdash$ is a τ - e.r. and $\nvdash \sigma$, then there is a maximal τ - e.r. $\vdash^*$ such that $\vdash \subseteq \vdash^*$ and $\nvdash^* \sigma$.

PROOFS. The quickest way is to use Zorn's lemma. For (3) we proceed as follows. Let $Z = \{ \vdash' \mid \vdash \subseteq \vdash', \sigma \vdash' \emptyset$ and $\vdash'$ is a non-trivial τ - e.r.$\}$. If Z can be shown to be non-empty then Zorn's lemma gives the desired result. In order to show that Z is non-empty it suffices to find one τ - e.r. $\vdash'$ such that $\vdash \subseteq \vdash'$, $\sigma \vdash' \emptyset$ and $\vdash'$ is non-trivial. Observe that if $\vdash^*$ is defined by $\Gamma \vdash^*$ iff $\Gamma \vdash \Sigma, \sigma$, then $\vdash'$ is a non-trivial a.e.r. and $\sigma \vdash^* \emptyset$. If τ is such that the a.e.r. $\vdash'$ can be closed under the τ-rules without becoming trivial, then the τ-closure of $\vdash^*$ is the required τ - e.r.

Łukasiewicz' Entailment Relations

An e.r. $\vdash$ which satisfies the following additional rules will be known as Łukasiewicz entailment relation (Ł.e.r.) - the two way rules are written with double lines and $\alpha \to \Gamma = \{\alpha \to \gamma \mid \gamma \in \Gamma\}$, correspondingly for $\Gamma \to \alpha$.

(COM) $\alpha \to (\beta \to \gamma) \vdash \beta \to (\alpha \to \gamma)$

$$(\to_1) \quad \frac{\alpha, \Gamma \vdash \Sigma}{\alpha \to \Gamma \vdash \alpha \to \Sigma} \quad (\Sigma \neq \emptyset)$$

$$(\to_2) \quad \frac{\Sigma \vdash \alpha, \Gamma}{\Gamma \to \alpha \vdash \Sigma \to \alpha} \quad (\Sigma \neq \emptyset)$$

$$(\wedge) \quad \frac{\Gamma, \alpha, \beta \vdash \Sigma}{\Gamma, \alpha \wedge \beta \vdash \Sigma}$$

$$(\vee) \quad \frac{\Gamma \vdash \Sigma\ \alpha, \beta}{\Gamma \vdash \Sigma\ \alpha \vee \beta}$$

$$(\Lambda_1) \quad \frac{\Gamma \vdash \Sigma, \alpha_{ab\ldots}}{\Gamma \vdash \Sigma, \Lambda x \alpha^{x}_{ab\ldots}}$$ the parameter a not occurring in the conclusion.

$$(\Lambda_2)\ \frac{\Gamma, \alpha^{t}_{ab\ldots} \vdash \Sigma}{\Gamma, \Lambda x \alpha^{x}_{ab\ldots} \vdash \Sigma}$$

$$(V_1)\ \frac{\Gamma, \alpha_{ab\ldots} \vdash \Sigma}{\Gamma, Vx\alpha^{x}_{ab\ldots} \vdash \Sigma},$$ the parameter a not occurring in the conclusion.

$$(V_2)\ \frac{\Gamma \vdash \Sigma, \alpha^{t}_{ab\ldots}}{\Gamma \vdash \Sigma, Vx\alpha^{x}_{ab\ldots}}.$$

REMARKS. The propositional part determining Łukasiewicz' many-valued propositional logic is shown in Scott 1974. The rules that we have chosen for the quantifiers are the usual classical ones. An immediate consequence of them is that

$$\Lambda x(\beta \vee \alpha_x) \vdash \beta \vee \Lambda x \alpha_x$$

and thus any Kripke style semantics for Ł.e.r's will probably involve non-empty constant domains.

We state, in the form of a theorem, some of the schemas (and derived rules) that hold in all Ł.e.r's.

THEOREM. If $\vdash$ is an Ł.e.r., then

(1) $\vdash (\alpha \to \alpha)$ (2) $\vdash \alpha \to (\beta \to \alpha)$ (3) $\dfrac{\alpha \vdash \beta}{\vdash (\alpha \to \beta)}$

(4) $\dfrac{\alpha \vdash (\beta \to \gamma)}{\beta \vdash (\alpha \to \gamma)}$ (5) $\vdash (\alpha \to \beta), (\beta \to \alpha)$.

TWO VALUED SEMANTICS FOR Ł.E.R.

Valuations. A function V mapping Snt into $\{t,f\}$ (where $t = 0$, and $f = 1$) is an L-valuation iff V is invariant under alphabetic equivalence, i.e. if $\sigma \approx \sigma'$, then $V(\sigma) = V(\sigma')$.

Models and Semantics. Given a language $L(\mathbb{C})$ then a $\mathbb{C}$-model is a non-empty set of $L(\mathbb{C})$-valuations. A non-empty set $\{A \mid \exists \mathbb{C}'_{\mathbb{C}' \supseteq \mathbb{C}}$ (A is a $\mathbb{C}'$ model)$\}$ will be called an <u>$L(\mathbb{C})$-semantics</u>.

A model $\mathcal{A} = \{\mathcal{V}_i \mid i \in I\}$ is a model of the sentence α, in symbols: $\mathcal{A}(\alpha) = t$, iff $\forall i_{i \in I}(\mathcal{V}_i(\alpha) = t)$. Correspondingly for sequents consisting of sentences.

A semantic $\underset{\sim}{S}$ for L is adequate for the e.r. $\vdash$ iff for every sentence-sequent $\Gamma \Rightarrow \Sigma$ (that is a sequent consisting exclusively of sequents)

$$\underline{\text{if}}\ \Gamma \vdash \Sigma\ ,\ \underline{\text{then}}\quad \forall \mathcal{A}_{\mathcal{A} \in S}[\mathcal{A}(\Gamma \Rightarrow \Sigma) = t]\ .$$

An adequate semantics which also satisfies the converse will be called a complete semantics for $\vdash$.

Canonical Valuations. Suppose that $\mathcal{T} = \langle A,B \rangle$ is a total $\vdash$-theory, where $\vdash$ is a non-trivial e.r. Let $\mathcal{V}_{\mathcal{T}}$ be the function such that

$$\mathcal{V}_{\mathcal{T}}(\alpha) = t \quad \text{for} \quad \alpha \in A$$

$$\mathcal{V}_{\mathcal{T}}(\beta) = f \quad \text{for} \quad \beta \in B\ .$$

It can easily be verified that $\mathcal{V}_{\mathcal{T}}$ is a valuation. $\mathcal{V}_{\mathcal{T}}$ is called the canonical valuation associated with $\mathcal{T}$. We let $\mathcal{M}(\vdash) = \{\mathcal{V} \mid$ is a total $\vdash$-theory$\}$. $\mathcal{M}(\vdash)$ is the canonical model of $\vdash$.

A model $\mathcal{A}$ such that for all sentence-sequents S of the language of an e.r. $\vdash$

$$\vdash S \ \text{ iff } \ \mathcal{A}(S) = t$$

is called a universal model for $\vdash$.

THEOREM. If $\vdash$ is a non-trivial τ - e.r. then $\mathcal{M}(\vdash)$ is a universal model for $\vdash$.

PROOF. Assume that $\vdash$ is a non-trivial τ - e.r.

Part 1. Suppose that $\Gamma \Rightarrow \Sigma$ is an s-sequent (i.e. a sentence-sequent), $\Gamma \vdash \Sigma$ and that $\mathcal{T} = \langle A,B \rangle$ is a total $\vdash$-theory such that $\Gamma \subseteq A$. Then since $\mathcal{T}$ is consistent, $\Sigma \not\subseteq B$; and hence since $\mathcal{T}$ is total there must be a sentence $\sigma \in \Sigma \cap A$. But then $\mathcal{V}(\sigma) = t$ and thus $\mathcal{V}(\Gamma \Rightarrow \Sigma) = t$.

Part 2. Let $\Gamma \Rightarrow \Sigma$ be an s-sequent such that $\Gamma \nvdash \Sigma$. Let $\mathcal{Z} = \{\langle A,B \rangle \mid \langle A,B \rangle$ is a consistent $\vdash$-theory, $\Gamma \subseteq A$ and $\Sigma \subseteq B\}$. Then $\mathcal{Z} \neq \emptyset$ and Zorn's lemma may be used to obtain a maximal element $\mathcal{T} \in \mathcal{Z}$. It can then be verified that $\mathcal{T}$ is total and $\mathcal{T}(\Gamma \Rightarrow \Sigma) = f$.

Canonical Semantics. Given that $\vdash$ is a non-trivial τ - e.r. on the language $L(\mathbb{C})$. Then $\vdash$ is complete iff the addition of any s-sequent from the language $L(\mathbb{C})$ renders it trivial.

DEFINITION. If $\vdash$ is a non-trivial τ - e.r. on a countable language and given that $\mathbb{C}^*$ is an infinite set of individual constants such that $\mathbb{C} \cap \mathbb{C}^* = \emptyset$. Then the C*-canonical semantics for $\vdash$, $\underset{\sim}{S}_{\vdash}(\mathbb{C}^*)$ [or simply: $\underset{\sim}{S}_{\vdash}$ if there is no risk of confusion] consists of all canonical models of the form $M(\vdash^*)$ where $\vdash \subseteq \vdash^*$ and $\vdash^*$ is a complete τ - e.r. with respect to a language $L(\mathbb{C} \cup \mathbb{C}')$, where $\mathbb{C}' \subseteq \mathbb{C}^*$.

Corresponding to our last theorem we have the following

THEOREM. If $\vdash$ is a non-trivial τ - e.r. then $\underset{\sim}{S}_{\vdash}$ is a complete semantics for $\vdash$.

PROOF. First of all observe that if $\vdash^*$ is a complete extension of $\vdash$ in the language $L(C \cup C^*)$, then every valuation in $M(\vdash^*)$ is an extension of a valuation in $M(\vdash)$. Thus we conclude that $\underset{\sim}{S}$ is an adequate semantics for $\vdash$. Furthermore since $M(\vdash)$ is a universal model for $\vdash$ we obtain that $\underset{\sim}{S}_{\vdash}$ is also complete for $\vdash$.

MODELS AND SEMANTICS FOR Ł.E.R. We shall now restrict ourselves to Łukasiewicz entailment relations. Of course all that has been said above about e.r. applies to Ł.e.r. However the total theories in (complete extensions of) an Ł.e.r. have some further properties. In fact by considering such total theories one is able to deduce how the connectives and quantifiers affect the valuations in a given canonical model. From such considerations one obtains a plausible definition for a quantificational model structure (q.m.s.) for Ł.e.r. This was done for propositional part of the Ł.e.r in D'Ottaviano and López-Escobar 1982. The conclusion obtained there (already suggested in Scott 1974) was that the model-structures for propositional Ł.e.r's should be of the form $A = \{V_i \mid i \in I\}$ where I is a subset of an interval [0,m] in an ordered abelian group such that

(1) V_i are monotone in the index, i.e. if $i \leq j$ and $V_j(\sigma) = t$ then $V_j(\sigma) = t$,

(2) For each σ, the set $\{i \in I \mid V_i(\sigma) = t\}$ if not empty then it has a smallest element,

(3) $V_i(T) = t$, $V_i(\bot) = f$ for all $i \in I$,

(4) $V_i(\alpha \wedge \beta) = t$ iff $V_i(\alpha) = t$ and $V_i(\beta) = t$,

(5) $V_i(\alpha \vee \beta) = t$ iff $V_i(\alpha) = t$ or $V_i(\beta) = t$,

(6) $V_i(\alpha \to \beta) = t$ iff $\forall_{j \in I}$ (If $V_j(\alpha) = t$ then $V_{i+j}(\beta) = t$).

For the propositional calculus it sufficed to consider maximal extensions of the Ł.e.r. In the case of the predicate calculus we need some means to handle the quantifiers. Hence we propose the following definition:

DEFINITION. An Ł.e.r. $\vdash$ on a language $L(\mathbb{C})$ is ε-like iff the following two conditions are satisfied:

(1) To every sentence $\bigvee x\alpha_x$ there corresponds an individual constant $c \in \mathbb{C}$ such that $\bigvee x\alpha_x \vdash \alpha_c$.

(2) To every sentence $\bigwedge x\alpha_x$ there corresponds an individual constant $d \in \mathbb{C}$ such that $\alpha_d \vdash \bigwedge x\alpha_x$.

The following is more or less part of the folklore:

THEOREM. If $\vdash$ is a non-trivial Ł.e.r. on a denumerable language $L(\mathbb{C})$, then there is a denumerable set $\mathbb{C}^*$ and a non-trivial extension $\vdash^*$ of $\vdash$ on the language $L(\mathbb{C} \cup \mathbb{C}^*)$ which is ε-like.

Applying Zorn's lemma one can show that

THEOREM. Given a non-trivial Ł.e.r. $\vdash$ on a language $L(C)$, then there is a denumerable infinite set $\mathbb{C}^*$ of new individual constants such that $\vdash$ is the intersection of all the ε-like complete extensions of $\vdash$ on the language $L(\mathbb{C} \cup \mathbb{C}^*)$.

The appropriate semantics for a Ł.e.r. $\vdash$ is then $S_{\vdash}^{\mathbb{C}^*} = \{M(\vdash^*) \mid \exists \mathbb{C}' \subseteq \mathbb{C}^* [\vdash \subseteq \vdash^*$ and $\vdash^*$ is a non-trivial, ε-like complete extension of $\vdash$ on the language $L(\mathbb{C} \cup \mathbb{C}')]\}$. And just as before it can be shown that if $\vdash$ is a non-trivial Ł.e.r. then $S_{\vdash}^{\mathbb{C}^*}$ is a complete semantics for $\vdash$.

By analysing the total theories in ε-like complete extensions of a non-trivial Ł.e.r. we arrive at the following definition of a quantificational model structure (q.m.s.) for an Ł.e.r on a language $L(\mathbb{C})$.

DEFINITION. A quantificational model structure (q.m.s.) for an Ł.e.r. on the language $L(\mathbb{C})$ consists of a non-empty set of relational structures $K = \{A_i \mid i \in I\}$ where A_i is a subset of the interval [0,1] of the rational numbers subject to the following conditions:

((1)) $0 \in I$,

((2)) There is a set U such that for $i \in I$, A_i is of the form $\langle U, P_i, c_i \rangle_{c \in \mathbb{C}}$ where $P_i \subseteq U \times U$ and $c_i \in U$ (in other words, the A_i are relational structures with constant universe U).

((3)) There is a concept of satisfaction $\models$ between the relational structures A_i and sentences of the language $L(U)$ such that

(i) $A_i \models Puv$ iff $\langle u,v \rangle \in P_i$,

(ii) $A_i \models (\alpha \wedge \beta)$ iff $A_i \models \alpha$ and $A_i \models \beta$,

(iii) $A_i \models (\alpha \vee \beta)$ iff $A_i \models \alpha$ or $A_i \models \beta$,

(iv) $A_i \models (\alpha \to \beta)$ iff $\forall_{j \in I}$ (if $A_j \models \alpha$ then $A_{i+j} \models \beta$),

(v) $A_i \models \top$,

(vi) $A_i \not\models \bot$,

(vii) $A_i \models \bigvee x \alpha_x$ iff $\exists u \in U(A_i \models \alpha^u_x)$,

(viii) $A_i \models \bigwedge x \alpha_x$ iff $\forall u \in U(A_i \models \alpha^u_x)$.

((4)) If $i \leq j$, then $P_i \subseteq P_j$ and $\forall c \in \mathbb{C}\ (c_i = c_j)$.

((5)) For each sentence σ, the set $\{i \in I \mid A_i \models \sigma\}$ is either empty or has a first element.

It can be shown, by induction on the complexity of the sentence σ, that:

if $i \leq j$ and $A_i \models \sigma$ then $A_j \models \sigma$.

We say that the q.m.s. $K = \{A_i \mid i \in I \subseteq [0,1]\}$ is a model of the sentence σ, in symbols: $K \models \sigma$, iff $A_o \models \sigma$ (which in view of the above is equivalent to $(\forall i \in I[A_i \models \sigma])$.

THE COMPLETENESS OF THE Q.M.S. FOR Ł.E.R.

The soundness of the q.m.s. for Ł.e.r. is similar to the soundness for the propositional Ł.e.r. and thus we shall not dwell any further on it. The completeness is shown, loosely speaking by transforming the models in the canonical semantics into q.m.s. Actually the models in the canonical semantics have too many valuations in them, because they are indexed by total theories and in general we have uncountably many total theories. We encountered similar problems in the propositional part, and solved it there by considering total theories which were determined by single sentences. In that way we ended with models having only denumerable many valuations in them.

Thus suppose we are given a non-trivial Ł.e.r. $\vdash$ on a denumerable language $L(\mathbb{C})$ and suppose $\vdash^*$ is an ε-like complete extension of $\vdash$ w.r.t. to the language $\mathbb{L} = L(\mathbb{C} \cup \mathbb{C}^*)$, where $\mathbb{C}^*$ is a denumerable set of new individual constants. Define, for sentences $\alpha,\beta \in Snt_{\mathbb{L}}$.

$$\alpha \leq \beta \quad \text{iff} \quad \beta \vdash^* \alpha$$

$$\alpha < \beta \quad \text{iff} \quad \alpha \leq \beta \quad \text{and not } (\beta \leq \alpha)$$

$$\alpha \equiv \beta \quad \text{iff} \quad \alpha \leq \beta \quad \text{and} \quad \beta \leq \alpha \ .$$

Since $\vdash^*(\alpha \to \beta), (\alpha \to \beta)$ and $\vdash^*$ is complete we conclude that for sentences α,β either $\alpha \leq \beta$ or $\beta \leq \alpha$. Furthermore since $\vdash^*$ is ε-like we obtain that for each sentence $\bigvee x\alpha_x$ there corresponds an individual constant d such that

(I) $\bigvee x\alpha_x \equiv \alpha_d$,

and similarly to each sentence $\bigwedge x\alpha_x$ there corresponds an individual constant c such that

(II) $\bigwedge x\alpha_x \equiv \alpha_c$.

Next, to each sentence $\alpha \in \mathbb{L}$, let $T_\alpha = \langle A_\alpha, B_\alpha \rangle$ where $A_\alpha = \{\beta | \beta \in Snt_{\mathbb{L}}$ and $\beta \leq \alpha\}$, $B_\alpha = Snt_{\mathbb{L}} - A_\alpha$. It can be shown that T_α is a total $\vdash^*$-theory. Furthermore if $\alpha \equiv \beta$ then $T_\alpha = T_\beta$. Let $I = Snt_{\mathbb{L}}/\equiv$. Then I is a linearly ordered denumerable set with a first element. <u>The reduced canonical model</u> is $N(\vdash^*) = \{V_{T_\alpha} \mid \alpha/\equiv\ \in I\}$ and the <u>reduced canonical semantics</u> is the collection of reduced canonical models. Almost immediate is that the reduced canonical semantics is complete for $\vdash$.

A reduced canonical model $N(\vdash^*)$ can be made into a collection $K' = \{A_\alpha | \alpha/\equiv\ \in I\}$ of relational structures by letting

$$A_\alpha = \langle \mathbb{C} \cup \mathbb{C}^*, \{\langle c', c''\rangle | V_{T_\alpha}(Pc'c'') = t\}, c\rangle_{c \in \mathbb{C}} \ .$$

Thus, in order to make K into a q.m.s. we have to show how to interpret I as a subset of the unit interval of rationals.

Using Scott 1974 and D'Ottaviano 1982 one can show that there is a function F from $Sent_{\mathbb{L}}$ into the interval [0,1] of an ultraproduct $\mathbb{Q}^*$ of the rationals such that

(1) If $\alpha \equiv \beta$ then $F(\alpha) = F(\beta)$,

(2) $F(\bot) = 1$, $F(T) = 0$,

(3) $F(\alpha \wedge \beta) = \max(F(\alpha), F(\beta))$,

(4) $F(\alpha \vee \beta) = \min(F(\alpha), F(\beta))$,

(5) $F(\alpha \rightarrow \beta) = F(\beta) - F(\alpha)$,

But then using (I) and (II) we obtain that

(6) $F(\bigvee x\alpha_x) = \mathrm{Min}\{F(\alpha_x^{c'}) \mid c' \in \mathbb{C} \cup \mathbb{C}^*\}$,

(7) $F(\bigwedge x\alpha_x) = \mathrm{Max}\ F(\alpha_x^{c'}) \mid c' \in \mathbb{C} \cup \mathbb{C}^*\}$.

Now using the fact that $\mathbb{Q}^*$ is a non-standard model of the rationals we may define $G(\alpha)$ = the rational closest to $F(\alpha)$. The conditions (1) - (7) apply equally well to G instead of F.

Let $J = G[Snt_{\mathbb{L}}] \subseteq [0,1] \subseteq \mathbb{Q}$, and for each $j \in J$ define $A_j = A_{\alpha}$ where $G(\alpha) = j$. Then using (1)-(7) one can verify that $K = \{A_j \mid j \in J \subseteq [0,1]\} = K'$ is a q.m.s.

Thus the reduced canonical models can be considered as q.m.s. Using the completeness for the reduced canonical semantics we obtain the desired completeness for q.m.s.

RELATION BETWEEN Ł.E.R'S AND PARACONSISTENT LOGICS

Poly-Valent Logics and Ł.E.R'.s

It was the investigation of modal notions of possibility and necessity that led Łukasiewicz to introduce the poly-valent propositional logics L_n , $n = 2,3,\ldots\aleph_0$ (see Łukasiewicz 1930). It may be worthwhile to observe that at first Łukasiewicz considered the bivalent, tri-valent and $\aleph_0$-valent logics to be the philosophically important ones (later on he also included $n = 4$ amongst his favorite cases).

We can recover the idea of n truth values from the more abstract

Ł.e.r's in the following way:

DEFINITION. *Let* $\aleph_o > n \geq 2$ *and* $\vdash$ *be a non-trivial Ł.e.r. Then* $\vdash$ *is an n-valued Ł.e.r. (or simply an n-Ł.e.r.) iff the following conditions are satisfied in the reduced canonical semantics* $S_{\sim\vdash}$ *for* $\vdash$:

(1) *Every model* $N(\vdash^*)$ *in* $S_{\sim\vdash}$ *contains at most (n-1)-valuations.*

(2) *There is at least one model in* $S_{\sim\vdash}$ *with (n-1)-valuations.*

Since every $N(\vdash^*)$ is countable we shall say that $\vdash$ (on a countable language) is $\aleph_o$-valued iff there is at least one model in the reduced canonical semantics which contains infinitely many valuations.

The following theorem can be obtained from D'Ottaviano 1982b.

THEOREM. *Let* $\aleph_o > n \geq 3$ *and* $\vdash$ *be a non-trivial Ł.e.r. Then* $\vdash$ *is n-valued iff it contains the following schema:*

$$\vdash \alpha_o, (\alpha_o \rightarrow \alpha_1), \ldots, (\alpha_{n-2} \rightarrow \alpha_{n-1})$$

where $\alpha_o, {}_1, \ldots, \alpha_{n-1}$ *are arbitrary formulas.*

For the case $n = 3$ we may use the more familiar scheme: $\vdash \alpha, (\alpha \rightarrow \neg\alpha)$, where $\neg\alpha$ is an abbreviation for $(\alpha \rightarrow \bot)$; the form given by Wajsberg 1931 was $\vdash((\alpha \rightarrow \neg\alpha) \rightarrow \alpha) \rightarrow \alpha)$ which is equivalent $\vdash \alpha, (\alpha \rightarrow \neg\alpha)$.

The Modal Operator ∇ For Possibility in 3-Ł.E.R.

Łukasiewicz introduced the 3-valued propositional logic L_3 because we wanted to be able to define the modal operator of possibility ∇ through the use of truth tables. It was later shown by A. Tarski that in L_3, ∇ could be defined by:

$$\nabla\alpha = (\neg\alpha \rightarrow \alpha).$$

Now a q.m.s. for a 3-Ł.e.r. $\vdash$ may be assumed to be of the form $K = \{A_o, A_{1/2}\}$, or pictorially:

$A_{1/2}$
|
A_o

where $A_o \subseteq A_{1/2}$. A straightforward computation shows that for $K = \{A_o, A_{1/2}\}$,

$A_i \models \nabla\sigma$ iff $A_{1/2} \models \sigma$, for $i = 0,1/2$.

Then using the completeness theorem one can verify that the following are derived rules of inference for 3-Ł.e.r's:

(1) $$\frac{\Gamma \vdash \Sigma,\ \vee x \nabla \alpha_x}{\Gamma \vdash \Sigma, \nabla \vee x \alpha_x}$$ (2) $$\frac{\Gamma \vdash \Sigma,\ \wedge x \nabla \alpha_x}{\Gamma \vdash \Sigma,\ \nabla \wedge x \alpha_x}$$

(3) $$\frac{\Gamma \vdash \Sigma,\ \wedge x \alpha_x}{\Gamma \vdash \Sigma,\ \neg \vee x \geq \alpha_x}$$ (3) $$\frac{\Gamma \vdash \Sigma,\ \neg \vee x \alpha_x}{\Gamma \vdash \Sigma,\ \wedge x \geq \alpha_x}$$

(5) $$\frac{\Gamma \vdash \Sigma,\ \vee x \alpha_x}{\Gamma \vdash \Sigma,\ \neg \wedge x \neg \alpha_x}$$ (6) $$\frac{\Gamma \vdash \Sigma,\ \neg \wedge x \alpha_x}{\Gamma \vdash \Sigma\ ,\ \vee x > \alpha_x}$$

The Paraconsistent Logic $\mathbb{J}_3$

In D'Ottaviano 1982 a paraconsistent logic $\mathbb{J}_3$ is introduced whose basic symbols are $\vee$, $\neg$, ∇ and $\vee$ (see also D'Ottaviano and da Costa 1970). Other connectives and quantifiers are introduced by definition and it is shown that $\mathbb{J}_3$ and L_3 are related. One important difference between $\mathbb{J}_3$ and L_3 is that in $\mathbb{J}_3$ there are two distinguished truth values, while in L_3 there is only one.

In D'Ottaviano 1982 a model theory is developed for $\mathbb{J}_3$. The $\mathbb{J}_3$-structures (in the similarity type we are using) are systems of the form $\mathcal{B} = \langle U,p,c\rangle_{c\in\mathbb{C}}$ where p is a function from $U \times U$ into $\{0,1/2,1\}$, where 0, 1/2 are the distinguished truth values. Clearly there is a natural way of transforming a $\mathbb{J}_3$-structure $\mathcal{B} = \langle U,p,c\rangle_{c\in\mathbb{C}}$ into a q.m.s. $K = \{A_o,A_{1/2}\}$ and vice-versa, $[A_i = \langle U,\{\langle u,u'\rangle \mid p(u,u') = i\}, c\rangle_{c\in\mathbb{C}}]$.

Let $\vdash_{\mathbb{J}_3}$ be the provability predicate in J_3 (i.e. the Hilbert type axiomatization given in D'Ottaviano 1982 and let $\Vdash^{(3)}$ be the 3-Ł.e.r. on the corresponding language. Then using the completeness of $\vdash_{\mathbb{J}_3}$ and $\Vdash^{(3)}$ one obtains the following

THEOREM. <u>Let σ be any sentence of the language $L(\mathbb{C})$. Then</u>

$$\vdash_{\mathbb{J}_3} \sigma \quad \text{iff} \quad \Vdash^{(3)} \nabla\sigma .$$

PROOF. Part 1. Suppose that σ is provable in $\mathbb{J}_3$ and that, for reductio ad absurdum, suppose that $\nabla\sigma$ (actually its translation) were not provable in the 3-Ł.e.r. $\Vdash^{(3)}$. Then using the completeness of $\Vdash^{(3)}$ we obtain a $K = \{A_o, A_{1/2}\}$ such that $A_o \nVdash \nabla\sigma$. But then $A_{1/2} \nVdash \sigma$. Let B be the associated $\mathbb{J}_3$-structure. Then $B(\sigma) = 1$ and thus using the soundness of $\vdash_{\mathbb{J}_3}$ we would obtain that $\nvdash_{\mathbb{J}_3} \sigma$, contrary to our assumptions.

Part 2. Analogous, this time using the completeness of $\vdash_{\mathbb{J}_3}$ and the soundness of $\Vdash^{(3)}$.

The shortcoming of our last theorem is that it deals only in the theorems and in general the notion of deducibility is much richer (and thus more interesting). Hence the following:

DEFINITION. Suppose $\alpha_o,\ldots,\alpha_k$ are sentences of $\mathbb{J}_3$. Then we set $\alpha_o,\ldots,\alpha_{k-1} \models_{\mathbb{J}_3} \alpha_k$ iff for every $\mathbb{J}_3$-structure B, if B assigns distinguished values to $\alpha_o,\ldots,\alpha_{k-1}$ then it also assigns a distinguished value to α_k.

Then the method of proof used above can be used to give the following theorem.

THEOREM. If $\alpha_o,\ldots,\alpha_k$ are sentences of the language $L(\mathbb{C})$, then the following conditions are equivalent:

(i) $\alpha_o,\ldots,\alpha_{k-1} \models_{\mathbb{J}_3} \alpha_k$.

(ii) $\nabla\alpha_o,\ldots,\nabla\alpha_{k-1} \Vdash^{(3)} \nabla\alpha_k$.

REMARKS. In a certain sense both theorems above characterize the paraconsistent logic $\mathbb{J}_3$ (which was introduced as a solution to Jaskowski's problem - see D'Ottaviano 1982 -). Or more specifically, ∇ acts as a faithful mapping of $\mathbb{J}_3$ into the more familiar tri-valent Łukasiewicz logic L_3. Furthermore, there are Kripke style models for L_3 and consequently the mapping can be used to define Kripke style models for $\mathbb{J}_3$.

Another observation that can be extracted from the above theorem: it is possible to develop paraconsistent logics with the conditional rather than negation. For example, one could introduce a paraconsistent logic $\mathbb{J}_n$, $n > 3$, by letting the theorems of $\mathbb{J}_n = \{\sigma \mid \vdash^{(n)} \nabla\sigma\}$, where $\Vdash^{(n)}$ is the n-Ł.e.r. on the corresponding language.

Furthermore it may be possible to characterize other paraconsistent logics using other kinds of conditionals. We propose to do that in forthcoming papers.

$\mathbb{J}_3$-THEORIES OF EQUALITY

In view of the last theorems a way to study $\mathbb{J}$-theories of equality is to consider the 3-Ł.e.r. theories of equality. And the simplest way to do the latter is to first consider equality in arbitrary Ł.e.r.'s. In this paper we shall only consider the pure theories of equality and to more or less conform with tradition we shall write "$a \doteq b$" instead of "Pab".

Definitions

(1) Eq <u>is the sentence:</u> $\Lambda x(x = x) \wedge \Lambda x \Lambda y(x \doteq y \rightarrow y \doteq x) \wedge \Lambda x \Lambda y \Lambda z\ (x \doteq y \wedge y \doteq z \rightarrow x \doteq z)$.

(2) Edt <u>is the sentence:</u> $\Lambda x(x \doteq x) \wedge \Lambda x \Lambda y(x \doteq y \rightarrow y \doteq x) \wedge \Lambda x \Lambda y \Lambda z\ (x \doteq y \rightarrow (y \doteq z \rightarrow x \doteq z)$

The following comes as no surprise:

THEOREM. <u>Suppose</u> $K = \{A_i \mid i \in I \subseteq [0,1]\}$ <u>is a q.m.s. which is a model of Eq and where for each</u> i, $A_i = \langle U, R_i \rangle$. <u>Then each</u> R_i <u>is an equivalence relation on</u> U <u>and if</u> $i \leq j \in I$ <u>then</u> $R_i \subseteq R_j$.

On the other hand with Edt the situation is quite different.

THEOREM. <u>Given that</u> $K = \{A_i \mid i \in I \subseteq [0,1]\}$ <u>is q.m.s. which is a model of Edt and where for each</u> i, $A_i = \langle U, R_i \rangle$. <u>Then there is an equivalence relation</u> $\equiv$ <u>on</u> U <u>and a metric on</u> $U/\equiv$ <u>such that for each</u> i, $R_i = \{\langle u', u'' \rangle \mid \rho(u'/\equiv, u''/\equiv) \leq i\}$.

PROOF. First of all observe that R_o is indeed an equivalence relation on U. Thus for $\equiv$ we way take R_o. Next let d be the function such that for $u', u'' \in U$

$$d(u', u'') = \min(\{i \mid \langle u', u'' \rangle \in R_i\} \cup \{1\}).$$

We complete the proof in 6 steps.

<u>Step 1</u>. If $v' \equiv u'$ and $v'' \equiv u''$ then $d(v', v'') = d(u', u'')$.

In order to prove step 1 it suffices to verify that if $\langle u',u''\rangle \in R_i$ and $\langle u',v'\rangle \in R_o$ then $\langle v',u''\rangle \in R_i$. And the latter follows from the interpretation of the conditional.

Step 2. $d(u',u'') = d(u'',u')$. This follows from $A_o \models \Lambda x \Lambda y(x \doteq y \rightarrow y \doteq x)$.

Step 3. $d(u',u') = 0$.

Step 4. If $u' \equiv u''$ then $d(u',u'') = 0$.

Step 5. $d(u',u'') \leq d(u',v) + d(v,u'')$. This follows from the additive character of the conditional and the fact that $A_o \models (u' \doteq v \rightarrow (v \doteq u'' \rightarrow u' \doteq u''))$

Step 6. Define ρ by $\rho(u/\equiv, u'/\equiv) = d(u,u')$.

The following is even simpler:

THEOREM. If $\langle U,\rho\rangle$ is a metric space where the range of $\rho \subseteq [0,1]$, $R_i = \{\langle u,u'\rangle \mid \rho(u,u') \leq i\}$ and $K = \{\langle U,R_i\rangle \mid i \in$ range $(\rho) \cap Q\}$, then K is a model of Edt.

The above theorems, although rather obvious, nevertheless bring out the fact that if interesting mathematics is going to be done in paraconsistent logics then it will not suffice to just change the logics. Commitments will have to be made as to basic properties (formulated in the language of the paraconsistent logic) that the mathematical objects are to satisfy.

REFERENCES

1. Arruda, A.I., A survey of paraconsistent logic, Mathematical Logic in Latin America, North-Holland Publ. Co., Amsterdam 1980.
2. Da Costa, N. and D'Ottaviano, I.M.L., Sur un problème de Jaśkowski, C.R. Acad. Sc. Paris 270A, 1970, pp. 1349-1353.
3. D'Ottaviano, I.M.L., Sobre uma Teoria de Modelos Trivalente, Ph.D. Thesis, Universidade Estadual de Campinas, 1982.
4. D'Ottaviano, I.M.L. and López-Escobar, E.G.K., The conditional and transmission of errors, VI Encontro Brasileiro de Lógica, Fortaleza, 1982.
5. Łukasiewicz, J., Untersuchungen über den Aussangenkalkül, C.R. Soc. Sci. Lett. Varsovie 23, 1930, pp. 39-50.
6. Mostowski, A., On rules of proof in the Pure Functional Calculus of first order. The Journal of Symbolic Logic, vol. 16, 1951, pp. 107-111.

7. Scott, D.S., Completeness and axiomatizability, Tarski Symposium, Berkeley, California, 1974.

8. Wajsberg, M., Aksjomatyzacja trōjwartosciowego rachunku zdān. C.R. Soc. Sci. Lett. Varsovie, vol. 24, 1931, pp. 126.

DEUX RELATIONS DÉNOMBRABLES, LOGIQUEMENT ÉQUIVALENTES POUR LE SECOND ORDRE, SONT ISOMORPHES (MODULO UN AXIOME DE CONSTRUCTIBILITÉ)

Roland Fraïssé/ Departément de Mathématique, Université de Provence à Marseille, Marseille, France

Rappelons que deux relations dénombrables, i.e. dont la base est un ensemble dénombrable, et plus particulièrement deux ordinaux dénombrables, peuvent être logiquement équivalents, en calcul logique du premier ordre avec identité, sans être isomorphes. Par exemple, les ordinaux $a.\omega^{\omega} + r$ où r est un ordinal $< \omega^{\omega}$ et a un ordinal non nul, sont logiquement équivalents pour toutes valeurs non nulles de a; voir par exemple [F 1974] p. 26.

En logique du second ordre, avec des formules de longueur finie, mais dont les quantifieurs portent non seulement sur des individus, mais aussi sur des prédicats, l'isomorphie de deux relations dénombrables logiquement équivalentes a été conjecturée par[F 1950]Il a été prouvé par [M 1973], avec l'aide de l'axiome de constructibilité, ou plus économiquement avec la seule constructibilité des ensembles hérédénombrables, que deux ordinaux dénombrables logiquement équivalents dans le second ordre, sont identiques; autrement dit deux bonordres équivalents sont isomorphes.

Le but du présent article est de généraliser ce résultat aux relations dénombrables quelconques.

A l'origine de la question, on savait, au moins expérimentalement, que l'isomorphie conjecturée ne pouvait résulter des seuls axiomes de Zermelo-Fraenkel. En effet considérons par exemple les relations binaires, et partons d'une numérotation de l'ensemble des formules du second ordre à un prédicat binaire libre: numérotation aisément définissable dans Z-F.

Alors pour chacune de ces relations binaires R, nous voyons que la théorie complète du second ordre de R est codée par un ensemble infini d'entiers, à savoir l'ensemble des numéros des formules qui vérifient R. Autrement dit, la théorie de R est codée par un nombre réel. Donc si, pour chaque ordinal dénombrable A, la théorie complète de A est distincte des théories complètes des autres ordinaux dénombrables, alors elle est codée par un réel qui correspond au seul A. Donc à la ω_1-suite des ordinaux dénombrables correspond la ω_1-suite des réels codant leurs théories du second ordre.

Or il est connu qu'avec la seule théorie ZF, on ne sait pas nommer une ω_1-suite de réels, c'est-à-dire définir une telle suite avec, en plus de l'existence, un énoncé d'unicité. Impossibilité prouvée par exemple par [L 1970] en supposant consistante ZF plus l'axiome du cardinal inaccessible. De plus G. Sacks (article non publié) obtient cette impossibilité en supposant seulement ZF consistante: il construit, par forcing, une ZF-appartenance qui ne comprend qu'une infinité dénombrable de bonordres dénombrables constructibles; or la possibilité de nommer, avec unicité, une ω_1-suite de réels donnerait, pour chaque intervalle initial de cette suite, un bonordre dénombrable et constructible; référence citée par [M 1973].

Le présent article reprend, avec addition de résultats annexes et refonte de la rédaction, le préprint [F 1977] qui exigeait une lecture préalable de [F 1975] (dont les définitions sont reprises ici). Le résultat principal a été retrouvé indépendamment et de manière différente par [A 1979].

RAPPELS SUR LE RANG FONDAMENTAL ET LE RANG DE CONSTRUCTION

Partons de l'axiomatique Z-F avec l'axiome de fondement: tout ensemble a admet un élément disjoint de a. Rappelons que cet axiome équivaut au schema de fondement suivant: si une condition C vérifie C(0) (où 0 désigne le vide), et si C(x) pour tout élément x de a entraine C(a), alors C est vraie pour tout ensemble. Par cet axiome, toute suite

x_i telle que $x_{i+1} \in x_i$, est finie.

Rappelons qu'un ordinal est un ensemble transitif et totalement ordonné pour l'appartenance. Alors à chaque ensemble a est associé par induction un ordinal appelé son rang fondamental: le rang du vide 0 est 0; pour a non vide, le rang de a est égal à Sup(rang x plus 1) où x parcourt a.

Rappelons que le rang fondamental d'un ordinal a est a lui-même.

Arrivons-en à une rapide définition de l'ensemble constructible et de son rang de construction, par l'intermédiaire des formules logiques ordinalées.

Considérons une formule logique du calcul du premier ordre avec identité; notre formule comprendra un unique prédicat binaire d'appartenance, noté ε. De plus à chaque quantifieur nous attribuons, en plus de son indice x, un ordinal α : au lieu de $\forall_x$, nous écrirons $\forall_x^\alpha$ et nous lirons: "quel que soit x de rang de construction $< \alpha$". De même pour les quantifieurs $\exists$. Une formule ainsi définie sera dite ordinalée.

Par définition inductive, attribuons le rang de construction 0 au seul ensemble 0.

Soit α un ordinal positif; supposons définis les ensembles ayant un rang de construction $< \alpha$; et supposons qu'ils forment un ensemble. Attribuons alors le rang de construction α à un ensemble a non encore affecté d'un rang, lorsque α est le plus petit ordinal vérifiant (1) et (2) ci-après:

(1) les éléments de a ont des rangs de construction $< \alpha$;

(2) il existe une formule P(x) de prédicat binaire ε avec un seul individu libre x, ordinalée avec des ordinaux au plus égaux à α; substituons alors ε par l'appartenance entre ensembles de rangs $<\alpha$; chaque quantifieur ordinalé par un $\beta \leq \alpha$, étant relativisé à l'ensemble des ensembles de rangs $< \beta$; alors la valeur de vérité de P(x) sera + lorsque x est substitué par un élément de a, et-lorsque x est substitué par un ensemble de rang $<\alpha$ qui n'appartient pas à a.

La formule ordinalée précédente sera dite une formule de construction de a. Enfin un ensemble a sera constructible lorsqu'il existe un ordinal α et une formule de construction définie comme précédemment.

Notons qu'il résulte de la définition, que les ensembles de rang α forment un ensemble, dont le cardinal est au plus celui des suites finies d'ordinaux $\leq \alpha$, multiplié par le dénombrable, puisque les formules logiques (sans ordinaux) sont elles-mêmes en infinité dénombrable. Donc pour α infini et modulo l'axiome du choix, les ensembles constructibles de rang $\leq \alpha$ sont en nombre égal au cardinal de α.

Notons encore, par induction, que le rang de construction d'un ensemble constructible est supérieur ou égal à son rang fondamental.

Un ensemble est dit héréfini (abréviation de "héreditairement fini") lorsqu'il est de rang fondamental fini. Equivalemment, lorsque la cloture transitive de cet ensemble est finie; la cloture de a est le plus petit surensemble transitif de a; ou encore l'ensemble des termes x_i des suites finies telles que $x_1 \in a$ et pour chaque entier i, $x_{i+1} \in x_i$.

Pour chaque entier p, les héréfinis de rang p sont en nombre fini; donc l'ensemble de tous les héréfinis est dénombrable.

Chaque héréfini est constructible, le rang de construction étant égal au rang fondamental; inversement tout constructible de rang fini est héréfini.

Rappelons que chaque ordinal α est constructible de rang égal à α.

L'ensemble des constructibles de rang $< \alpha$ est lui-même constructible de rang α.

L'ensemble des constructibles de rang α est lui-même constructible de rang $\alpha + 1$.

La réunion, l'intersection de deux constructibles de rang α, β est de rang au plus égal au maximum de α, β. Si a est de rang de construction α, son complément par rapport à l'ensemble de tous les constructibles de rangs $< \alpha$, est constructible de rang α.

Si a est constructible de rang α et si $\beta \leq \alpha$, alors l'ensemble des éléments de a qui ont un rang $< \beta$, est constructible et de rang $\leq \alpha$.

Tout ensemble fini de constructibles a_i est constructible, avec un rang égal au maximum des rangs des a_i, plus 1.

L'ensemble réunion d'un constructible de rang α, est constructible de rang $\leq \alpha$.

L'ensemble des parties constructibles d'un constructible de rang α, est constructible.

Mais ici, si nous voulons limiter efficacement le rang de cet ensemble, il faut faire appel à un énoncé beaucoup plus difficile que les trivialités précédentes.

<u>Théorème de la partie constructible</u> (voir par exemple [C 1966] p. 95). Soit a un ensemble constructible de rang α. Alors pour chaque partie constructible b de a, le rang de construction β de b verifie Card $\beta \leq$ Card α.

Corollaire. Soit b constructible, a_i les éléments de b, et α_i le rang de construction de a_i. Alors le rang β de b vérifie Card β = = Card Sup(α_i + 1).

L'axiome de constructibilité affirme que tout ensemble est constructible. Avec notre définition par formules ordinalées, il est évident que l'axiome de constructibilité entraîne l'axiome du choix. Il suffit de bien-ordonner les suites finies d'ordinaux et les formules (qui, sans leurs indices ordinaux, sont en infinité dénombrable).

Pour voir l'utilité du théorème de la partie constructible, notons qu'il en résulte immédiatement que l'axione de constructibilité entraîne l'axiome, ou hypothèse, du continu. En effet, l'ensemble N des entiers est identique à l'ordinal ω donc constructible de rang ω. Donc si chaque partie de N est constructible, son rang étant dénombrable par le théorème de la partie constructible, chaque partie de N est définie par une formule ordinalée dont les indices ordinaux sont finis ou dénombrables. Or, modulo l'axiome du choix, ces formules ordinalées sont en nombre égal à ω_1. Le lecteur étendra cette preuve au cas géneral, déduisant l'axiome, ou hypothèse du continu généralisée, de l'axiome de constructibilité.

ENSEMBLE HEREDENOMBRABLE; CODAGE D'UN HEREDENOMBRABLE PAR UN GRAPHE: THEORIE DE L'APPARTENANCE SUR LES HEREFINIS ET SUR LES HEREDENOMBRABLES

Définissons un ensemble hérédénombrable (abréviation de "héréditairement dénombrable") par les conditions suivantes:

(1) l'ensemble vide 0 est hérédénombrable;

(2) tout ensemble fini ou dénombrable d'éléments hérédénombrables, est hérédénombrable;

(3) tout hérédénombrable est obtenu à partir du vide par le (2) précédent; plus rigoureusement, l'ensemble des hérédénombrables est l'intersection des ensembles U tels que 0 appartient à U et tels que tout ensemble fini ou dénombrable inclus dans U, appartient à U.

D' apres (3), si une propriété P est vraie pour 0, et si, quand P est vraie pour chaque élément d'une ensemble x fini ou dénombrable P est vraie pour x, alors P est vraie pour chaque ensemble hérédenombrable; il suffit de voir que l'ensemble U des hérédénombrables qui vérifient P, satisfait à la condition (3).

Comme exemples d'hérédénombrables, citons l'ensemble ω des entiers; plus généralement, citons chaque ordinal fini ou dénombrable (preuve par induction). Citons encore chaque ensemble héréfini, et l'ensemble de tous les héréfinis, puisqu'il est dénombrable.

Tout élément d'un hérédénombrable, ou toute partie d'un hérédénombrable, est hérédénombrable.

Tout ensemble hérédénombrable est fini ou dénombrable.

1.1. Un ensemble a est hérédénombrable si et seulement si la cloture transitive de a est finie ou dénombrable.

□ Supposons a hérédénombrable, et supposons que la cloture transitive de a soit de cardinal infini non dénombrable. Nous pouvons toujours supposer que, pour chaque élément x de a, la cloture de x est au plus dénombrable. Sinon nous aurions une suite infinie ascendante par appartenance: suite dont chaque terme comprend comme élément le terme suivant, contre l'axiome de fondement. La cloture transitive de a est la réunion de a, qui est au plus dénombrable, et des clotures transitives des éléments de a, qui sont chacune au plus dénombrable et sont au plus en infinité dénombrable.

Inversement, supposons que la cloture transitive de a soit au plus dénombrable; donc a est fini ou dénombrable. Supposons que a ne soit pas hérédédombrable. Nous pouvous toujours supposer que chaque élément de a est hérédémombrable: sinon nous aurions une suite infinie ascendante par appartenance. Donc a est hérédénombrable, en tant qu'ensemble au plus dénombrable formé d'éléments hérédénombrables. □

1.2. Appelons H_0 l'ensemble du seul ensemble vide 0. Soit i un ordinal non nul, fini ou dénombrable; supposons défini H_j pour chaque $j < i$, et appelons H_i l'ensemble des éléments et des parties finies ou dénombrables de la réunion des H_j $(j < i)$. Nous avons donc $H_i \subseteq H_{i'}$, lorsque $i \leq i'$.

Pour chaque ordinal fini ou dénombrable i, l'ensemble H_i est celui des ensembles hérédénombrables de rangs fondamentaux $< i$. La différence $H_i - (\cup H_j)(j < i)$ étant l'ensemble des hérédénombrables de rang fondamental i; c'est encore la différence $H_i - H_{i-1}$ lorsque i est un ordinal successeur.

Il en résulte que l'ensemble des héréfinis est la réunion des H_i pour les ordinaux finis i; l'ensemble des hérédénombrables est la réunion des H_i pour les ordinaux i finis ou dénombrables. Ce dernier point résulte du fait que tout ensemble dénombrable d'ordinaux dénombrables, est majoré par un ordinal dénombrable.

Il en résulte encore que tout hérédénombrable est de rang fondamental fini ou dénombrable. Mais, alors qu'un ensemble est héréfini si et seulement si son rang fondamental est fini, l'ensemble H_ω, formé de tous les ensembles d'héréfinis, est de rang fondamental $\omega + 1$, et il est continupotent, donc n'est pas hérédénombrable. Par exemple, il comprend comme élément chaque ensemble d'entiers.

L'ensemble des hérédénombrables est continupotent.

□ Nous voyons qu'il est au moins continupotent puisqu'il comprend comme élément chaque ensemble d'entiers. De plus il est la réunion des ensembles H_i précédemment définis. Or les ordinaux finis ou dénombrables i qui servent d'indices aux H_i sont en infinité aleph 1, donc inférieure ou égale au continupotent. D'autre part chaque H_i est fini pour les i finis, et continupotent à partir de $i = \omega$, puisque formé de toutes les parties finies ou dénombrables, plus les éléments, de la réunion des H_i $(j < i)$. □

<u>1.3.</u> Un ensemble constructible est hérédénombrable si et seulement si son rang de construction est fini ou dénombrable.

□ Soit a un ensemble constructible hérédénombrable. Raisonnons par l'absurde en supposant que son rang de construction est infini non dénombrable. Nous pouvons toujours supposer que, pour chaque élément de a, le rang de construction est au plus dénombrable. Sinon, en remontant d'un ensemble de rang infini non dénombrable à un élément de rang infini non dénombrable, nous aurions une suite infinie ascendante par appartenance, contre l'axiome de fondement. Mais puisque a est hérédénombrable, il est dénombrable, donc est dénombrable le plus petit ordinal > rangs de construction des éléments de a; d'après le corollaire du théorème de la partie constructible, le rang de construction de a est au plus dénombrable.

Inversement, soit a un ensemble constructible dont le rang de construction α est fini ou dénombrable. Raisonnons par l'absurde, en supposant a non hérédénombrable, et α minimum. Donc chaque élément de a, ayant un rang de construction $< \alpha$, est hérédénombrable.

Pour prouver que a est hérédénombrable, donc pour obtenir la contradiction, il suffit de prouver que a est fini ou dénombrable. Puisque chaque élément de a est de rang de construction $< \alpha$, chacun est défini par une formule ordinalée d'ordinal maximum $< \alpha$. Il suffit de remarquer que les suites finies d'ordinaux $< \alpha$, donc aussi les formules ordinalées correspondantes, sont au plus en infinité dénombrable. □

En conséquence, si nous admettons l'axiome de constructibilité, les hérédénombrables sont exactement les ensembles ayant un rang de construction fini ou dénombrable.

Si nous n'admettons pas cet axiome, notons que toute partie d'un ensemble de rang de construction dénombrable, est hérédénombrable. Par exemple tout ensemble d'entiers, qu'il soit ou non constructible , est hérédénombrable. Notons encore que tous les hérédénombrables ne sont pas obtenus comme des parties d'un constructible de rang dénombrable: par exemple, partons d'un ensemble a supposé hérédénombrable et non constructible; le singleton $\{a\}$ est encore hérédénombrable et non constructible, et ce singleton n'est inclus dans aucun ensemble constructible.

1.4. La relation binaire d'appartenance, basée sur l'ensemble des héréfinis, vérifie les axiomes de Zermelo Fraenkel, excepté l'axiome de l'infini; de plus elle vérifie l'axiome de choix et l'axiome de finitude, exprimant que tout ensemble est fini.

Elle vérifie même l'axiome de constructibilité, les seuls rangs de construction étant les entiers.

La relation binaire d'appartenance, basée sur l'ensemble des hérédénombrables, vérifie les axiomes de Zermelo-Fraenkel, excepté l'axiome de la puissance; de plus elle vérifie l'axiome exprimant que tout ensemble est fini ou dénombrable; l'axiome du choix est vérifié. En effet, étant donné un ensemble hérédénombrable a dont les éléments sont disjoints, la réunion de a est dénombrable, donc en bijection avec l'ensemble des entiers. Représentons par $\in$ cette appartenance.

Pour préciser les énoncés précédents, représentons l'axiome du choix par la formule suivante, ou $\mathcal{D}xy$ est l'abréviation de $\neg\exists_t(t \in x \wedge t \in y)$, qui prend la valeur + lorsque x et y sont disjoints; formule exprimant que, pour tout ensemble a dont les éléments sont non vides et deux à deux disjoints, il existe un ensemble de choix b comprenant un élément et un seul commun à b et à chaque élément de a:

$$(\forall_a (\forall_x \in a \Longrightarrow \exists_t \in x) \wedge (\forall_{x,y} (x \in a \wedge y \in a) \Longrightarrow \mathcal{D}xy))$$

$$\Longrightarrow \exists_b (\forall_x \; x \in a \Longrightarrow \exists_t ! \; t \in x \wedge t \in b);$$

où $\exists_t ! \; P(t)$ est l'abréviation de $\exists_t P(t) \wedge \forall_{u,v} (P(u) \wedge P(v) \Longrightarrow u = v)$.

Représentons comme suit l'axiome de finitude, que nous préciserons en disant que a est fini lorsque tout ensemble b de parties de a comprend un élément y minimal par inclusion, en ce sens qu'il n'existe aucun élément de b strictement inclus dans y (notons $x \subseteq a$ l'abréviation de $\forall_t \; t \in x \Longrightarrow t \in a$, et $x \subset y$ l'abréviation de $x \subseteq y \wedge x \neq y$):

$$\forall_{a,b} \; (\exists_x \; x \in b \wedge (\forall_x \; x \in b \Longrightarrow x \subseteq a))$$

$$\Longrightarrow (\exists_y \; y \in b \wedge \neg \exists_x \; x \in b \wedge x \subset y)).$$

Enfin, pour préciser une formule exprimant qu'un ensemble a est dénombrable, c'est-à-dire équipotent à l'ensemble des entiers, il suffit de rappeler que "a est entier" s'exprime par la conjonction de "a est fini" et "a est un ordinal (ensemble transitif et totalement ordonné par appartenance)"; enfin que "a et b sont équipotents" s'exprime par l'existence d'un ensemble c de couples (x,y) où x appartient à a et y appartient à b, avec la condition de bijectivité: pour tout x de a, il existe un y de b et un seul avec (x,y) appartenant à c , et inversement en échangeant a et b .

1.5. Graphe codant un ensemble hérédénombrable; ensemble hérédénombrable récursif.

Soit a un ensemble hérédénombrable; nous avons vu en 1.1 que la clôture transitive de a est finie ou dénombrable; il en est de même de la clôture transitive du suivant de a, soit $a \cup \{a\}$, qui est encore hérédénombrable. Associons donc à l'ensemble a la relation binaire d'appartenance basée sur la clôture transitive du suivant de a. Si cette clôture est finie, autrement dit si a est héréfini, étendons-la à une base dénombrable en ajoutant une infinité dénombrable d'éléments, la relation prenant la valeur + pour tout couple dont l'un au moins des termes est un élement ajouté. Enfin transformons la relation par isomorphie, de façon à ce qu'elle soit basée sur l'ensemble des entiers naturels. Nous appellerons graphe de a l'une quelconque de ces relations, deux à deux isomorphes, basée sur les entiers naturels.

Par exemple, un graphe de l'ensemble vide est une relation prenant la valeur - seulement pour un couple de la forme (u,u) où u est un entier; un graphe de l'ensemble 1 = {0} est une relation prenant la valeur - pour deux couples réflexifs (u,u), (v,v), ainsi que pour (v,u), l'ensemble 0 devant être appliqué sur u et 1 sur v.

Etant donné un graphe A, il existe un entier u et un seul tel que A(x,u) = - pour tout x vérifiant A(x,x) = - : cet u est le transformé de l'ensemble vide 0. Il existe un entier v et un seul tel que A(v,x) = - pour tout x vérifiant A(x,x) = -: ce v est le transformé de l'ensemble donné a, et nous l'appellerons le sommet du graphe. Enfin, étant donné deux éléments antiréflexifs u,v, vérifiant A(u,u) = A(v,v) = - , nous dirons que u est un antécédent immédiat de v lorsque A(u,v) = + ; qu'il est un antécédent de v lorsqu'il existe une suite finie allant de u jusqu'à v avec valeur + pour A à chaque étape.

Nous voyons qu'une relation A basée sur l'ensemble des entiers est un graphe si et seulement si elle est fondée: tout suite d'antécédents consécutifs est finie; extensionnelle: deux éléments ayant mêmes antécédents immediats sont identifiques; et si elle admet un sommet. De plus si elle est antiréflexive (cas général) ou si elle admet une restriction finie antiréflexive (représentant un ensemble héréfini), la relation prenant valeur + pour tout couple dont l'un des termes est en dehors de l'ensemble fini considéré.

Incidemment, on définit un hérédénombrable récursif lorsqu'il existe un graphe récursif qui le représente, au sens usuel des relations récursives basées sur l'ensemble des entiers.

Information bibliographique: le codage des hérédénombrables par graphes, ainsi que la construction des formules du second ordre $\equiv^*$ et $\in^*$ de 3.3 et 3.4 ci-dessous, remontent à [Z 1971] .

BONORDRE FORMALISABLE DES CONSTRUCTIBLES EN CALCUL LOGIQUE DU PREMIER ORDRE

2.1. Considérons, comme précédemment, des formules logiques du premier ordre avec identité; encore avec un unique prédicat binaire ε que nous substituerons par l'appartenance. Nous n'ordinalons plus ces formules, mais nous autorisons plusieurs individus libres x,y,... que nous substituerons chacun par un ensemble.

Définissons la valeur absolue (+ ou -) prise par une telle formule P lorsque chaque individu libre est substitué par un ensemble. Il s'agit simplement de la valeur usuelle de vérité, mais étendue à tous les ensembles: les difficultés logiques provenant de l'inexistence d'un ensemble de tous les ensembles.

Donc lorsque P se réduit à $x = y$ et que x,y sont substitués par deux ensembles a,b, la valeur absolue sera + ssi a est identique à b. Lorsque P se réduit à $x \in y$, valeur absolue + ssi a appartiente à b. Le lecteur passera de manière évidente d'une formule P à sa négation; de deux formules P,Q à leur conjonction ou à leur disjonction.

Lorsque P est de la forme $\forall_x Q$, valeur absolue + ssi la formule Q prend la valeur absolue + quel que soit l'ensemble substitué à x. Définition analogue avec le quantifieur $\exists$.

Aucune difficulté logique ne se présente avec la valeur absolue, tant que nous considérons un nombre fini de formules. Ou même avec une infinité de formules, tant que le nombre des alternances de quantifieurs $\forall$ et $\exists$ est borné par un entier fixé. Par contre, l'introduction de la valeur absolue pour toute formule constituerait non une définition, mais un schéma de définition assimilable à un renforcement de nos axiomes de départ (qui sont Z-F plus la constructibilité de certains ensembles), sans garantie d'éviter la contradiction.

Le lecteur vérifiera aisément que chaque axiome Z-F prend la valeur absolue + .

Cela est une simple répercussion de la métathéorie dans laquelle nous nous exprimons (et qui part des axiomes Z-F) sur la notion de valeur absolue concernant chaque axiome Z-F codé, ou représenté.

Le lecteur vérifiera aisément que, si une formule P déduit une formule Q au sens formel usuel, et si P prend la valeur absolue + , alors Q prend la valeur absolue +. En conséquence, si une formule P se déduit des axiomes Z-F, donc d'un nombre fini d'axiomes Z-F en raison du théorème classique de compacité, alors P prend la valeur absolue +.

Aucune difficulté logique ne se présente encore, avec la valeur absolue, dans le cas où nous nous restreignons à une base fixe; par exemple, dans les paragraphes suivants où nous nous restreignons à l'appartenance définie sur l'ensemble des hérédénombrables, constructibles, puis supposés constructibles par axiome. En ce cas, la valeur absolue se réduit à la valeur de vérité usuelle, sur les hérédénombrables.

Dans ce qui suit, nous prendrons toujours, avec la valeur absolue, l'une au moins des deux précautions: ne considérer qu'une ou un nombre fini de formules; ou bien travailler sur l'ensemble fixe des hérédénombrables, muni de la relation d'appartenance.

Le reste du présent paragraphe sera consacré à la description sommaire des formules qui conduisent aux trois formules: R (dite "formule-rang"); C (dite "formule de constructibilité"); et D (dite "formule de bonordre").

(1) La formule-rang $R(x,t)$ prendra la valeur absolue + lorsque α, substitué à t, est un ordinal, et a, substitué à x, est un ensemble constructible de rang α. Les quantifieurs de R seront relativisés chacun à un ensemble, soit à l'ordinal α, soit à l'ensemble des constructibles de rangs $< \alpha$, soit à l'ensemble des formules ordinalées d'indices ordinaux $\leq \alpha$. Ainsi, lorsque a et α sont fixés, la valeur absolue de $R(a,\alpha)$ se réduira à une valeur de vérité usuelle.

(2) La formule de constructibilité sera $C(x) = \exists_t R(x,t)$; elle prendra donc la valeur absolue + ou - pour a selon que a est ou non constructible.

(3) La formule de bonordre $D(x,y)$ comprendra deux individus libres x,y.

Puisqu'il est impossible de parler d'un bonordre sur tous les ensembles constructibles, disons que D définira un tel bonordre chaque fois qu'on la restreindra à un ensemble de constructibles. Plus précisément:
des axiomes Z-F se déduira la formule de réflexivité $\forall_x C(x) \Longrightarrow D(x,x)$;
de ce qui précède, il résulte alors que D prendra la valeur absolue + lorsque x et y sont substitués par un même ensemble constructible;
de Z-F se déduira la formule de transitivité:

$$\forall_{x,y,z}(C(x) \wedge C(y) \wedge C(z) \wedge D(x,y) \wedge D(y,z)) \Longrightarrow D(x,z);$$

de ce qui précède, il résulte que, si a,b,c sont constructibles et si $D(a,b)$ et $D(b,c)$ ont la valeur absolue +, alors $D(a,c)$ est de valeur absolue + ;
de Z-F se déduira la formule d'antisymétrie: $\forall_{x,y}C(x) \wedge C(y) \wedge D(x,y) \wedge D(y,x)) \Longrightarrow x = y$;

de ce qui précède, il résulte que, si a,b sont constructibles et si D(a,b) et D(b,a) ont la valeur absolue +, alors a et b sont identiques; de Z-F se déduira la formule de comparabilité: $\forall_{x,y}(C(x) \wedge C(y)) \Rightarrow D(x,y) \vee D(y,x))$; de ce qui précède, il résulte que si a,b sont constructibles, ou bien D(a,b) ou D(b,a) est de valeur absolue + ; de Z-F se déduira une formule d'existence du minimum:

$$\forall_u(\exists_x\ x \in u) \wedge (\forall_x\ x \in u \Rightarrow C(x)) \Rightarrow \exists_x(x \in u \wedge \forall_y(y \in u \Rightarrow D(x,y));$$

de ce qui précède, il résulte que, pour tout ensemble non vide de constructibles, l'un d'eux, soit m, donne la valeur absolue + pour D(m,a) avec chacun des autres.

Approfondissons maintenant notre description des formules R et D .

L'association à un ensemble constructible, de son rang, utilise une formule ordinalée, que nous écrirons sous forme prénexe. La longueur de cette formule sera le nombre de termes du préfixe (suite des quantifieurs placés en tête), ou encore égal au nombre des individus de la partie libre, ou matrice, diminué de 1: en effet un seul individu reste libre dans une formule de construction telle que nous l'avons définie en RAPPELS . Nous pourrons toujours numéroter les individus dans l'ordre où ils figurent au préfixe, réservant le dernier numéro pour l'unique individu libre.

Il existe alors une formule de codage rang-longueur L(x,t,u) qui prendra la valeur absolue + lorsque, x étant substitué par un constructible a et t substitué par le rang α de a, le troisième individu u est substitué par la plus courte longueur possible d'une formule de construction de a avec le rang α. Notons que R(x,t) se réduit à $\exists_u L(x,t,u)$.

Continuons notre codage. Étant donné une formule ordinalée définissant a, de rang α et de longueur minimum h, le préfixe de cette formule associe à chaque $i < h$, d'une part le symbole d'un quantifieur $\forall$ ou $\exists$, d'autre part un indice ordinal $\beta \leq \alpha$. Ordonnons lexicographiquement les suites de longueur h formées de 0 à la place de chaque $\forall$ et 1 à la place de chaque $\exists$; ce qui nous donne un entier, rang de la suite des quantifieurs. De plus ordonnons lexicographiquement les suites de longueur h formées d'ordinaux $\leq \alpha$; le codage le plus commode est l'association, à une telle suite $\beta_0,\beta_1,\ldots,\beta_{h-1}$, de l'ordinal

$\gamma = \beta_{h-1}(\alpha + 1)^{h-1} + \dots + \beta_1(\alpha + 1) + \beta_0$: la décomposition cantorienne de γ suivant les puissances décroissantes de $\alpha + 1$, d'exposants $h-1, h-2, \dots, 1, 0$.

Il existe alors une formule de codage rang-longueur-quantifieurs $K(x,t,u,v,w)$ qui prendra la valeur absolue + lorsque, x étant substitué par un constructible a et t par son rang α et u par la plus petite longueur possible h d'une formule de construction de rang α ; alors de plus v est substitué par le rang lexicographique de la suite des quantifieurs, et w par l'ordinal codant γ les plus petits possibles.

Pour achever notre codage, nous avons besoin de la notion d'opérateur libre, qui précise la notion usuelle de formule libre, ou sans quantifieur.

Disons qu'une suite finie d'entiers est canonique lorsqu'elle est de la forme $u_0, u_1, \dots, u_h$ avec $u_0 = 0$, puis $u_1 = 0$ ou 1, puis en général pour chaque entier $i \leq h$, lorsque u_i est au plus égal au maximum des valeurs u_j $(j < i)$ augmenté de 1. Nous voyons que toute suite finie v de termes $v_0, v_1, \dots, v_h$ est isomorphe à une suite canonique et une seule, appelée la transformée canonique de v, par l'isomorphisme f tel que $f(v_0) = 0$, puis $f(v_1) = 0$ ou 1 selon que v_1 est identique ou distincte de v_0, est ainsi de suite.

Etant donné l'entier longueur h, associons-lui un opérateur $(h+1)$-aire, ou d'arité $h+1$ (pour tenir compte de l'individu qui restera libre); avec un unique prédicat binaire ε. Cet opérateur se définit comme une fonction qui associe la valeur de vérité + ou - à chaque système formé d'une suite canonique u de $h+1$ termes, et d'une relation binaire antiréflexive basée sur l'ensemble des valeurs $\{u_0, u_1, \dots, u_h\}$; une relation binaire est elle-même une fonction A qui prend la valeur + ou - pour chaque couple ordonné de ces u_1 $(i \leq h)$. Par exemple, l'opérateur libre usuellement représenté par la formule $z \varepsilon x \wedge y = z$, que nous reécrirons $x_2 \varepsilon x_0 \wedge x_1 = x_2$, prend la valeur - pour tous les systèmes débutant par la suite canonique $(0,0,1)$ ou par la suite $(0,1,0)$ ou encore par $(0,1,2)$; suites dont les deux derniers termes sont distincts, ce qui entraine la valeur - pour $x_1 = x_2$. L'opérateur prend la valeur - encore pour la suite $(0,0,0)$ puisque nous ne considérons que des relations antiréflexives, comme est l'appartenance. Enfin notre opérateur prend la valeur + pour les systèmes formés de la suite $(0,1,1)$ et d'une relation A telle que $A(1,0) = +$, de manière à donner la valeur + pour

$x_2 \varepsilon x_0$.

Ordonnons lexicographiquement les suites canoniques d'une longueur donnée h+1. Puis ordonnons lexicographiquement, dans l'ensemble des valeurs de notre suite canonique, les couples ordonnés; puis les relations qui prennent une valeur + ou - pour chaque couple. Finalement nous ordonnons lexicographiquement les systèmes (suite canonique plus relation), et enfin nous obtenons un ordre lexicographique des opérateurs puisque ce sont des fonctions prenant la valeur + ou - pour chaque système de la longueur donnée h+1. Le codage est maintenant terminé, puisque deux ensembles constructibles ayant la même formule ordinalée de construction, sont identiques.

Il existe alors une formule de codage rang-longueur-quantifieurs-opérateurs, que nous noterons $J(x,t,u,v,w,o)$, qui prendra la valeur absolue + lorsque x est substitué par un constructible a, et t par le rang α de a, et u par la plus petite longueur h d'une formule de construction de rang α, et v par le plus petit codage de la suite des quantifieurs, et w par le plus petit ordinal γ codant la suite des indices ordinaux, et enfin o par le plus petit entier codant l'opérateur libre qui achève de définir la formule ordinalée de construction de a.

Evidemment la formule-rang R se réduit à J précédée des quantifieurs $\exists$ pour u,v,w et o. Quant à la formule D de bonordre, le lecteur la construira très aisément à partir de J, en exprimant la condition que le constructible a est antérieur au constructible b lorsque: soit le rang de a est strictement inférieur au rang de b; soit ces rangs sont égaux mais la longueur minimum d'une formule de construction de a est inférieure à celle de b; soit ces longueurs sont égales mais le numéro de codage de la suite des quantifieurs, choisi minimum, est plus petit pour a que pour b ; soit ces numéros sont égaux mais l'ordinal codant minimum γ est plus petit pour a que pour b ; soit enfin que ces ordinaux sont égaux mais que l'entier minimum codant l'opérateur libre pour a est plus petit que pour b .

2.2. Rappelons que, par 1.3, utilisant le théorème de la partie constructible, un ensemble constructible est hérédénombrable ssi son rang de construction est $< \omega_1$.

Il en résulte que l'ensemble des hérédénombrables constructibles est constructible, et de rang ω_1, au moyen de la simple formule de construction $x = x$ affectée l'ordinal ω_1 .

Il existe un bonordre de l'ensemble des hérédénombrables constructibles, et une formule D(x,y) du calcul du premier ordre avec identité, comprenant un unique prédicat binaire ε et deux individus libres x,y, de sorte que: si nous substituons ε par l'appartenance basée sur l'ensemble des hérédenombrables constructibles, alors D prend la valeur + ssi l'ensemble hérédénombrable constructible substitué à x est antérieur à celui substitué à y, modulo le bonordre considéré.

Ils'agit ici de la valeur de vérité usuelle, et non plus de la valeur absolue.

□ Signalons d'abord que la formule "u est un ordinal", autrement dit "u est transitif et totalement ordonné par appartenance", avec l'interprétation considérée dans l'énoncé, prend la valeur de vérité + ssi u est substitué par un ordinal fini ou dénombrable. L'énoncé resulte alors de nos remarques du 2.1, en prenant D identique à la formule D du 2.1: la valeur absolue qui était prise par D se réduit ici à la valeur de vérité sur l'ensemble des hérédénombrables constructibles, et le bonordre, qui était défini par D sur un ensemble quelconque de constructibles, concerne ici ceux qui sont hérédénombrables. □

TRADUCTION DE LA FORMULE DE BONORDRE DU PREMIER ORDRE, EN UNE FORMULE DU CALCUL LOGIQUE DU SECOND ORDRE, SANS PREDICAT D'APPARTENANCE

3.1. Dans ce qui suit, nous introduirons d'abord une formule du calcul du second ordre notée $I(\iota)$, comprenant seul predicat binaire libre ι, et qui exprime que la relation binaire substituée à ι est isomorphe à l' ordinal ω; ou est un ω-bonordre. Plus précisement ι est reflexif, transitif, antisymétrique; deux éléments quelconques de la base sont comparables modulo ι; de plus ι est bienordonnant: pour toute relation unaire, représentée par un prédicat unaire ξ lié par le quantifieur $\forall$, s'il existe un élément de la base vérifiant ξ, il en existe un qui soit minimum modulo ι ; enfin pour tout élément autre que le minimum, il existe, modulo ι, un prédécesseur immédiat.

3.2. Reprenons le codage de chaque ensemble hérédénombrable a au moyen d'un graphe, qui est une relation binaire basée sur l'ensemble des entiers naturels, et isomorphe à la relation d'appartenance basée sur la cloture transitive du suivant de a (ou à l'extension à valeurs + d'une telle relation, lorsque a est héréfini); voir 1.5.

En rapport avec ce codage, construisons une formule $G(\iota,\xi)$ du second ordre, à deux prédicats binaires libres ι et ξ, telle que, si ι est substitué par la comparaison des entiers naturels, G prend la valeur + si et seulement si ξ est substitué par un graphe; autrement dit G exprime que ξ est antiréflexive, fondée, extensionnelle, et admet un sommet, au sens de 1.5. Ou encore G prend la valeur + pour tous couples sauf certains pris dans une partie finie de la base, la restriction à cette partie finie étant antiréflexive, fondée, extensionnelle et admettant un sommet.

3.3. Associons, à chaque identiteur x = y, ou formule atomique d'identité du calcul logique du premier ordre, une formule du second ordre que nous noterons $\xi \equiv^* \eta$. Cette formule exprimera l'isomorphie entre la relation binaire substituée au prédicat binaire ξ (que nous considérerons comme le transformé de l'individu x), et la relation substituée au prédicat binaire η (transformé de l'individu y). Plus précisément, la formule exprimera l'existence d'une relation binaire qui permute la base et qui transforme la première relation en la deuxième. En particulier, si nous conjonctons $G(\iota,\xi)$, $G(\iota,\eta)$ et $\xi \equiv^* \eta$, et si nous substituons au prédicat ι la comparaison des entiers naturels, alors la conjonction considerée exprime que ξ et η sont deux graphes isomorphes, donc représentant un même ensemble hérédénombrable.

3.4. Associons, à chaque opérateur $x \varepsilon y$, ou formule atomique d'appartenance du calcul du premier ordre, une formule du second ordre que nous noterons $\xi \in^* \eta$. Cette formule, à trois prédicats binaires libres ι,ξ et η, sera la conjonction de $G(\iota,\xi)$, $G(\iota,\eta)$ qui expriment que ξ et η sont deux graphes (une fois ι substitué par la comparaison des entiers); et d'une troisième formule exprimant que le graphe substitué à ξ est obtenu en partant du graphe substituté à η, en prenant un élément u, antécédent immédiat du sommet du graphe η, puis en prenant ξ isomorphe à la restriction du graphe η aux éléments antécédents de u; si de plus, ces derniers éléments sont en nombre fini, le graphe ξ est complété par des valeurs +, puisqu'il est basé sur l'ensemble des entiers.

Il résulte alors de 1.5 que $\in^*$ code l'appartenance entre ensembles hérédénombrables, eux-mêmes codés par ξ et η.

Notons que, dans la construction des formules G et $\in^*$, le prédicat ι est notamment utilisé pour exprimer la finitude de certain ensembles.

Ainsi, dans G, pour exprimer que ξ est soit antiréflexive, soit de valeurs + sauf sur un ensemble fini où elle est antiréflexive, fondée et extensionnelle; la finitude pourra s'exprimer en identifiant l'ensemble fini avec les entiers antérieurs, pour ι, à l'un d'eux. De même, dans la formule $\xi \in^* \eta$, la condition d'antécédence d'un élément u à un élément v, pour le graphe η, s'exprimera par l'existence d'un entier r et d'une permutation de la base qui transforme chaque entier $i \leq r$ en un élément u_i, avec $u_0 = u$, $u_r = v$ et $\eta(u_i, u_{i+1}) = +$ pour chaque $i < r$; la consécutivité $j = i+1$ s'exprimant de façon évidente à l'aide du prédicat ι.

3.5. A l'aide de la formule D du premier ordre, du 2.1 ci-dessus, et des formules I, G, $\equiv^*$ et $\in^*$ précédentes, et en utilisant l'axiome de constructibilité des hérédénombrables, nous allons obtenir une formule du second ordre à trois prédicats binaires libres ι, ξ, η, soit $D^*(\iota,\xi,\eta)$, vérifiant ce qui suit:

si ι est substitué par la comparaison des entiers naturels, D^* vaut + si et seulement si ξ et η sont substitués par deux graphes représentant deux ensembles hérédénombrables, et si l'ensemble représenté par ξ est antérieur à l'ensemble représenté par η, dans le bonordre défini par la formule D de 2.1 sur l'ensemble des hérédénombrables.

□ A chaque indice x, libre ou lié, de la formule D de 2.1, associons un prédicat binaire ξ. A chaque identiteur $x = y$ figurant comme sous-formule dans D, associons une formule $\xi \equiv^* \eta$ construite comme en 3.3 ci-dessus, les prédicats binaires ξ, η étant transformés des indices x,y. A chaque opérateur $x \in y$ figurant comme sous-formule dans D, associons une formule $\xi \in^* \eta$ comme en 3.4. Conservons les connections qui figurent dans D; enfin à chaque quantifieur $\forall_x$ figurant dans D, associons le quantifieur $\forall$ portant sur un prédicat binaire ξ, transformé de l'indice x, et relativisé aux graphes; plus précisément, remplaçons $\forall_x$ par $\forall_\xi G(\iota,\xi) \Rightarrow$ où G est la formule construite en 3.2 ci-dessus; de même, remplaçons $\exists_x$ par $\exists_\xi G(\iota,\xi) \wedge$.

Nous obtenons au lieu de D, une formule du second ordre à trois prédicats binaires libres ι,ξ,η. Conjonctons-la avec les trois formules $I(\iota)$, $G(\iota,\xi)$ et $G(\iota,\eta)$ de 3.1 et 3.2, qui exigent, pour prendre la valeur +, que ι soit substitué par la comparaison des entiers naturels (ou une relation isomorphe), et que ξ et η soient substitués par des graphes représentant des hérédénombrables: nous obtenons alors la formule D^* de l'énoncé. □

3.6. Dans ce qui suit, nous noterons $H(x,\iota,\xi)$ une formule du second ordre comprenant un seul individu libre x et deux prédicats binaires libres ι et ξ. Cette formule, une fois ι substitué par la comparaison des entiers naturels, vaudra + si et seulement si, x étant substitué par un élément de la base, qui se trouve être un entier naturel du fait de la comparaison ι, le prédicat ξ est substitué par l'un quelconque des graphes (isomorphes entre eux) qui représentent l'entier x. Pour préciser la construction de la formule H, rappelons que le graphe de l'entier x est, à l'isomorphie près, la relation binaire d'appartenance, ou équivalemment l'ordre strict (<) entre les entiers $0,1,\ldots,x-1$, x, complété par la valeur + pour tout couple d'entiers dont l'un au moins est $> x$. D'après ce qui précède, la formule $\exists_x H(x,\iota,\xi)$, où ι est substitué par la comparaison des entiers, vaut + si et seulement si ξ est un graphe d'un entier.

3.7. Soit n un entier positif, fixe dans ce qui suit, et que nous appellerons l'arité. Convenons de définir le n-uple d'entiers $(a_0,\ldots,a_{n-1})$ comme étant l'ensembles des paires $\{0,\{\{a_0\}\}\},\{1,\{\{a_1\}\}\},\ldots,\{n-1,\{\{a_{n-1}\}\}\}$. Aucune confusion n'est possible entre un rang $i(0 \leq i < n)$, qui est un entier et le bi-singleton $\{\{a_i\}\}$, puisque le seul entier qui soit un singleton est $1 = \{0\}$, et ce n'est pas un bi-singleton. Sur le modèle de la formule H, construisons une formule $H_n(x^0,x^1,\ldots,x^{n-1},\iota,\xi)$, du second ordre, comprenant n individus libres $x^i(0 \leq i < n)$, et deux prédicats binaires libres ι et ξ. Cette formule, une fois ι substitué par la comparaison des entiers naturels, vaudra + si et seulement si, chaque x^i étant substitué par un élément de la base, qui se trouve être un entier naturel a_i du fait de la comparaison ι, predicat ξ est substitué par l'un quelconque des graphes (isomorphes entre eux) qui représentent le n-uple $(a_0,\ldots,a_{n-1})$, défini comme ci-dessus par des bi-singletons. D'après ce qui précède, la formule $\exists_{0,\ldots,n-1} H_n(x^0,\ldots,x^{n-1},\iota,\xi)$, où ι est substitué par la comparaison des entiers, vaut + si et seulement si ξ est un graphe d'un n-uple d'entiers.

Il en résulte que la formule suivante, où ι est substitué de même, que nous noterons $G_n(\iota,\alpha)$ vaut + si et seulement si α est un graphe d'une relation n-aire basée sur les entiers naturels: $\forall_\xi(\xi \in^* \alpha \Rightarrow \exists_{0,\ldots,n-1} H_n(x^0,\ldots,x^{n-1},\iota,\xi)$. Conjonctions cette formule avec la formule suivante; $\exists_\xi H_n(x^0,\ldots,x^{n-1},\iota,\xi) \wedge \xi \in^* \alpha$; alors ι

étant substitué par la comparaison des entiers, les individus libres $x^0,\dots,x^{n-1}$ étant substitués par les entiers $a_0,\dots,a_{n-1}$, la conjonction obtenue vaudra + si et seulement si α est substitué par le graphe d'une relation n-aire valant + pour le n-uple $(a_0,\dots,a_{n-1})$

3.8. Construisons maintenant une formule qui exprimera, ι étant substitué par la comparaison, que les relations binaires substituées aux prédicats α et β sont des graphes représentant deux relations n-aires isomorphes.

Cette formule, que nous noterons $Is(\iota,\alpha,\beta)$ exprimera l'existence d'une permutation notée $\overline{\omega}$ (plus exactement, substituant un prédicat binaire $\overline{\omega}$), telle que, pour tous $x^0,\dots,x^{n-1}$, nous ayons $\exists_\xi H_n(x^0,\dots,x^{n-1},\iota,\xi)$ si et seulement si'il existe $y^0,\dots,y^{n-1}$ et une relation binaire η telle que:
$\overline{\omega}(x^0,y^0) \wedge \dots \wedge \overline{\omega}(x^{n-1},y^{n-1}) \wedge H_n(y^0,\dots,y^{n-1},\iota,\eta) \wedge \eta \in^* \beta$.

Cette isomorphie entre relations n-aires se traduit, sur les graphes, par une équivalence plus large que l'isomorphie entre α et β : rappelons en effet que l'isomorphie des graphes correspond à l'identité des ensembles hérédénombrables qu'ils representent, donc ici à l'identité des relations. Nous voyons ainsi que, même si nous avons n=2, nous ne pouvons pas identifier une relation binaire basée sur les entiers, avec son graphe représentatif; le travail précédent est donc à faire même dans ce cas.

PREUVE FINALE

Nous sommes maintenant en mesure de prouver que deux relations n-aires dénombrables logiquement équivalentes pour le second ordre, sont isomorphes.

□ Étant donné une relation n-aire R, associons-lui un ω-bonordre I de la base $|R|$; à chacun des bonordres I, correspond une interprétation des éléments de la base $|R|$ comme entiers naturels, donc une interprétation de R comme relation basée sur l'ensemble de ces entiers. Ainsi, à chaque ω-bonordre I, correspond un graphe et un seul (à l'isomorphie près) qui représente une isomorphe de R sur les entiers naturels. Choisissons I de façon à obtenir le graphe minimum dans le bonordre des graphes défini par la formule D* de 3.5. Appelons A le graphe correspondant (défini à l'isomorphie près). Nous voyons qu'en notant ρ le prédicat n-aire substituable par R, et notant α le prédicat binaire substituable par A, et naturellement ι celui substituable par I, la formule du second ordre suivante, des prédicats libres ι, ρ, α, une fois ρ substitué par R, prendra la valeur + si et seulement si ι est substitué par le ω-bonordre I qui rend minimum (dans le bonordre défini par D*) le graphe représentant R, rapportée aux entiers définis par l'ordre I; et si de plus α est substitué par ce graphe minimum A;

$I(\iota) \wedge (\forall_{0,\ldots,n-1}\ \rho x^0 \ldots x^{n-1} \Longleftrightarrow (\exists_\xi H_n(x^0,\ldots,x^{n-1},\iota,\xi) \wedge \xi \in^* \alpha)) \wedge G_n(\iota,\alpha)$ $\wedge\ \forall_\beta\ Is(\iota,\alpha,\beta) \Longrightarrow D^*(\iota,\alpha,\beta)$ (rappelons que $G_n(\iota,\alpha)$ exprime que α est le graphe d'une relation n-aire (voir 3.7).

Soit maintenant S une relation n-aire dénombrable et non isomorphe à R. Comme nous l'avons fait pour R, associons à S le ω-bonordre que nous noterons encore I, qui rend minimum (dans le bonordre défini par D*) le graphe représentant S, rapportée aux entiers définis par l'ordre I. Alors en confondant l'ordre I associé à R et celui associé à S (qui sont tous deux isomorphes à l'ordinal ω), les relations R et S ainsi redéfinies sur les entiers, sont évidemment distinctes, puisque non isomorphes. Fixons les idées en désignant par $(a_0,\ldots,a_{n-1})$ un n-uple d'entiers qui donne la valeur + à R et la valeur − à S. Soit b le maximum des entiers a_i $(0 \leq i < n)$. Notons σx (x égale zéro) l'abréviation de $\forall_y \urcorner\, \iota xy$; notons γxy (y consecutif à x) l'abréviation de $\iota xy \wedge x \neq y \wedge \forall_x(\iota zx \vee \iota yz)$. Alors pour obtenir une formule du second ordre et du seul prédicat libre P, valant + pour R et − pour S, il suffit de reprendre la formule ci-dessus, de la conjoncter avec la formule suivante:

$$\exists_{0,\ldots,b}(\sigma x^0 \wedge \gamma x^0 x^1 \wedge \ldots \wedge \gamma x^{b-1} x^b \wedge \rho x^{a_o} \ldots x^{a_{n-1}}).$$

puis de faire précéder cette conjonction du quantifieur $\exists$ portant sur les deux prédicats ι et α. □

REMARQUE FINALE

Si nous nous plaçons dans Z-F plus choix, sans constructibilié des hérédénombrables, alors la formule D* de 3.5 ne concerne que les graphes constructibles. En ce cas nous prouvons seulement que deux relations basées sur les entiers, logiquement équivalentes pour le second ordre et constructibles, sont isomorphes. Il faut entendre "constructible" au sens strict, non à l'isomorphie près. Par exemple un ordinal dénombrable α n'est "constructible" que s'il existe une relation binaire de bonordre A de type α, définie sur les entiers et constructible; autrement dit si le couple de relations (I,A) où I est l'ordre de entiers, est constructible.

BIBLIOGRAPHIE

1. Ajtai, M. Isomorphism and higher order equivalence. Annals of Math. Logic, vol. 16, 1979, 181-203.

2. Cohen, P. The independence of the continuum hypothesis. Proceedings Nat. Acad. Sc. vol. 50 et 51, 1963, 1143-1148 et 105-110.

3. Fraïssé, R. Une nouvelle classification des systèmes de relations. Comptes rendus, vol. 230, 1950, 1022-1024.

   Course of Mathematical Logic, Model Theory, trad. David Louvish, Dordrecht (Reidel), vol. 2, 1974, 192 p.

   Cours de Logique Mathématique, Récursivité et Constructibilité, Paris (Gauthier-Villars), vol. 3, 1975, 138 p.

   Publications du Département de Mathématiques de l'Université de Lyon, vol. 14-2, 1977, p. 41-62.

4. Levy, A. Definability in set theory, Logic and Foundations of Set Theory, Amsterdam (North-Holland), 1970.

5. Marek, W. Consistance d'une hypothèse de Fraïssé sur la définissabilité dans un langage du second ordre. Comptes rendus, vol. 276(A), 1973, p. 1147-1150 et 1169-1172.

6. Zbierski, P. Models for higher order arithmetics. Bulletin Acad. Pol. Sc. (Math. Astr. Phys.), vol. 19, 1971, p. 557-562.

A GENERALIZED ŁUKASIEWICZ-STYLE IDENTITY LOGIC

Siegfried Gottwald/Department of Philosophy, Karl Marx University, Leipzig, German Democratic Republic

ABSTRACT

It is an old problem in Łukasiewicz first order logic, at least for the case of finitely many generalized truth values, to consider first order logic with identity. With the usual many-valued generalizations of identity axioms of classical first order logic one gets only trivial interpretations of the identity symbol, i.e. only such interpretations which allow only two generalized truth values at all for the identity relation.

We propose a modification of the Leibnizian axiom with the effect that we get for a restricted class of interpretations non-trivial identity relations and an axiomatization theorem for the class of all restrictedly valid sentences.

INTRODUCTION

We consider here many-valued first order logics which have as set of truth values a suitable subset of the real interval [0,1], which have the Łukasiewicz implication described by the truth value function seq with

$$seq(s,t) = \min\{1, 1 - s + t\},$$

and the negation described by the truth value function non with

$$non(s) = 1 - s$$

as sentential connectives, and for which generalization and existential quantification are characterized by the infimum resp. supremum.

For the case of only finitely many truth values $k/m-1$ with $0 \leq k < m$, $m \in \mathbb{N}$, $\mathbb{N}$ the set of natural numbers, these logics were extensively studied by H. Thiele [10], and different axiomatizations are known (cf. e.g. [7], [8]).

But, usually, one is discussing these first order logics as logics without identity only. This is mainly caused by the fact the enlargement of the language by an identity symbol $\equiv$ and of the set of axioms by many-valued "translations" of the usual identity axioms.

(i) $x \equiv x$

(ii) $x \equiv y \longrightarrow (A(x) \longrightarrow A(y))$

result only in interpretations where the identity symbol is interpreted as the usual two-valued identity; cf. e.g. [10], p. 111.

From the intuitive point of view this is not a satisfactory situation, because, if one allows more than two truth values it seems appropriate to allow also interpretations of the identity symbol which really have as truth values with respect to some individuals also non-trivial ones, i.e. truth values different of "truth" and "falsity", i.e. of 1 and 0.

Recently, Ch. G. Morgan has considered such first order logics with identity. In [4] he is also forced to consider only trivial interpretations. But in [5] he discusses a more general notion of non-trivial many-valued identity relations.

We were led to the same problem of non-trivial identity relations by considerations on fuzzy set theory [1], [2], [3]. Our main points there have been to discuss first an inclusion of fuzzy sets which itself was "fuzzy" in the sense of sometimes having a truth value different of 1 and 0, and later on by this fuzzy inclusion we defined as usual a "fuzzy identity" of fuzzy sets which really could have non-trivial truth values. With this, the problem arose to construct a cumulative system of fuzzy sets in such a way that always "almost identical" fuzzy sets should behave in "almost the same way". This was realized by using the language of Łukasiewicz first order logic - with some special sentential connectives only - as a natural tool in connection with the generalized membership values of fuzzy sets. Thus, this system of fuzzy sets was something like an interpretation of (a fragment of) Łukasiewicz first order logic with identity which had a non-trivial identity relation.

Out of this the ideas of a development of Łukasiewicz first order logic with identity allowing non-trivial identity relations by changing axiom (ii) and restricting the class of interpretations arose which will be described here.

In comparison to Morgan [5] it should be noted that he uses for the approach in [5] in an essential way the J_k-connectives of Rosser/Turquette [8] which are described by truth functions j_k with

$$j_k(s) = \begin{cases} 1 , & \text{if } s = k \\ 0 , & \text{if } s \neq k \end{cases}$$

and hence in some sense realize an essential element of two-valuedness inside these considerations of many-valuedness.

Instead, in most papers on fuzzy sets theory these J_K-connectives are not used. Of course, the Łukasiewicz implication and the Łukasiewicz negation, which are normally used there, suffice to define all the J_k-connectives (for fixed number of truth values), but the definitions depend of the number of truth values. Thus, we are interested in an axiomatization of many valued identity logic which does not (directly) use these J_k-connectives.

The principle of construction for such an axiomatic system is as in [5] the addition of "identity axioms" to one of the usual systems for first order many valued predicate calculus. Therefore we consider only the Łukasiewicz logics $\mathbb{L}_m$ with finitely many truth values because $\mathbb{L}_{\aleph_1}$ with truth values set the whole real interval [0,1] was proved to be non-axiomatizable by Scarpellini [9].

FUNDAMENTAL NOTIONS

We are considering first order logics $\mathbb{L}_m$ with finite sets of truth values $W_m = \{k/m-1 \mid 0 \leq k < m\}$ for every integer $m > 2$. We suppose to have in each case a language consisting of

(1) an infinite set $V = \{x_j \mid j \geq 0\}$ of individual variables;

(2) a (finite or infinite) set $\mathbb{C} = \{k_1, k_2, \ldots\}$ of sentential connectives together with an arity function $\alpha_{\mathbb{C}} : \mathbb{C} \longrightarrow \mathbb{N}$, and binary connectives $\rightarrow$ and $\neg$ not contained in $\mathbb{C}$;

(3) a (finite or infinite) set $\mathbb{P} = \{P_1, P_2, \ldots\}$ of predicate symbols together with an arity function $\alpha_{\mathbb{P}} : \mathbb{P} \longrightarrow \mathbb{N}$, and a special binary predicate symbol $\equiv$ not contained in $\mathbb{P}$;

(4) the quantifiers $\forall$ and $\exists$ for generalization and existential quantification;

(5) parentheses.

We suppose also to have fixed some $\mathbb{L}_m$ in advance; and all the further considerations will be with respect to the corresponding truth value set W_m.

The notion of well-formed formula (wff) is defined inductively as usual. We omit the details.

Any $\mathbb{L}_m$-interpretation $\mathcal{A}$ has to specify a set A of elements, its support, has to map every sentential connective to a truth function of corresponding arity in W_m, and has to map every predicate symbol to a mapping (of corresponding arity) from A into W_m.

The interpretations we consider have to satisfy some very natural restrictions which later on will slightly be generalized. Therefore we call them <u>identity interpretations</u>. Such an identity interpretation consists of

(1) a nonempty set A of individuals, the support of $\mathcal{A}$;

(2) the function $\mathrm{seq} : W_m \times W_m \longrightarrow W_m$ defined by $\mathrm{seq}(s,t) = \min\{1, 1-s+t\}$ as interpretation of the implication $\rightarrow$;

(3) the function $\mathrm{non} : W_m \longrightarrow W_m$ defined by $\mathrm{non}(s) = 1 - s$ as interpretation of the negation sign $\neg$;

(4) an identity function $i : A \times A \longrightarrow W_m$ with properties

 (i) $i(a,a) = 1$

 (ii) $i(a,b) \leq \mathrm{seq}(i(a,c), i(b,c))$ for all individuals a,b,c;

(5) an $\alpha_{\mathbb{C}}(k)$-ary function f_k from W_m into W_m for every connective $k \in \mathbb{C}$;

(6) an $\alpha_{\mathbb{P}}(P)$-ary function g_P from A into W_m for every predicate symbol $P \in \mathbb{P}$, which in every argument has the property

$$i(a,b) \leq \mathrm{seq}(g_P(\ldots,a,\ldots), g_P(\ldots,b,\ldots))$$

for all individuals a,b.

We consider the truth value 1 as the only positively designated one. In accordance with the definition of a wff we define the truth value $val_{A,C}(H)$ of any wff H with respect to an interpretation A and any valuation $C : V \to A$ in the usual way inductively by using as a definitional instance e.g.

$$val_{A,C}(\forall x_j H) = \inf_{a \in A} val_{A,C^{[j/a]}}(H)$$

with $C^{[j/a]} : V \to A$ defined by $C^{[j/a]}(x_i) = C(x_i)$ for every $i \neq j$ and $C^{[j/a]}(x_j) = a$.

A wff H is called A-valid iff $val_{A,C} H = 1$ for all $C : V \to A$. And a wff H is called valid, denoted: $\models_{ID} H$, iff it is A-valid for all identity interpretations A.

ALTERATION OF IDENTITY AXIOMS

The main point of the following approach is in the syntactic direction a change of the many-valued versions of the identity axioms.

Of course, the first axiom

(I1) $\forall x(x \equiv x)$

will not be considered for alteration. So we have to discuss the second one of the identity axioms

(x) $x \equiv y \to (H(x) \to H(x/y))$.

Here H(x) is a wff which may have x among its free variables, and H(x/y) results of H(x) by substituting all free occurrences of x in H(x) by y (and making an appropriate change of bounded variables if some of the free occurrences of x in H(x) are in the scope of a quantifier binding y).

Our aim is a weakening of this axiom. This we get if we assume that in (x) the formula H(x/y) is choosen in a more specialized manner: H(x/y) has to result of H(x) by substituting <u>at most one</u> free occurrence of x in H(x) by y (and by suitable changes of bounded variables).

To put it loosely: if we suppose only once that $x \equiv y$, then we should use this information with respect to H(x) also only once.

The fact that we did not include the sign $\equiv$ into P forces us now to state

(I2) $\forall x \forall y \forall z(x \equiv y \rightarrow (x \equiv z \rightarrow y \equiv z))$

as an axiom additionally.

But more care is necessary because almost nothing is supposed for the sentential connectives in $\mathbb{C}$. To see this, let us assume to have an unary connective k' and the interpretation function $f_{k'}$ with

$$f_{k'}(s) = \max\{t \in W_m \mid t \leq s^2\},$$

and consider the following wff H_o which is an instance of the weakened version of (x) we like to start with:

$$(\forall)(x_1 \equiv x_2 \rightarrow (k'(x_1 \equiv x_1) \rightarrow k'(x_1 \equiv x_2))).$$

Suppose now to have an interpretation A and a valuation C such that

$$val_{A,C}(x_1 \equiv x_2) = 0.5 ,$$

then it follows that

$$val_{A,C} H_o \leq 0.75 .$$

Therefore we will simplify (x) not only by assuming that at most one free occurrence of x in $H(x)$ was changed to y, but also by assuming that $H(x)$ is a predicate formula $P(x_1,\dots,x_r)$ with $P \in \mathbb{P}$, $r = \alpha_{\mathbb{P}}(P)$ only. Hence we add to the generalized identity axioms (I1), (I2) the schema

(I3) $(\forall)(x \equiv y \rightarrow (P(x_1..x..x_r) \rightarrow P(x_1..y..x_r)))$

for each $P \in \mathbb{P}$ and $r = \alpha_{\mathbb{P}}(P)$, and for each position of x in the r-tuple $(x_1,\dots,x_r)$. Here $(\forall)$ denotes many valued generalization of all free variables.

The generalization of this idea is straightforward. We use the following notational conventions:

(a) by $H(x/n/y)$ we denote a formula which results of $H(x)$ by substituting at most n of the free occurrences of x in $H(x)$ by y (and by suitable changes of bounded variables);

(b) for iterated implications we define the wff $G \underset{n}{\rightarrow} H$ inductively for $n \geq 1$ by the rules

- the wff $G \underset{1}{\rightarrow} H$ is the wff $G \rightarrow H$;
- the wff $G \underset{n+1}{\rightarrow} H$ is the wff $G \rightarrow (G \underset{n}{\rightarrow} H)$.

Now our weakened version of (x) can also be formulated as the schema

$$(I3^*) \quad (\forall)(x \equiv y \xrightarrow[n]{} (H(x) \longrightarrow H(x/n/y)))$$

with $H(x)$ as predicate formula only.

It is easy to get (I3) as special case of (I3*), but conversely it is also easy to derive (13*) from (13) immediately by usual means of Łukasiewicz logic. Therefore it is enough to consider (I1), (I2), (I3). By Id we denote the set of wff {(I1), (I2), I3)}.

As a side remark we mention the fact that a conjunction & of many valued logic can be defined as

$$H_1 \,\&\, H_2 =_{df} \neg(H_1 \longrightarrow \neg H_2) ,$$

having as corresponding truth value function the function et given by

$$et(s,t) = \max\{0, s+t-1\}$$

and yielding the formulas

$$H_1 \longrightarrow (H_2 \longrightarrow H_3) , \qquad H_1 \,\&\, H_2 \longrightarrow H_3$$

as semantically equivalent ones, i.e. such that both represent the same truth function. With this notation (I2), (I3) are equivalent with resp.

$$(\forall) \ (x \equiv y \,\&\, x \equiv z \longrightarrow y \equiv z) ,$$

$$(\forall) \ (x \equiv y \,\&\, P(x_1..x..x_r) \longrightarrow P(x_1..y..x_r)).$$

With one new notation more introduced through

$$(+) \quad H^1 =_{df} H , \qquad H^{n+1} =_{df} H^n \,\&\, H$$

also (I3*) can be formulated equivalently as

$$(\forall) \ ((x \equiv y)^n \,\&\, H(x) \longrightarrow H(x/n/y)).$$

THE COMPLETENESS THEOREM

We assume that Ax is any set of axioms or axiom schemes and Abl any - syntactically defined - consequence relation such that for (Ax, Abl) there holds true a completeness theorem with respect to the class of all valid sentences of $\mathbb{L}_m$ (in the language without identity symbol); e.g. Ax, Abl may be choosen as in [10].

We write $\vdash H$ iff H is Abl-derivable from Ax. We write $\Gamma \vdash H$ iff H is Abl-derivable from $Ax \cup \Gamma$, Γ any set of wff.

Extending the language by the identity symbol, we allow for all expressions of Ax and all those that appear in the definitional rules for Abl to be formulated in this extended language. Then we write for any set Γ of wff.

$$\vdash_{ID} H \quad \text{iff} \quad H \text{ is Abl-derivable from } \underline{Ax} \cup \underline{Id}$$
$$\text{iff} \quad \underline{Id} \vdash H$$

and

$$\Gamma \vdash_{ID} H \quad \text{iff} \quad H \text{ is Abl-derivable from } \underline{Ax} \cup \Gamma \cup \underline{Id}$$
$$\text{iff} \quad \Gamma \cup \underline{Id} \vdash H$$

And we write also

$\Gamma \models_{ID} H$ iff H is A-valid in all identity interpretations A for which all the wff of Γ are A-valid.

From the definition of identity interpretation by straightforward calculations it now follows for each $\mathbb{L}_m$-interpretation A: all the identity axioms (I1), (I2), (I3) are A-valid iff A is an identity interpretation.

Furthermore, the set Id is consistent, i.e. there does not exist any wff E such that $\underline{Id} \vdash E \ \& \ \neg E$. This fact may be proved as in [5] using from Rosser/Turquette [8] the equivalence of consistency and satisfiability of any set of wff. (Of course, all the necessary logical operators satisfy the standard conditions of [8]).

Hence we get like Morgan [5] the

Completeness Theorem: For every set Γ of wff:

$$\Gamma \models_{ID} H \quad \text{iff} \quad \Gamma \vdash_{ID} H .$$

It is not necessary to check all the details here (cf. [5]).

A GENERALIZATION

Up to now, the main point of our approach was presented in the identity axioms (I3) resp. (I3*) and the notion of identity interpretation. And as the notion of identity interpretation may appear as a natural one, the final result - the completeness theorem - looks also satisfactory.

Discrepancies between our and the traditional two-valued approach are likely to arose out of the behaviour of the sentential connectives of our many valued logic. But also another source can be imagined:

If we try to define from a given predicate symbol $P \in \mathbb{P}$ with arity $n = \alpha_{\mathbb{P}}(P)$ a new $(n-1)$-ary one Q by identification of two of the variables, e.g. as

$$Q(x_1,\dots,x_{n-1}) =_{df} P(x_1,\dots,x_{n-1},x_{n-1})$$

then the function $g_Q : A^{n-1} \longrightarrow W_m$ describing the interpretation of the predicate symbol Q may for a considered identity interpretation $\mathcal{A}$ with support A fail to satisfy condition (6) in the definition of identity interpretation.

To surmount this difficulty it seems appropriate again to change (I3). The leading point comes from (I3*). For every predicate symbol $P \in \mathbb{P}$ we suppose to be given not only its arity $n = \alpha_{\mathbb{P}}(P)$, but also a n-tuple $(v_1^P,\dots,v_n^P)$ and consider the axiom schema

$$\text{(I4)} \quad (\forall)((x \equiv y)^{v_i^P} \;\&\; P(x_1..x..x_n) \longrightarrow P(x_1..y..x_n))$$

with x resp. y in the i-th place of P and $(x \equiv y)^{v_i^P}$ as defined in (+).

Condition (6) for identity interpretations was the essential one for having schema (I3) valid for all identity interpretations. Therefore with (I3) we now have to change this condition also.

The new formulation needed is easiest to describe in analogy to (I4). First we remember that the truth function et associated with connective & was given by

$$et(s,t) = \max\{0\ ,\ s+t-1\}.$$

The finite iteration of this function needed for describing the truth value H^k as a function of the truth value of H is given by

$$konj_k(s) = \max\{0,\ k\cdot s - (k-1)\}.$$

By an <u>extended identity interpretation</u> we now understand a $\mathbf{L}_m$-interpretation which satisfies (1),...,(5) as before and consists also of

(7) an $\alpha_{\mathbb{P}}(P)$-ary function g_P from A into W_m and an $\alpha_{\mathbb{P}}(P)$-tuple $(v_1^P,\dots,v_n^P)$, $n = \alpha_{\mathbb{P}}(P)$, for every predicate symbol $P \in \mathbb{P}$, which for each $1 \leq i \leq \alpha_{\mathbb{P}}(P)$ with respect to the i-th argument has the property

$$konj_{v_i^P}(i(a,b)) \leq seq(g_P(..,a,..),g_P(..,b,..))$$

for all individuals a,b.

As in the last section now we are able to prove the completeness of the logical calculus given by Ax + (I1) + (I2) + (I4) and Abl-derivability with respect to the class of all extended identity interpretations.

The way of proof and most of the details are as in the last section. Again we omit the details.

A LOOK AT FUZZY SETS

As a starting point for the problem discussed in this paper we mentioned work on fuzzy sets. Therefore now we will have a look back at this topic.

Our main concern is the system developed e.g. in [1], [2]. As in these papers and also by Morgan/Pelletier [6] we enlarge the usual notion of fuzzy set of Zadeh [11] in such a way that we distinguish systems of fuzzy sets with respect to each one of our sets W_m and of [0,1] as sets of generalized membership values.

Zadeh [11] and most papers on fuzziness consider only [0,1] as set of generalized membership values. For this, the infinite valued logic $\mathbb{L}_{\aleph_1}$ is providing the appropriate language, treating the generalized membership values of fuzzy sets a generalized truth values of a many valued membership relation. But the set of valid sentences of $\mathbb{L}_{\aleph_1}$ is not axiomatizable and therefore the formal treatment of the theory of that kind of fuzzy sets has to be in some sense incomplete.

Yet for most applications a finite set of generalized membership values, i.e. a system of fuzzy sets developed in one of the systems $\mathbb{L}_m$ will suffice. This was argued for by Morgan/Palletier [6] and is in accordance with points of view of numerical mathematics and computer science. From this point of view our considerations on identity logic $\mathbb{L}_m$ are related to fuzzy sets theory.

Such ideas have been leading ones already in [1], [2]. The resulting system was a cumulative structure of fuzzy sets which represents an identity interpretation - with a proper class as support, but this does not matter here - for a $\mathbb{L}_m$-language with two binary many valued predicates: g for the membership relation and $\subset$ for rank relation. ($\subset$ is de facto only binary).

If we start with the corresponding $\mathbb{L}_m$-language with identity symbol we have with respect to this intrepretation as basic results:

$$\models a \equiv b \rightarrow (c \in a \rightarrow c \in b)$$
$$\models a \equiv b \rightarrow (a \in c \rightarrow b \in c)$$
$$\models a \equiv b \rightarrow (a \subset c \rightarrow b \subset c)$$
$$\models a \equiv b \rightarrow (c \subset a \rightarrow c \subset b) .$$

Extensionality becomes the property

$$\models \forall x(a \in x \rightarrow b \in x) \rightarrow a \equiv b ,$$

the separation axiom now is the schema

$$\models \forall a \exists b \forall x(x \in b \leftrightarrow x \in a \wedge H(x))$$

with

$$H_1 \wedge H_2 =_{df} H_1 \,\&\, (\neg H_1 \,\&\, H_2)$$

as the "usual" min-conjunction and $H(x)$ anyone of the possible wff of our language with the essential collecting property

$$x \equiv y \,\&\, H(x) \rightarrow H(y).$$

And replacement can be formulated as the schema

$$\models \forall x \forall y \forall z(F(x,y) \wedge F(x,z) \rightarrow y \equiv z)$$
$$\rightarrow \forall a \exists b \forall y(y \in b \leftrightarrow \exists x(x \in a \wedge F(x,y)))$$

for every wff $F(x,y)$ such that

$$\models x \equiv y \,\&\, F(z,y) \rightarrow F(z,x).$$

Furthermore, the empty set axiom becomes

$$\models \exists a \forall x(\neg x \in a) ,$$

and the pairing axiom is valid as

$$\models \forall a \forall b \exists c \forall x(x \in c \leftrightarrow (x \equiv a \vee x \equiv b))$$

with max-disjunction $\vee$ defined e.g. by

$$H_1 \vee H_2 =_{df} \neg(\neg H_1 \wedge \neg H_2).$$

Many valued generalizations of the axioms of union and power set are valid in this interpretation consisting of a cumulative universe of fuzzy sets e.g. in the following forms (cf. [1]):

$$\models \forall a \exists b \forall x(x \in b \leftrightarrow H(a,x))$$

with for the axiom of union

$$H(a,x) : \exists z(x \in z \,\&\, z \in a)$$

or also the corresponding wff with $\wedge$ instead of &, with for the axiom of power set

$$H(a,x) : \forall y(y \in x \rightarrow y \in a) \ \& \ x \subset a$$

in case of nonexistence of urelements in the intrepretation and otherwise with

$$H(a,x) : \forall y(y \in x \rightarrow y \in a) \ \& \ x \subset a \ \& \ \exists z(z \sqsubset a)$$

using

$$z \sqsubset a =_{df} z \subset a \ \& \ \neg a \subset z.$$

The axiom of infinity, finally, is valid e.g. in the form

$$\models \exists a(\exists x(x \in a) \ \& \ \forall x(x \in a \rightarrow \exists y(y \in a \ \& \ x \sqsubset y))).$$

The validity of all these many valued versions of ZF-axioms for the cumulative system of fuzzy sets of [1], [2] is supporting the authors claim that the many valued predicate calculi with identity considered here are, apart from the interest they have in themselves, useful for general considerations of fuzzy sets theory.

REFERENCES

1. S. Gottwald, A cumulative system of fuzzy sets, in: Set Theory Hierarchy Theory (W. Marek et al. eds.), Lect. Notes Math., Berlin-Heidelberg-New York, vol. 537, 1976, pp. 109-119.

2. S. Gottwald, Set theory for fuzzy sets of higher level, Fuzzy Sets Syst. 2 (1979), 125-151.

3. S. Gottwald, Fuzzy uniqueness of fuzzy mappings, Fuzzy Sets Syst. 3 (1980), 49-74.

4. Ch. G. Morgan, A theory of equality for a class of many-valued predicate calculi, Ztschr. math. Logik Grundl. Math. 20 (1974), 427-432.

5. Ch. G. Morgan, Similarity as a theory of graded equality for a class of many-valued predicate calculi, in: Proc. Intern. Symp. Multiple-Valued Logic, Bloomington, Ind., 1975, New York 1975, pp. 436-449.

6. Ch. G. Morgan and F.J. Pelletier, Some notes concerning fuzzy logics, Linguist. Philos. 1 (1977), 79-97.

7. N. Rescher, Many-Valued Logic, New York etc. 1969.

8. J.B. Rosser and A.R. Turquette, Many-Valued Logics, Amsterdam 1952.

9. B. Scarpellini, Die Nichtaxiomatisierbarkeit des unendlichwertigen Prädikatenkalküls von Łukasiewicz, Journal Symbolic Logic 27 (1962), 159-170.

10. H. Thiele, Theorie der endlichwertigen Łukasiewiczschen Prädikatenkalküle der ersten Stufe, Ztschr. math. Logik Grundl. Math. 4 (1958), 108-142.

11. L.A. Zadeh, Fuzzy sets, Information and Control 8 (1965), 338-353.

UNE CLASSE D'ANNEAUX MONADIQUES SANS CONSTANTE

Marcel Guillaume/Departément de Mathématiques Pures, Université de Clermont-Ferrand II, Clermont-Ferrand, France

Théorème: Si X est un ensemble infini et si A est une sous-algèbre booléienne de parties de X, à laquelle appartiennent toutes les parties dénombrables (infinies ou non) de X, l'algèbre monadique libre sur A n'a pas de constante.

Nous empruntons notre terminologie à Halmos (par exemple [1]) et Jurie ([2]). Ce dernier a défini l'algèbre monadique libre sur une algèbre de Boole A comme (à isomorphisme près) triplet (S, ∃, u) tel que S est une algèbre de Boole, ∃ un quantificateur ([1]) sur S et u un homomorphisme de A vers S, possédant la propriété que pour tout autre triplet (S', ∃', u') il existe un unique homomorphisme monadique f de S vers S', tel que fu = u'.

Dans le même travail, Jurie a montré que la construction de l'algèbre monadique libre sur A pouvait s'effectuer en deux temps.

Le premier temps consiste dans le construction de ce qu'il a nommé l'hémimorphisme ([1]) unitaire libre sur A. A isomorphisme près, celui-ci est, comme hémimorphisme, application h de A vers une algèbre de Boole C, telle que h(0) = 0 et h(p v q) = h(p) v h(q) pour tous p,q ∈ A, comme unitaire ([2]), tel que h(1) = 1, et comme libre ([2]), tel que pour tout couple (C',h') composé d'une algèbre de Boole C' et d'un hémimorphisme unitaire h' de A vers C', il existe un unique homomorphisme f de C vers C' tel que $fh = h'$.

Jurie a établi ([2]) que C est alors nécessairement engendrée par l'image de h, et produit une représentation de h, lorsque A est atomique. Si X est l'ensemble des atomes de A - et pour $a \in A$, si $at(a)$ est l'ensemble des atomes $x \in X$ tels que $x \leq a$ - si Y est l'ensemble des parties finies non vides de X, l'application h de A vers l'algèbre $P(Y)$ des parties de Y, définie pour tout $a \in A$ par

$$h(a) = \{J \in Y | J \cap at(a) \neq \phi\}$$

est un hémimorphisme unitaire complet.

Si on lui donne pour but la sous-algèbre C_A engendrée par $h[A]$ dans $P(Y)$, c'est (à isomorphisme près) l'hémimorphisme unitaire libre sur A.

Les éléments de C_A sont des réunions (bornes supérieures) finies d'intersections (bornes inférieures) finies d'éléments de $h[A]$. Ces intersections sont de la forme $h(a_1) \cap \ldots \cap h(a_n) \cap (Y - h(a))$, où $a_1,\ldots,a_n$, $a \in A$ et où $Y - b$ dénote le complémentaire de b.

Remarque: $h(a_1),\ldots,h(a_n)$ et $Y - h(a)$ ne sont pas des parties quelconques de Y; ce sont des intervalles "généralisés" pour l'ordre induit sur Y par l'inclusion. Nous entendons par là que si $J, K, L \in Y$ sont tels que $J \subset L \subset K$, et si $J \cap at(a) \neq \phi \neq K \cap at(a)$, on a aussi $L \cap at(a) \neq \phi$ (la seconde condition est même superflue); et si $J \cap at(a) = \phi = K \cap at(a)$, on a aussi $L \cap at(a) = \phi$ (la première condition est superflue); or il est clair que toute famille d'intervalles a pour intersection un intervalle (peut-être vide).

Ainsi C_A se compose de réunions finies d'intervalles de Y.

Si $c \in P(Y)$ nous dirons d'une chaîne strictement croissante $J_o \subset J_1 \subset \ldots \subset J_{2p}$ d'éléments de Y telle que $J_o,J_2,\ldots,J_{2p} \in c$ et $J_1,J_3,\ldots,J_{2p-1} \notin c$, qu'elle est caractéristique de c, et qu'elle est - comme telle - de longueur $p+1$. Si l'on définit la longueur de c comme égale au maximum des longueurs de ses chaînes caractéristiques, nous voyons que toute $c \in C_A$ est de longueur finie. □

Revenant à la construction de l'algèbre monadique libre sur A, Jurie a montré dans [2] qu'on l'obtenait (sans hypothèse d'atomicité sur A) à partir de l'hémimorphisme unitaire libre $h : A \longrightarrow C$ sur A en construisant l'algèbre monadique libre sur h.

Il a défini au même endroit l'algèbre monadique libre sur un hémimorphisme h d'une algèbre de Boole A vers une algèbre de Boole C comme (à isomorphisme près) quadruplet $(S, \exists, u, v)$, tel que S est une algèbre de Boole, $\exists$ un quantificateur sur S, et u et v des homomorphismes (booléiens), respectivement, de A vers S et de C vers S, tels que $v[C] \subset \exists[S]$, $\exists u = vh$, et que pour tout autre quadruplet analogue $(S', \exists', u', v')$ il existe un unique homomorphisme monadique f de S vers S' tel que $fu = u'$ et $fv = v'$.

Dans le même travail Jurie a établi que S est alors nécessairement engendrée par la réunion des images de u et de v - cette dernière coïncidant alors avec celle du quantificateur - que v est injectif si et seulement si h est unitaire - et que $u^{-1}(0) = h^{-1}(0)$.

Si h est l'hémimorphisme unitaire libre sur A, et $(S, \exists, u, v)$ l'algèbre monadique libre sur h, le triplet $(S, \exists, u)$ est algèbre monadique libre sur A ([2]).

Nous sommes alors dans un cas où v est injectif, et si A est atomique, comme on voit aussitôt que $h(a) = \{J \in Y | J \cap at(a) \neq \phi\} = \phi$ si et seulement si $a = 0$, l'homomorphisme u est injectif aussi. Jurie a également donné dans [2] une représentation de $(S, \exists, u)$, mais nous n'en aurons pas besoin dans ce qui suit, où nous exploiterons le résultat suivant de [2] :

Si $h : A \longrightarrow C$ est un hémimorphisme unitaire de l'algèbre de Boole A vers l'algèbre de Boole C, et $(S, \exists, u, v)$, algèbre monadique libre sur h, il existe une bijection canonique de l'ensemble des constantes de l'algèbre monadique $(S, \exists)$ sur l'ensemble des homomorphismes f de A vers C tels que $f \leq h$ (c'est-à-dire tels que $f(a) \leq h(a)$ pour tout $a \in A$).

Nous sommes dans cette situation lorsque h est hémimorphisme unitaire libre sur A, et $(S, \exists, u, v)$ algèbre monadique libre sur h; et donc $(S, \exists, u)$, algèbre monadique libre sur A. Nous aurons démontré le théorème annoncé au début de ce travail lorsque nous aurons montré que, dans les conditions de son énoncé, il ne peut exister d'homomorphisme f de A vers C_A, tel que l'on ait $f \leq h$.

Dans ce qui suit, nous supposons que X est un ensemble infini, que Y est l'ensemble des parties finies non vides de X, et que $A = P(X)$; si $x \in X$ nous assimilons l'atome $\{x\}$ à x, et si $a \subset X$, nous assimilons $at(a)$ à a. On a de la sorte

$h(a) = \{J \in Y | a \cap J \neq \phi\}$ pour tout $a \in A$.

Il nous sera utile d'introduire

$h^o(a) = y - h(x - a) = \{J \in Y | J \subset a\}$.

Si f est un homomorphisme de A vers $P(Y)$ on a $f \leq h$ si et seulement si on a $h^o \leq f$.

<u>Lemme</u>: Il y a une bijection canonique entre les homomorphismes f de A vers $P(Y)$, tels que $h^o \leq f$, et les fonctions de choix sur Y: tout homomorphisme tel est induit ([3]) par une fonction de choix.

<u>Démonstration</u>: Si f est un homomorphisme de l'espèce, si $J = \{x_1,\ldots,x_k\} \in Y$, comme on a $J \subset J$, on a

$J \in h^o(J) \subset f(J) = f(x_1) \cup \ldots \cup f(x_k)$

- ce qui montre, en passant, que f est injectif - et il y a un $x_J \in J$ et un seul, tel que $J \in f(x_J)$, parce que les $f(x)$ pour $x \in X$ sont deux à deux disjoints, et forment une partition de Y.

Puisqu'on a $J \in f(x_J)$ pour tout $J \in Y$, si χ dénote la fonction de choix $J \longrightarrow x_J$, on a $J \in \chi^{-1}(x)$ si et seulement si $x = x_J$ si et seulement si $J \in f(x)$, et donc $f(x) = \chi^{-1}(x)$, pour tout $x \in X$ - ce qui montre, en passant, que f est complet.

Réciproquement si $\chi : Y \longrightarrow X$ est une fonction de choix, $f = \chi^{-1}$ est un homomorphisme tel que pour tout $J \in Y$ on a $J \in f(x)$ si et seulement si $\chi(J) = x$, ce qui exige $x \in J$ et donc

$f(x) \subset h(x) = \{J \in Y | x \in J\}$ pour tout $x \in X$,

d'où résulte que pour tout $a \in A$, on a $f(a) \subset h(a)$, puisque f et h sont l'un et l'autre complets, c'est-à-dire respectent toutes les bornes supérieures. □

Soit $\Phi_o = (J_o, J_1 \ldots, J_p, \ldots)$ une chaîne infinie strictement croissante (pour l'inclusion) d'éléments de Y, et $\chi : Y \longrightarrow X$ une fonction de choix et $f = \chi^{-1}$ un homomorphisme associés.

Il est possible que $\chi[\Phi_o]$ soit un ensemble infini. Un exemple en est fourni lorsque $X = N$ par l'application $\chi_M : J \longrightarrow \text{Max } J$; on voit facilement que pour aucune chaîne Φ_o l'ensemble $\chi_M[\Phi_o]$ ne saurait être infini.

Nous procédons maintenant à la démonstration de notre théorème.

Si $\chi[\Phi_o]$ est infini, et si nous posons $x_p = \chi(J_p)$ pour tout $p \in N$, nous aurons $J_p \in f(x_p)$ pour tout $p \in N$, mais l'énumération

$\chi[\Phi_o] = \{x_o, x_1, \ldots, x_p, \ldots\}$

comportera des répétitions.

Soit $i_o = 0 < i_1 < \ldots < i_p < \ldots$ une suite infinie croisssante d'entiers telle que l'énumération

$\chi[\Phi_o] = \{x_{i_o}, x_{i_1}, \ldots, x_{i_p}, \ldots\}$

ne comporte plus de répétition; il est facile de déterminer une telle suite.

On aura $J_{i_o} \subset J_{i_1} \subset \ldots \subset J_{i_p} \subset \ldots$, et si

$a = \{x_{i_o}, x_{i_2}, \ldots, x_{i_{2p}}, \ldots\} \in A$

on aura $J_{i_o}, J_{i_2}, \ldots, J_{i_{2p}}, \ldots \in f(a)$, et

$J_{i_1}, J_{i_3}, \ldots, J_{i_{2p+1}}, \ldots \notin f(a)$,

de telle sorte que $f(a) \notin C_A$ puisque la longueur de $f(a)$ n'est pas finie: la présence de a dans A exclut que l'image de f soit incluse dans C_A.

Si $\chi[\Phi_o]$ est fini, il existe un x_o, et une suite infinie Φ_o^1 extraite de Φ_o (par exemple $\Phi_o \cap f(x_o)$), telle que $\Phi_o^1 \subset f(x_o)$ et $\{x_o\} \notin \Phi_o^1$.

Considérons alors $\Phi_1 = \Phi_o^1 - \{x_o\}$ (qui se compose des éléments de Y obtenus de ceux de Φ_o^1 en leur enlevant x_o).

Si $\chi[\Phi_1]$ est infini, on raisonne sur Φ_1 comme plus haut sur Φ_o.

Si $\chi[\Phi_1]$ est fini, il existe un $x_1 \neq x_o$, et une suite infinie Φ_1^2 extraite de Φ_1 (par exemple $\Phi_1 \cap f(x_1)$) telle que $\Phi_1^2 \subset f(x_1)$ et $\{x_1\} \notin \Phi_1^2$.

Nous définissons alors, pour un raisonnement ultérieur:
Φ_o^2 comme la suite extraite de Φ_o^1, telle que $\Phi_1^2 = \Phi_o^2 - \{x_o\}$ et $\{x_o, x_1\} \notin \Phi_o^2$: on a $\Phi_1^2 \subset f(x_o)$;
$\Phi_2 = \Phi_1^2 - \{x_1\} = \Phi_o^2 - \{x_o, x_1\}$.

Si $\chi[\Phi_2]$ est infini, on raisonne sur Φ_2 comme plus haut sur Φ_o.

Si $\chi[\Phi_2]$ est fini, il existe un $x_2 \neq x_o$, $x_2 \neq x_1$, et une suite infinie Φ_2^3 extraite de Φ_2 (par exemple $\Phi_2 \cap f(x_2)$), telle que

$\Phi_2^3 \subset f(x_2)$ et $\{x_2\} \notin \Phi_2^3$.

Nous définirons alors:

Φ_1^3 , extraite de Φ_1^2, telle que $\Phi_2^2 = \Phi_1^3 - \{x_1\}$ et $\{x_1,x_2\} \notin \Phi_1^3$: on a $\Phi_1^3 \subset f(x_1)$;

Φ_o^3 , extraite de Φ_o^2, telle que $\Phi_1^3 = \Phi_o^3 - \{x_o\}$, donc telle que $\Phi_2^3 = \Phi_o^3 - \{x_o , x_1\}$ et $\{x_o , x_1 , x_3\} \notin \Phi_o^3$: on a $\Phi_o^3 \subset f(x_o)$;

$\Phi_3 = \Phi_2^3 - \{x_2\} = \Phi_1^3 - \{x_1 , x_2\} = \Phi_o^3 - \{x_o , x_1 , x_2\}$,

et raisonnerons comme auparavant sur Φ_o ; et ainsi de suite, par une récurrence que l'on pourra facilement formaliser si nécessaire.

Ou bien à l'une des étapes de cette récurrence, on aura atteint Φ_p telle que $X[\Phi_p]$ soit infini et introduit à partir d'elle un $a \in A$ tel que $f(a)$ soit de longueur infinie, donc tel que $f(a) \notin C_A$. Ou bien - comme l'exemple du minimum sur N montre que cela peut se produire - on aura obtenu une suite infinie de Φ_p telle que tous les $X[\Phi_p]$ soient finis, et une suite infinie sans répétition $(x_n)_{n \in N}$ d'éléments de X tels que pour tous $n,p \in N$ tels que $p \leq n$, on ait $\Phi_p^n \subset f(x_p)$.

Considérons maintenant $a = \{x_o , x_1,\ldots, x_{2p},\ldots\}$ et $q \in N$. Le $(2q + 2)^e$ élément K_{2q+2} de Φ_o^{2q+2} a ses termes dans $f(x_o)$, on a $K_{2q+2} \in f(a)$.

Soit $K_{2q+1} = K_{2q+2} - \{x_o\}$: il appartient à Φ_1^{2q+2}, inclus dans $f(x_1)$; on a donc $K_{2q+1} \notin f(a)$.

Soit $K_{2q} = K_{2q+1} - \{x_1\} = K_{2q+2} - \{x_o , x_1\}$: il appartient à Φ_2^{2q+2} , inclus dans $f(x_2)$; on a donc $K_{2q} \in f(a)$. Etc...

Pour $p \leq 2q+2$ nous aurons $K_{2q+2-p}=K_{2q+2}-\{x_o,x_1,\ldots,x_{p-1}\} \in \Phi_p^{2q+2}$, inclus dans $f(x_p)$: on a donc $K_{2q+2-p} \in f(a)$ si p est pair, et $K_{2q+2-p} \notin f(a)$ si p est impair.

Ainsi, la chaîne $K_o \subset K_1 \subset K_2 \subset \ldots \subset K_{2q+2}$ est caractéristique de $f(a)$, et de longueur $q + 2$.

L'ensemble des longueurs des chaînes caractéristiques de $f(a)$ n'est pas borné; la longueur de $f(a)$ est donc infinie; à nouveau nous obtenons que $f(a) \notin C_A$.

Ceci achève de montrer qu'aucun homomorphisme f de A dans P(Y) tel que $h^{o} \leq f \leq h$ n'a son image incluse dans C_A.

Comme les $a \in A$ tels que $f(a) \notin C_A$ construits précédemment à partir de la fonction de choix associée à f ont toujours été dénombrables, nous pouvons tenir le même raisonnement en remplaçant A par un sous-algèbre de P(X) comportant toutes les parties dénombrables de A: nous "appauvrirons" C_A en éléments, mais garderons dans A des éléments qui "font sortir de C_A" l'image de tout homomorphisme de A vers P(Y) tel que $f \leq h$. Ainsi notre théorème est-il complètement établi.

BIBLIOGRAPHIE

1. P.R. Halmos, Algebraic Logic I. Monadic Boolean algebras. Compositio Mathematica 12, 1955, pp. 217-249.
2. P.F. Jurie, Anneaux monadiques libres, hémimorphismes unitaires libres. Bull. Soc. Math. France 98, 1970, pp. 36-46.
3. R. Sikorski, Boolean Algebras, Ergeb. der Math. und ihrer Grenzgeb. Neue Folge, Band 25, Springer, Berlin - Göttingen - Heidelberg - New York 2e éd., 1964, x + 237 pp.

ELEMENTARY EXTENSION AND THE HYPERREAL NUMBERS

William S. Hatcher/Department of Mathematics, Laval University, Quebec, Canada

ELEMENTARY EXTENSION REVISITED

Let τ be a type of finitary operations and relations, and let L be the first-order language with equality of type τ. We identity constant symbols with nullary operation symbols. Let g be an interpretation of L in some nonempty set A, defining thereby a structure $\mathcal{A} = \langle A,g \rangle$ of type τ. Let Z be any formula of L all of whose free and bound variables occur in the list $x_1,\ldots,x_n$, $n \geq 1$ (we say that n is <u>appropriate</u> for Z). Then, we let $g_n(Z) = \{\langle a_1,\ldots,a_n \rangle \in A^n \mid \langle a_1,\ldots,a_n \rangle \models_{\mathcal{A}} Z\}$. Where n is understood, we write simply g(Z). We define, in a similar manner, the function $g_n(t) : A^n \longrightarrow A$ where t is any term of L all of whose variablès occur in the list $x_1,\ldots,x_n$, $n \geq 1$. Again, we write g(t) when n is understood (and <u>appropriate</u>). Obviously, $g_n(t) = g_{n+1}(t = x_{n+1})$ whenever n is appropriate for the term t.

The sets of the form g(Z) for Z some formula of L are, of course, just the g-definable sets (over the base set A). We let $\mathcal{D}(A)$ represent the collection of all such sets. If, further, we let S denote the set of formulas of L and $S_n \subset S$ the subset of formulas Z for which n is appropriate, $n \geq 1$, then $g_n : S_n \longrightarrow\!\!\!\!\rightarrow \mathcal{D}(A^n)$, where $\mathcal{D}(A^n) \subset \mathcal{P}(A^n)$ is the set of g-definable subsets of A^n. Thus, $\mathcal{D}(A) = \bigcup_{n=1}^{\infty} \mathcal{D}(A^n)$. In particular, the interpretations of the non-logical symbols of L are all definable sets since $g(K) = g_n(Kx_1 \ldots x_n)$ where K is any n-ary predicate symbol of L, $n \geq 1$, and $g(f) = g_{n+1}(fx_1 \ldots x_n = x_{n+1})$ where f is any n-ary operational symbol of L, $n \geq 0$.[1] Also, for f an n-ary operational symbol of L, $n \geq 1$, $g(f) = g_n(t)$ where t is the term $fx_1 \ldots x_n$.

Any formula Z or term t of L has a smallest appropriate $n \geq 1$. We call n the <u>index</u> of Z or of t. If n is the index of Z, then for all $m \geq n$, and for all $\langle a_1,\ldots,a_m \rangle \in A^m$, $\langle a_1,\ldots,a_m \rangle \models_{\mathcal{A}} Z$ iff $\langle a_1,\ldots,a_n \rangle \models_{\mathcal{A}} Z$. Similarly, for n the index of the term t of L, for $m \geq n$, and for $\langle a_1,\ldots,a_m \rangle \in A^m$, $g_m(t)(a_1,\ldots,a_m) = g_n(t)(a_1,\ldots,a_n)$.

Suppose, now, that $\mathcal{B} = \langle B, g^* \rangle$ is a structure of type τ defined by another interpretation g^* of L in the nonempty set B. To say that $\mathcal{B}$ is an elementary extension of $\mathcal{A}$, $\mathcal{A} < \mathcal{B}$, is to say that $A \subset B$ and that, for every formula Z of L, for every n which is appropriate for Z, and for every sequence $\vec{a} \in A^n$, $\vec{a} \models_{\mathcal{A}} Z$ if and only if $\vec{a} \models_{\mathcal{B}} Z$. When $A \subset B$, this latter condition is equivalent to the condition $g^*(Z) \cap A^n = g(Z)$ since $\vec{a} \models_{\mathcal{A}} Z$ iff $\vec{a} \in g(Z)$ and $\vec{a} \models_{\mathcal{B}} Z$ iff $\vec{a} \in g^*(Z)$. Moreover, since A and B are g-definable and g^*-definable, respectively, by any logically true sentence Z of L, the condition $g^*(Z) \cap A^n = g(Z)$ for all Z and appropriate n implies that $A \subset B$.

In other words, to say that $\mathcal{A} < \mathcal{B}$ is to say that, for all formulas Z of L and all appropriate n, $g^*(Z) \subset B^n$ is a superset of $g(Z)$ which is identical with $g(Z)$ on A^n. We say that $g^*(Z)$ <u>extends</u> $g(Z)$. In fact, it is obviously sufficient that $g_n^*(Z)$ extend $g_n(Z)$ where n is the index of Z. This gives rise to the following proposition:

<u>Proposition 1</u>. Let L be any first-order language with equality of type τ and let $\mathcal{A} = \langle A, g \rangle$ and $\mathcal{B} = \langle B, g^* \rangle$ be two structures of type τ deriving from interpretations g and g^* of L. Then $\mathcal{A} < \mathcal{B}$ iff, for every formula Z of L, $g_n^*(Z) \cap A^n = g_n(Z)$, where n is the index of Z.

Proof. Where $m \geq n$ is appropriate for a formula Z of L, and for $\langle a_1,\ldots,a_m \rangle \in A^m$, $\langle a_1,\ldots,a_m \rangle \models_{\mathcal{A}} Z$ iff $\langle a_1,\ldots,a_n \rangle \models_{\mathcal{A}} Z$ iff $\langle a_1,\ldots,a_n \rangle \models_{\mathcal{B}} Z$ iff $\langle a_1,\ldots,a_m \rangle \models_{\mathcal{B}} Z$, whenever the condition of the proposition is satisfied.

Given two structures $\mathcal{A}$ and $\mathcal{B}$ as above, if $\mathcal{A} < \mathcal{B}$, then for each $n \geq 1$ we can define a function $*n : \mathcal{D}(A^n) \longrightarrow \mathcal{D}(B^n)$ by the rule $*n(g_n(Z)) = g_n^*(Z).*n$ is well-defined on $\mathcal{D}(A^n)$ since $g_n^*(Z) = g_n^*(Y)$ whenever $g_n(Z) = g_n(Y)$ as follows from the fact that $\mathcal{A} < \mathcal{B}$. This gives the following commutative diagram for each $n \geq 1$, where each of the functions is surjective:

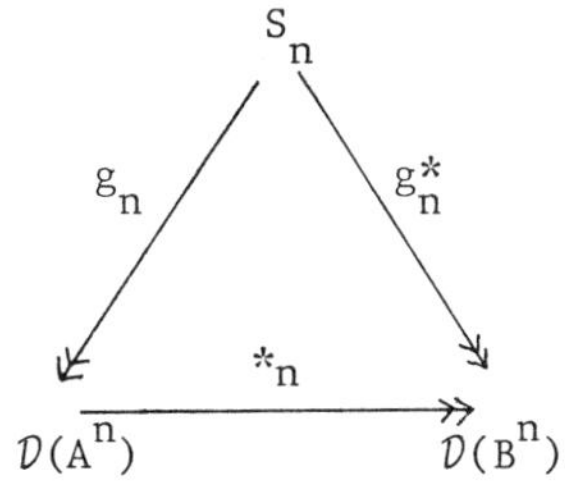

It is also useful to consider, for each $n \geq 1$, the set $I_n \subset S$ of formulas Z of L for which n is the index. Obviously, $S_1 = I_1$ and $I_{n+1} = S_{n+1} - S_n$ for $n \geq 1$. Since $I_n \subset S_n$, we notationally confuse each of g_n and g_n^* and its respective restriction to I_n. We thus also have the commutative diagram:

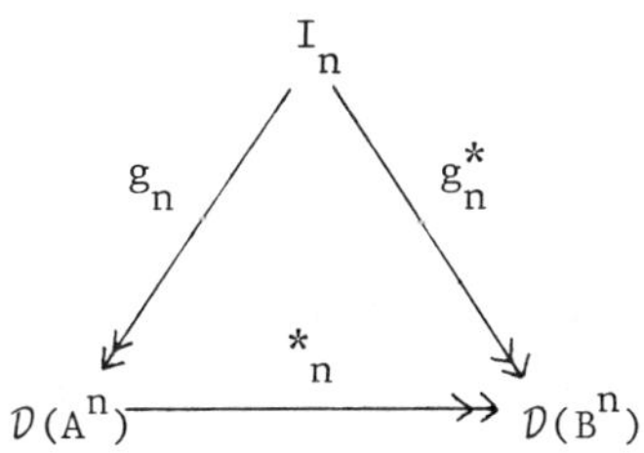

Even though $I_n \neq S_n$ in general, the restrictions of g_n and g_n^* to I_n are still surjective since every formula $Z \in S_n$ is logically equivalent to the formula $(Z \wedge x_n = x_n) \in I_n$ which, of course, defines the same relation on A^n as does Z.

The sets $\mathcal{D}(A^n)$ and $\mathcal{D}(A^m)$ are disjoint whenever $n \neq m$. Thus, any collection of functions $*n : \mathcal{D}(A^n) \longrightarrow \mathcal{D}(B^n)$ defines a single function $* : \mathcal{D}(\mathcal{A}) \longrightarrow \mathcal{D}(\mathcal{B})$ whose restriction to $\mathcal{D}(A^n)$ is $*n$. Conversely, any single arity-preserving function $* : \mathcal{D}(\mathcal{A}) \longrightarrow \mathcal{D}(\mathcal{B})$ defines, by restriction, a collection of functions $*n : \mathcal{D}(A^n) \longrightarrow \mathcal{D}(B^n)$. From now on, whenever we deal with an arity-preserving function $* : \mathcal{D}(\mathcal{A}) \longrightarrow \mathcal{D}(\mathcal{B})$, we will drop the subscript n when restricting $*$ to $\mathcal{D}(A^n)$, and we will write $g_n(Z)*$ (or $g(Z)*$) instead of $*(g_n(Z))$ (respectively, $*(g(Z))$) when applying $*$ to the argument $g_n(Z)$ (respectively, $g(Z)$). We also suppose that all $* : \mathcal{D}(A) \longrightarrow \mathcal{D}(B)$ considered are arity-preserving.

Suppose, then, that we are given two structures A and B and an arity-preserving $* : \mathcal{D}(A) \longrightarrow \mathcal{D}(B)$ as above. Under what conditions will $*$ determine B as an elementary extension of A? Clearly it will if, for each $Z \in I_n$, $g_n(Z)*$ extends $g_n(Z)$ and if, further, $*g_n = g_n^*$ (i.e., the second diagram above commutes), because these two conditions immediately give us the condition of Proposition 1. On the other hand, we have already seen above that the existence of a function $*$ satisfying these two conditions follows from the condition of Proposition 1. We thus have:

Proposition 2. Given L, A, and B as above, then $A \prec B$ iff (1) there is an arity-preserving function $* : \mathcal{D}(A) \longrightarrow \mathcal{D}(B)$; (2) for each definable set $X \subset A^n$, $X* \cap A^n = X$ (i.e., $X*$ extends X); (3) for all $n \geq 1$, $*g_n = g_n^*$ where g_n and g_n^* are regarded as having domain I_n.

Notice that, since g_n is surjective, the conjunction of conditions (1) and (3) of Proposition 2 is equivalent to the single condition that $\ker(g_n) \subset \ker(g_n^*)$, i.e., that $g_n^*(Z) = g_n^*(Y)$ whenever $g_n(Z) = g_n(Y)$.

AN ALGEBRAIC CHARACTERIZATION OF ELEMENTARY EXTENSION

Suppose, now, that we are given L, A, B and a function $*$ satisfying conditions (1) and (2) of Proposition 2. Then a necessary and sufficient condition that the function $*$ determine B as an elementary extension of A is that, for all $Z \in I_n$, $g_n(Z)* = g_n^*(Z)$, in other words, that $*$ satisfy condition (3) of Proposition 2. The two sets $\mathcal{D}(A)$ and $\mathcal{D}(B)$ have a rich algebraic structure. For example, for each $n \geq 1$, $\mathcal{D}(A^n)$ and $\mathcal{D}(B^n)$ are closed under set union, intersection, and complement, making $\mathcal{D}(A^n)$ (respectively, $\mathcal{D}(B^n)$) a Boolean subalgebra of $\mathcal{P}(A^n)$ (respectively, of $\mathcal{P}(B^n)$). Similarly, the operational part of $\mathcal{D}(A)$ (respectively, of $\mathcal{D}(B)$) contains each of the standard projections on A^n (respectively, on B^n), $n \geq 1$, and is closed under composition of operations, making it a subclone of the full clone $\mathcal{O}(A)$ (respectively $\mathcal{O}(B)$) of all operations on A (respectively, on B). What we now establish is that, for $*$ to satisfy condition (3) of Proposition (2), it is necessary and sufficient that $*$ preserve certain parts of the algebraic structure from $\mathcal{D}(A)$ to $\mathcal{D}(B)$.

Theorem 1. Given the language L and the structures A and B as above, and given $*$ satisfying conditions (1) and (2) of Proposition 2, suppose that for each operation symbol f and each relation symbol K of L, $g(f)* = g*(f)$[2] and $g(K)* = g*(K)$. Then, a necessary and sufficient condition that $*$ determine B as an elementary extension of A is that $*$ preserve the following data: (1) All standard projections π_i^n, $n \geq 1$; (2) all n-ary constant operations, $n \geq 1$, definable by constant symbols of L; (3) composition of term-defined operations; (4) equalizers (difference kernels) of n-ary term-defined operations, $n \geq 1$; (5) the Boolean operations from $\mathcal{D}(A^n)$ to $\mathcal{D}(B^n)$ for each $n \geq 1$; (6) Cylindrifications.

Proof. Necessity is straighforward, so we concentrate on sufficiency. We prove by induction on formulas that the conjunction of conditions (1) to (6) above implies that $*$ satisfies conditions (3) of of Proposition 2.

First, we observe that every atomic formula $Kt_1 \ldots t_n$ of L where the t_i are terms is logically equivalent to the following formula Z:

$$(\exists x_{m+1})\ldots(\exists x_{m+n})\left(\bigwedge_{i=1}^{n}(t_i = x_{m+i}) \wedge (\exists x_1)\ldots(\exists x_n)(Kx_1\ldots x_n \bigwedge_{i=1}^{n}(x_i = x_{m+i}))\right),$$

where m is the maximum of n and k, k being the smallest natural number appropriate for $Kt_1\ldots t_n$. Z involves atomic formulas either of the form $Kx_1\ldots x_n$ or $t_i = t_j$. Given the universal substitutivity of logical equivalence, every formula of L is thus equivalent to a formula involving only atomic formulas of one of these two forms. For this proof, we now suppose that L contains only formulas built up from atomic formulas of the form $Kx_1\ldots x_n$ where K is a non-logical predicate symbol of L or of the form $t_i = t_j$ where t_i and t_j are terms of L. The point is that the set of g-definable sets $\mathcal{D}(A)$, for $A = \langle A, g\rangle$ where g is some interpretation of L, is unaffected by this restriction on L. There is, therefore, no loss of generality. Notice, in particular, that Z has the same free variables as $Kt_1\ldots t_n$ in the above. We now proceed with our inductive proof through a series of lemmas.

Lemma 1. Where t is any term of L for which n is appropriate, $g_n(t)* = g_n*(t)$.

Proof. The proof is by induction on the structure of t. It t is the variable x_i, then $g(t) = \pi_i^n$ and $g(t)* = (\pi_i^n)* = g*(t)$ since $*$

preserves standard projections, and the ith standard projection is the interpretation of the variable x_i for $i \leq n$. If t is the constant a of L, then $g(t)$ is the n-ary constant function $f : A^n \longrightarrow A$ with constant value $g(a)$. Because $*$ preserves such constant functions, $g(t)^* : B^n \longrightarrow B$ is a constant function. Since $*$ also satisfies condition (2) of Proposition 2, $A \subset B$, and $g(t)^*$ is an extension of $g(t)$. $g(t)^*$ thus can only have constant value $g(a)$. Moreover, as we have already noted, the hypotheses of the theorem imply that $g(a)=g^*(a)$ where a is any constant symbol of L. Thus, $g(t)^* = g^*(t)$ by the definition of the interpretation function g^*.

Inductively, if $t = ht_1 \ldots t_m$ where h is an m-ary operational symbol of L, then $g(t) = g(h)(g(t_1),\ldots,g(t_m))$, by definition of $g(t)$, which implies $g(t)^* = g(h)^*(g(t_1)^*,\ldots,g(t_m)^*)$ since $*$ preserves the composition of term-defined functions. But this latter is $g^*(h)(g^*(t_1),\ldots,g^*(t_m)) = g^*(t)$ by the induction hypothesis and the definition of g^*. This proves the lemma.

<u>Lemma 2</u>. Where $Z \in I_n$ is atomic, $g_n(Z)^* = g_n^*(Z)$.

Proof. If Z is $Kx_1 \ldots x_n$, where K is an n-ary predicate symbol of L, then $g_n(Z) = g(K)$. Thus, $g_n(Z)^* = g(K)^* = g^*(K) = g_n^*(Z)$ by the hypotheses of the theorem. If Z is $t_1 = t_2$, where t_1 and t_2 are terms at least one of which has index n, then $g(Z) = \{\vec{a} \in A^n \mid g(t_1)(\vec{a}) = g(t_2)(\vec{a})\} = Eq(g(t_1),g(t_2))$. Since $*$ preserves equalizers of term-defined operations, $(Eq(g(t_1),g(t_2)))^* = Eq(g(t_1)^*,g(t_2)^*)$. This latter is $Eq(g^*(t_1),g^*(t_2))$ by Lemma 1, which is $g^*(t_1 = t_2)$ by definition of g^*. Since, as we have supposed above, all atomic formulas of L have one of these two forms, the lemma is proved.

<u>Lemma 3</u>. Where Z is of the form $\sim Y$ or $\vee YW$, $g_n(Z)^* = g_n^*(Z)$ where Z, and thus Y and/or W, have index n.

Proof. The result follows immediately from the fact that $*$ preserves the Boolean operations from $\mathcal{D}(A^n)$ to $\mathcal{D}(B^n)$. We illustrate with the case of negation: $g(\sim Y)^* = (A^n - g(Y))^* = B^n - g(Y)^* = g^*(\sim Y)$ by the definition of the functions g and g^* and by the fact that $*$ preserves complements. This establishes the lemma.

<u>Lemma 4</u>. Where Z is of the form $(\exists x_i)Y$ and n is the index of Z, $g_n(Z)^* = g_n^*(Z)$.

Proof. This follows immediately from the fact that $*$ preserves cylindrifications: $g(Z)^* = (C_i(g(Y))^* = C_i(g(Y)^*) = C_i(g^*(Y)) = g^*(Z)$.

The four lemmas establish that, for all formulas Z of L, whenever Z has index n, then $g_n(Z)^* = g_n^*(Z)$. Thus, condition (3) of Proposition 2 is satisfied and the theorem is established.

Having now characterized elementary extension in terms of preservation properties of the function * of Proposition 2 for the general case of any type τ of finitary structure, we turn to an examination of certain special cases. In particular, we will be interested in the case where the type is that of the full structure over the set A. We will see that considerable simplification in the preservation conditions on * can be achieved. We will also see that this case merits our attention, especially when A supports and ordered field structure.

FULL STRUCTURES

Let A be any nondegenerate set, i.e., a set with two distinct, distinguished elements denoted 0 and 1. Let L be any first-order language with equality having for its set of operation symbols precisely the set $\mathcal{O}(A)$ of all finitary operations over the set A. We let g be any interpretation of L in A which interprets each operation symbol as itself: $g(f) = f$, for all $f \in \mathcal{O}(A)$. The nonlogical predicate symbols of L are interpreted in A in any appropriate manner, but which remains of course fixed for the given interpretation g. As above, $\mathcal{A} = \langle A, g \rangle$ is the structure determined by this interpretation. We let $\mathcal{B} = \langle B, g^* \rangle$ be the structure resulting from the interpretation g^* of of L in any other nonempty set B.

Let $f : A^n \longrightarrow A$, $n \geq 1$, be given. We say that f is a <u>characteristic function</u> if $f(A^n) \subset \{0,1\}$. We say that f is a characteristic function <u>of</u> the relation $M = f^{-1}(\{0\}) \subset A^n$. Notice that M is also the equalizer of f and the function $h : A^n \longrightarrow A$ which is constantly 0. Since every n-ary relation $M \subset A^n$ has a characteristic function f which is an operation symbol of L, every such relation M is g-definable by the formula $fx_1 \ldots x_n = 0$. Thus, the set $\mathcal{D}(A)$ of all g-definable relations is just the set $\mathcal{R}(A)$ of all finitary relations over the set A.

The interpretation g^* of L in B induces a function * from operations f on A to operations on B by the definition $f^* = g^*(f)$.

This can be naturally extended to relations $M \subset A^n$ by defining $M^* = f^{*-1}(\{0^*\})$ where f is the characteristic function of M. Any function $* : \mathcal{R}(A) \longrightarrow \mathcal{D}(B)$ satisfying this latter condition is said to preserve ideals of characteristic functions. Notice that if such a * also sends a function which has constant value 0 to a function with constant value 0*, then this condition is a special case of preserving equalizers. We will now determine the precise conditions under which such a * function defines $\mathcal{B}$ as an elementary extension of $\mathcal{A}$. In particular, we will see that the strong condition (6) of Theorem 1 can be dropped altogether because the structure $\mathcal{A}$ involves the full operational structure over A.

Theorem 2. Let L, $\mathcal{A}$, and $\mathcal{B}$ be given as above. Let an arity-preserving function $* : \mathcal{R}(\mathcal{A}) \longrightarrow \mathcal{D}(\mathcal{B})$ be given such that $g(f)^* = f^* = g^*(f)$ for all operations $f \in \mathcal{O}(\mathcal{A})$ and $g(K)^* = g^*(K)$ for all nonlogical predicate symbols K of L. We suppose further that the algebraic part of $\mathcal{B}$ is an extension of the algebraic part of $\mathcal{A}$. Then, a necessary and sufficient condition that * determine $\mathcal{B}$ as an elementary extension of $\mathcal{A}$ is that * preserve the following data: (1) All standard projections; (2) All n-ary constant operations, $n \geq 1$; (3) operational composition; (4) ideals of characteristic functions; (5) the Boolean operations from $\mathcal{P}(A^n)$ to $\mathcal{D}(B^n)$ for each $n \geq 1$.

Proof. Again, necessity is straightfoward and we concentrate on proving sufficiency. We first establish that * satisfies condition (2) of Proposition 2. Since the algebraic part of $\mathcal{B}$ is an extension of the algebraic part of $\mathcal{A}$, $A \subset B$ and $g^*(a) = g(a) = a$ for all $a \in A$. Moreover, every function $f^* : B^n \longrightarrow B$ is an extension of $f : A^n \longrightarrow A$, $n \geq 1$. Since * preserves ideals of characteristic functions, for $M = f^{-1}(\{0\})$, $M^* = f^{*-1}(\{0\})$, where $f : A^n \longrightarrow A$, $n \geq 1$, is the characteristic function of M. Thus, $M^* \cap A^n = M$ for all $M \subset A^n$ since every such M has a characteristic function f which is the restriction to A^n of f*. In other words, each M* extends M, and condition (2) of Proposition 2 is established.

Hence, to complete the proof of Theorem 2, we have only to establish that * satisfies condition (3) of Proposition 2, namely, that $g(Z)^* = g^*(Z)$ for all formulas Z of L. As it turns out, a straightfoward inductive proof of this condition will not go through as it did for Theorem 1. We therefore conceive the following strategy: We observe that the relation $g(Z) \subset A^n$ has a characteristic function f. Thus $\mathcal{A} \models Z \equiv fx_1 \ldots x_n = 0$. We prove by induction on the structure of the formula Z that, in every such case, $\mathcal{B} \models Z \equiv fx_1 \ldots x_n = 0$. We also show that $g(W)^* = g^*(W)$ for W a formula of the form $t = 0$, $g(t)$ a characteristic function. We thus have, for all formulas Z of L, $g(Z)^* = g(W)^* = g^*(W) = g^*(Z)$, establishing condition (3) of Proposition 2 and thereby Theorem 2. We now see how the preservation conditions on * allow us to prove that $\mathcal{B} \models Z \equiv fx_1 \ldots x_n = 0$ whenever $\mathcal{A} \models Z \equiv fx_1 \ldots x_n = 0$ where f is the characteristic function of $g(Z)$.

Since * preserves standard projections, constant operations, and operational composition, and since, for n-ary operations f on A, $n \geq 0$, f* is an extension of f, we obtain Lemma 1 as before: $g_n(t)^* = g_n^*(t)$ whenever $n \geq 1$ is appropriate for the term t of L. We use this to prove:

Lemma 5. Let t be any term of L such that $g(t)$ is a characteristic function. Let Z be the formula $t = 0$ of L. Then $g(Z)^* = g^*(Z)$.

Proof. By definition, $g(Z) = g(t)^{-1}(\{0\})$. Thus, $g(Z)^* = (g(t)^{-1}(\{0\}))^* = g(t)^{*-1}(\{0\})$ since * preserves ideals of characteristic functions. But this latter is equal to $g^*(t)^{-1}(\{0\}) = g^*(Z)$ by the above (Lemma 1) and the definition of g^*.

In preparation for the next lemma, let us observe that, by definition of logical interpretation, $g(=)$ is the diagonal relation π_1^1. Since * preserves standard projections, $\pi_1^{1*} = \pi_1^1 = g^*(=)$, the latter equality again by the definition of the logical interpretation function g^*. Thus, $g(=)^* = g^*(=)$. Moreover, by the hypotheses of Theorem 2, $g(K)^* = g^*(K)$ where K is any nonlogical predicate symbol of L. Since = is the only logical predicate symbol, the condition $g(K)^* = g^*(K)$ therefore holds for all (logical or nonlogical) predicate symbols of L.

Lemma 6. Let Z be any atomic formula of L. If m is appropriate for Z, then $g_m(Z)^* = g_m^*(Z)$.

Proof. If Z is atomic, it is of the form $Kt_1 \ldots t_n$, $n \geq 1$, where K is a nonlogical or the logical (equality) predicate symbol of L and the t_i are terms of L for which m is appropriate. $g(K) \subset A^n$ has a characteristic function $f : A^n \longrightarrow A$. Thus:
(i) $\mathcal{A} \models Kx_1 \ldots x_n \equiv fx_1 \ldots x_n = 0$. Substituting the terms t_i in the appropriate manner yields: (ii) $\mathcal{A} \models Kt_1 \ldots t_n \equiv ft_1 \ldots t_n = 0$. Now, $g_n(Kx_1 \ldots x_n)^* = g(K)^* = g^*(K) = g_n^*(Kx_1 \ldots x_n)$ as established above, and

$g_n(fx_1\ldots x_n = 0)^* = g_n^*(fx_1\ldots x_n = 0)$ by Lemma 5. Applying this to (i) immediately yields (iii) $\mathcal{B} \models Kx_1\ldots x_n \equiv fx_1\ldots x_n = 0$ which, again by substitution of terms, yields: (iv) $\mathcal{B} \models Kt_1\ldots t_n \equiv ft_1\ldots t_n = 0$. Since $f = g(f)$ is a characteristic function, so is $g_m(ft_1\ldots t_n)$ and thus $g_m(ft_1\ldots t_n) = g(f)(g_m(t_1),\ldots,g_m(t_n))$. Letting $t = ft_1\ldots t_n$ we thus obtain $g_m(t = 0)^* = g_m^*(t = 0)$ by Lemma 5. Now, (ii) says that $g_m(Z) = g_m(t = 0)$ for any appropriate m, and thus, $g_m(Z)^* = g_m(t = 0)^* = g_m^*(t = 0) = g_m^*(Z)$, the last equality by (iv). Thus, $g_m(Z)^* = g_m^*(Z)$ and the lemma is established.

We now use Lemma 6 to attack directly an inductive proof of the following proposition: Let m be appropriate for the formula Z of L. Then $\mathcal{B} \models Z \equiv t = 0$ whenever $\mathcal{A} \models Z \equiv t = 0$, where t is $fx_1\ldots x_m$ and f is the characteristic function of $g_m(Z)$. We now suppose m to be chosen and fixed for the remainder of the proof. The induction is on the structure of the formula Z.

If Z is atomic, then $\mathcal{A} \models Z \equiv fx_1\ldots x_m = 0$ where f is the characteristic function of $g(Z)$. Thus $g(Z)^* = g(fx_1\ldots x_m = 0)^*$ which implies, by Lemma 6, that $g^*(Z) = g^*(fx_1\ldots x_m = 0)$, in other words, that $\mathcal{B} \models Z \equiv fx_1\ldots x_m = 0$.

We illustrate the case of the propositional connectives with negation. If Z is of the form $\sim Y$, then Z and Y have the same variables and so m is appropriate for Y. We let f be the characteristic function of $g(Y)$ and $f^\sim$ that of $g(Z)$ $(=A^m - g(Y))$. We thus have: $\mathcal{A} \models Y \equiv fx_1\ldots x_m = 0$ and also $\mathcal{A} \models Z \equiv f^\sim x_1\ldots x_m = 0$. Now, $g^*(Z) = g^*(\sim Y) = B^m - g^*(Y)$ by the definition of the interpretation function g^*. But, $B^m - g^*(Y) = B^m - g^*(fx_1\ldots x_m) = 0)$ by the induction hypothesis, and the latter equals $B^m - g(fx_1\ldots x_m = 0)^*$ by Lemma 6 and thus $(A^m - g(fx_1\ldots x_m = 0))^*$, since * preserves complements of subsets of A^m, which is $g(\sim fx_1\ldots x_m = 0)^* = g(f^\sim x_1\ldots x_m = 0)^*$ by the definition of g and of $f^\sim$. Finally, $g(f^\sim x_1\ldots x_m = 0)^* = g^*(f^\sim x_1\ldots x_m = 0)$ by Lemma 5. This establishes $g^*(Z) = g^*(f^\sim x_1\ldots x_m = 0)$, i.e. $\mathcal{B} \models Z \equiv f^\sim x_1\ldots x_m = 0$, which was to be proved.

The case of the other propositional connectives is treated in a similar manner. We now come to the last and most delicate case of all, that of quantified formulas.

Suppose, then, that Z is of the form $(\exists y)W$. If y has no free occurrence in W, then the result is immediate by the induction hypothesis.

We thus suppose that y has a free occurrence in W. To simplify the exposition, we suppose that y is the variable x_m. Letting f be the characteristic function of $g(Z)$, we thus have (1) $\mathcal{A} \models Z \equiv fx_1\ldots x_m = 0$. Letting d be the characteristic function of $g(W)$, we also have (2) $\mathcal{A} \models W \equiv dx_1\ldots x_m = 0$ from which we obtain (3) $\mathcal{B} \models W \equiv dx_1\ldots x_m = 0$ by induction hypothesis and therefore (4) $\mathcal{B} \models Z \equiv (\exists x_m)(dx_1\ldots x_m = 0)$ by logic.

We now let $h: A^{m-1} \longrightarrow A$ be a Skolem function for Z, i.e., a function such that, for all $\langle a_1,\ldots,a_m \rangle \in A^m$, $\langle a_1,\ldots,a_{m-1}, h(a_1,\ldots,a_{m-1})\rangle \models_A W$ if and only if $\langle a_1,\ldots,a_{m-1},a_m\rangle \models_A Z$. Such an h exists by the axiom of choice and is a symbol of L since every k-ary operation on A in an operation symbol of L, $k \leq 0$. Letting $tx_1\ldots x_{m-1}$ be the term $dx_1\ldots x_{m-1}hx_1\ldots x_{m-1}$, we thus obtain (5) $\mathcal{A} \models tx_1\ldots x_{m-1} = 0 \equiv fx_1\ldots x_m = 0$ by the definitions of d, h, and f and thus (6) $\mathcal{B} \models tx_1\ldots x_{m-1} = 0 \equiv fx_1\ldots x_m = 0$ by Lemma 6 and the fact m is appropriate for the atomic formula $tx_1\ldots x_{m-1} = 0$.

Now, by logic, $\mathcal{A} \models W \supset Z$, and so $\mathcal{A} \models dx_1\ldots x_m = 0 \supset tx_1\ldots x_{m-1} = 0$ by (2), (1) and (5). By Lemma 6, and by virtue of the fact that * preserves complements and unions, we then obtain
$\mathcal{B} \models dx_1\ldots x_m = 0 \supset tx_1\ldots x_{m-1} = 0$, and thus
$\mathcal{B} \models (\exists x_m)(dx_1\ldots x_m = 0) \equiv tx_1\ldots x_{m-1} = 0$ by logic (x_m is not free in $tx_1\ldots x_{m-1} = 0$). Applying (4) and then (6) to this latter, we finally obtain (7) $\mathcal{B} \models Z \supset fx_1\ldots x_m = 0$, which is half of what we must prove. Going the other way, we observe that $\mathcal{B} \models (tx_1\ldots x_{m-1} = 0 \supset (\exists x_m)(dx_1\ldots x_m = 0)$ by logic. Applying (6) and (4) to this, we obtain (8) $\mathcal{B} \models fx_1\ldots x_n \supset Z$.

The conjunction of (7) and (8) establishes the conclusion of the quantifier case hence the inductive proposition. Given, now, any formula Z of L, $g(Z)* = g(fx_1\ldots x_m = 0)* = g*(fx_1\ldots x_m = 0) = g*(Z)$ by Lemma 6 and the inductive proposition, where f is the characteristic function of $g(Z)$. This establishes condition (3) of Proposition 2 and hence Theorem 2.

Corollary 1. Let A be any nondegenerate set and B a superset of A. Let * be a mapping from $\mathcal{R}(A)$ to $\mathcal{R}(B)$ which extends every n-ary relation on A, $n \geq 1$, to an n-ary relation on B. We further suppose that * extends n-ary operations on A to n-ary operations on B, $n \geq 0$ (for $n = 0$, this means that * leaves singleton sets invariant: $\{a\}* = \{a\}$ for all $a \in A$). Let L be the full diagram language of A, i.e., the first-order language with equality having the operations and relations of

A for its nonlogical operation and relation symbols. Let $\mathcal{A} = \langle A,g \rangle$ be the structure on A determined by the interpretation g(K) = K where K is a nonlogical symbol of L, and let $\mathcal{B} = \langle B,g^* \rangle$ be the structure on B determined by the interpretation $g^*(K) = K^*$ for K a nonlogical symbol of L. Then a necessary and sufficient condition that $\mathcal{A} < \mathcal{B}$ is that * preserve the following data: (1) All standard projections; (2) all n-ary constant operations, $n \geq 1$; (3) operational composition; (4) ideals of characteristic functions; (5) the Boolean operations from $\mathcal{P}(A^n)$ to $\mathcal{P}(B^n)$ for each $n \geq 1$.

Proof. It is immediate that the structure $\mathcal{B}$ defined in the indicated manner from * satisfies all the hypothesis of Theorem 2.

The advantage in dealing with the full structure on the set A is that we can define * directly by its preservation properties without having previously defined either the structure $\mathcal{B}$ or the language L (though, of course, the superset B of A must be previously chosen). Moreover, the structure of the extension $\mathcal{B}$ of $\mathcal{A}$ is much richer and therefore more potentially useful. Finally, and significantly, every ultrapower extension B of the set A gives rise to an elementary extension $\mathcal{B}$ of the full structure on A since the ultrapower construction is defined independently of relations and operations on A. Thus, elementary extension of full structures exist in abundance (and, in particular proper ones exist in abundance if we assume the axiom of choice). In other words, Corollary 1 gives an algebraic characterization of elementary extension for a useful and abundant case.

FIELD STRUCTURES

We now suppose that A is a nondegenerate set which supports a field structure. We suppose that the distinguished elements 0 and 1 of A are the 0 and 1 of the field structure on A.

Theorem 3. Let $B \supset A$ be a superset of A which supports a field structure which is an extension of the field structure on A. Let * extend every n-ary relation and operation on A to an n-ary relation or operation on B, extending the field operations on A to the corresponding field operations on B. Then, the structure $\mathcal{B}$ defined on B by * is an elementary extension of the full structure $\mathcal{A}$ on A if and only if * preserves: (1) standard projections; (2) n-ary constant operations, $n \geq 1$;

(3) operational composition; (4) characteristic functions of relations. Moreover, if the field structure $\mathcal{A}$ on A is ordered, then the field structure $\mathcal{B}$ in an ordered field extension of $\mathcal{A}$.

Proof. By hypothesis, $\mathcal{B}$ is a field extension of the field structure on A. Moreover, for each $n \geq 1$, we can represent subsets of A^n by their characteristic functions and we can define representations of the Boolean operations on such subsets in terms of the field operations in combination with these characteristic functions. Since * preserves the field operations and characteristic functions of relations by hypothesis, * preserves the Boolean operations on $\mathcal{P}(A^n)$. Thus, the hypotheses of of Corollary 1 are satisfied and its conclusion follows: $\mathcal{B}$ is an elementary extension of $\mathcal{A}$. Thus, if $\mathcal{A}$ is ordered, $\mathcal{B}$ will be ordered by the extension of the order relation on A.

Let us note also that, since * preserves the field operations, and since 0 and 1 are the only idempotents of a field, the * of a characteristic function will always be a characteristic function. Hence, condition (4) of Theorem 3 is here equivalent to the weaker condition (4) of Corollary 1. Moreover, if we allow inhomogeneous composition of operations, then condition (2) can be dropped from Theorems 1,2, and 3 and from Corollary 1, because the n-ary constant function $f : A^n \longrightarrow A$, $n \geq 1$, with constant value $a \in A$, can then be defined as the composite $\pi_1^2(a, \pi_1^n)$. This latter is preserved by * whenever a is the interpretation g(h) of a constant symbol h of L.

If, in Theorem 3, A is the set R of real numbers, and if B = R* is a proper extension of R, then R* is a hyperreal field. In that case, the hypotheses of Theorem 3 amount to a set of objective axioms for the hyperreal numbers, i.e., axioms not involving any notions of formal logic whatever. This stands in contrast to the well-known axioms of Keisler (see [7] and [8]) which are partly objective and partly subjective. (In particular, the solution set axiom of Keisler depends significantly on a certain notion of a system of formulas and the attendant notion of the satisfaction of the system of formulas by a vector of values).

For the remainder of the paper, we consider the case of Theorem 3 where A = R is the set of real numbers and B = R* a proper superset of R. Theorem 3 determines R* as an elementary extension of R, but does not characterize R* up to isomorphism. As it turns out, there are some very natural algebraic conditions that can be added that allow us to determine R^* up to isomorphism. One such condition is expressed in the following theorem.

Theorem 4. Let R^* be an ordered field extension on R having order type η_1, and suppose that other hypotheses of Theorem 3 are satisfied. Then R^* is an elementary extension of the full structure on R and is unique up to an ordered field isomorphism. In fact, R^* is isomorphic to the field structure R^N/M where M is a maximal ideal containing the ideal F of sequences having finite support in the ring R^N of all sequences of real numbers.

Proof. Since the hypotheses of Theorem 3 are satisfied, R^* is an elementary extension of R as a field. It is thus real-closed and hence has a uniquely determined order structure. Since it is an ordered field extension of R, this unique order must be the *-extension of the order structure on R, making R^* an elementary extension of R as an ordered field. By virtue of the main theorem of [1], a real-closed field having order type η_1 is unique up to an ordered field isomorphism. Finally, the existence of a set with order type η_1 implies the continuum hypothesis which, in turn, implies that the field R^N/M, $F \subset M$, has order type η_1. But R^N/M is also an elementary extension of the reals (see [3]) and therefore real-closed. It is thus isomorphic to R^*, again by appeal to the main theorem of [1].

Let us note that the field R^N/M is an ultrapower of the reals (see [3]). Thus, Theorem 4 gives a set of objective conditions which determine a certain structure R^* as an ultrapower model of the full structure on the reals.

The paper [6] contains a more comprehensive study of the order structure of the hyperreal number system.

BIBLIOGRAPHICAL NOTE

The history of these results is as follows. The result of Corollary 1 of the present paper, applied to the case where A is the set R of real numbers, was announced in [2]. It is also the basis of [3], which was written and completed in the Spring of 1981. A full proof of the general case (i.e., Corollary 1 of the present paper) was published as Theorem 1 of [4]. Theorem 2 of [4] is a weaker form of Theorem 3 of the present paper. The sharper version appears here for the first time. Theorem 1 of the present paper was announced in [5], and the proof given here is the first published one. Theorem 2 of the present paper is new in its present form, but its proof is essentially a streamlined version of the proof of Theorem 1 of [4]. Theorem 4 of the present paper is essentially Theorem 3 of [4], and is also contained in a slightly different from in [6]. There are also some significant differences in approach between the present paper and [4] where notions of category theory play a greater importance than they do here.

NOTES

1. Here and elsewhere in this paper, we avoid unnecessary complications by identifying definable constants g(f), f a 0-ary operation symbol of L, with the singleton sets $\{g(f)\} = g_1(f = x_1)$ they define, thereby regarding them as unary relations (sets) over A.

2. For f a constant symbol, that means $\{g(f)\}^* = \{g^*(f)\}$. This immediately yields $g(f) = g^*(f)$ since * satisfies condition (2) of Proposition 2.

REFERENCES

1. P. Erdös, L. Gillman, and M. Henriksen, "An isomorphism theorem for real-closed fields", Ann. of Math., 61 (1955), pp.542-554.
2. W.S. Hatcher, "Objective axioms for the hyperreals", Abstracts A.M.S., 3 (1982), p. 282.
3. W.S. Hatcher, "Calculus is algebra", Amer. Math. Monthly, 89 (1982), pp. 362-370.
4. W.S. Hatcher, "Clone embeddings and the hyperreal numbers" in Recent Trends in Mathematics, Kurke, Mecke, Triebel, and Thiele, Ed., Teubner-Texte zur Mathematik 50, Leipzig, 1983, pp. 164-173.
5. W.S. Hatcher, "An algebraic characterization of elementary extension". Abstracts A.M.S., 4 (1983), p. 189.
6. W.S. Hatcher and C. Laflamme, "On the order structure of the hyperreal line", Z. Math. Logik Grundlag. Math., 29 (1983), pp. 197-202.
7. H.J. Keisler, Elementary Calculus, Prindle, Weber & Schmidt, Boston, 1976.
8. H.J. Keisler, Foundations of Infinitesimal Calculus, Prindle, Weber & Schmidt, Boston, 1976.

RELEVANT QUANTUM ARITHMETIC

Chris Mortenson/Philosophy Department, University of Adelaide, Adelaide, Australia

Robert K. Meyer/Department of Philosophy, Research School of Social Sciences, Australian National University, Canberra, Australia

INTRODUCTION

In 'Quantum Mathematics' [1], Michael Dunn showed that the addition of Peano arithmetic to a classical quantum (orthomodular) logic enables all instances of the distribution axiom $A \& (B \vee C) \rightarrow (A \& B) \vee C$ to be proved, thus collapsing (classical) quantum arithmetic to classical Peano arithmetic $P^{\#}$. He also showed that distribution fails in second order Peano arithmetic without extensionality, but holds when extensionality is added. He expressed the opinion that the first of these results should not be of concern to the quantum logician (on the grounds that arithmetic need not be infected with quantum problems), but that the last should (on the grounds that the language of quantum mechanics is set theory, while the result seems to show that extensional set theory must admit distribution).

We wish to raise here the question whether Dunn's result can be reproved relevantly, i.e. whether distribution can be recovered in arithmetic founded on a base logic of <u>relevant</u> orthomodular logic. It is not apparent from Dunn's proof method that it can, but then it is not apparent that it cannot. As a contribution to this problem, we show in the next section that in arithmetic based on relevant ortho-logic, OR, distribution cannot be proved. OR differs from what one would take to be the relevant analogue of orthomodular logic in that the latter is slightly stronger. While both lack distribution, OR lacks any version of the orthomodular law $A \mathbin{\&} (\sim A \vee (A \mathbin{\&} B)) \leq B$; so what happens when the latter is added remains open. In the final section, we ask what happens when inconsistent arithmetics are added to relevant quantum logics. Inconsistent arithmetics were studied in [2]. For inconsistent arithmetics to be mathematically interesting, a logic in which $(A \mathbin{\&} \sim A) \rightarrow B$ fails is necessary; and the usual relevant logics provide such logics. We adapt here the methods of [2] to show a Dunn-like result, that in the extensional fragment of such arithmetics, even with a logic as weak as OR, distribution is provable (and so a fortiori when orthomodularity holds as well). Hence there is no interesting effect on the extensional theorems of inconsistent relevant arithmetic when distribution is dropped from the base logic. It is an open question whether this holds true of the intensional theorems of inconsistent arithmetic as well.

DISTRIBUTION FAILS IN $OR^{\#}$

As Dunn notes, it is not entirely clear just what, from the point of view of Hilbert-style logics, quantum logic is. The logic OR is formed from R by dropping the distribution axiom (see e.g. [3], p. 341). The arithmetic based on OR, called $OR^{\#}$, is formed by adding following intensional forms of the Peano postulates to OR (see [2]). AXIOMS: $x = y \leftrightarrow x' = y'$, $x = y \rightarrow (x = z \rightarrow y = z)$, $x' \neq 0$, $x + 0 = x$, $x + y' = (x + y)'$, $x \times 0 = 0$, $x \times y' = (x \times y) + x$. Rule: $F0$, $Fx \rightarrow Fx' / \therefore (x)Fx$.

Now we can state the result of this section.

Proposition 1. Distribution is not provable in $OR^{\#}$.

Proof. The proof is quite easy. Take the ortholattice below, which we call BR; first discussed, as far as we know, in [4].

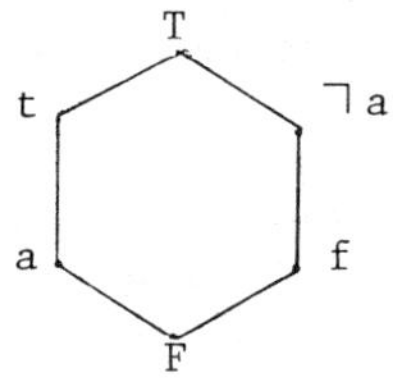

Conjunctions and disjunctions are lower and upper bounds respectively in BR, $F = \neg T$, $f = \neg t$. The $\rightarrow$ operation is given by:

$\rightarrow$	F	f	a	¬a	t	T
F	T	T	T	T	T	T
f	a	t	a	T	a	T
a	¬a	¬a	T	¬a	T	T
¬a	a	a	a	T	a	T
* t	F	f	a	¬a	t	T
* T	F	F	a	¬a	a	T

It is mechanical to verify that every theorem of OR holds in BR. Now we construct a model for arithmetic on BR as follows. The domain is the natural numbers with + , × and ' as in the standard model for arithmetic. The universal quantifier is interpreted as the lower bound of the set of values of its instantiations (i.e. generalized conjunction in the complete lattice BR). We assign the equations $n = n$, for each n, to the value t; all other equations are assigned to F. This induces the values f on $\neg n = n$, and T on $\neg n = m$ for distinct n,m. It is then standard (though nonfinitistic) to verify that every theorem of $OR^\#$ takes a designated value (t or T) in this model. But distribution fails, e.g. $t \mathbin{\&} (\neg a \vee a) \not\rightarrow (t \mathbin{\&} \neg a) \vee a$.

Note also that the rule γ $(A, \neg A \vee B / \therefore B)$ fails for $OR^\#$: it can be shown that $\vdash_{OR^\#} t$ and $\vdash_{OR^\#} \neg t \vee a$; but here not $\vdash_{BR} a$, so not $\vdash_{OR^\#} a$. This is of interest, since $OR^\#$ is (presumably!) consistent. Say that a theory is <u>prime</u> if whenever a disjunction is in the theory, at least one of the disjunts is. Now it is standard that if a theory consistent and prime, γ holds for it. Hence the failure of γ for $OR^\#$ must be due to a failure of primeness. And indeed this is so, for $\vdash_{OR^\#} \neg t \vee a$; but here neither $\vdash_{BR} \neg t$ nor $\vdash_{BR} a$, so neither $\vdash_{OR^{\#}} \neg t$ nor $\vdash_{OR^{\#}} a$. It is a common observation among paraconsistentist logicians that γ fails when a theory is inconsistent but nontrivial. We believe this to be the only extant example of an interesting theory in which γ fails because of failure of primeness.

Note finally that the above result does not circumvent Dunn's theorem relevantly, since BR is not an orthomodular lattice. Nor does what seems a natural relevant analogue of orthomodularity, $(A \mathbin{\&} (\sim A \vee (A \mathbin{\&} B))) \to B$, hold in BR.

INCONSISTENT QUANTUM ARITHMETIC

In [2], we investigate the inconsistent (but nontrivial) arithmetics which arise when an equation of the form $0 = n$ is added as a theorem to relevant arithmetic, $R^{\#}$. R is of course an inconsistency - tolerating logic, and the addition of such formulae does not lead to total collapse. Here we ask what happens when such equations are added to $OR^{\#}$, or indeed in any inconsistent extension thereof.

Definition. Let t be the sentence $0 = 0$ and f be $\neg t$. The arithmetical sentence A is a secondary equation of a theory iff $\vdash A \to t$. A is a secondary unequation iff $\vdash f \to A$.

Lemma. In $OR^{\#}$, if A and B are both secondary equations, then distribution holds of them, i.e. $\vdash A \mathbin{\&} (B \vee C) \to (A \mathbin{\&} B) \vee C$. Similarly, if B and C are secondary unequations, then $\vdash A \mathbin{\&} (B \vee C) \to (A \mathbin{\&} B) \vee C$.

Proof: First note that for any A, $\vdash (t \to A) \leftrightarrow A$ and $\vdash (A \to f) \leftrightarrow A$. (see, e.g. [5] or [6]). But also in OR, both $\vdash f \to \neg B \to . A \to f \to . A \to \neg B$ and $\vdash f \to \neg A \to . B \to f \to . B \to \neg A$. Assuming that A and B are secondary equations, then $\vdash f \to \neg B$ and $\vdash f \to \neg A$ by contraposition. So detatching $\vdash A \to f \to . A \to \neg B$ and $\vdash B \to f \to . B \to \neg A$. But since in OR, $\vdash (A \to \neg B) \leftrightarrow (B \to \neg A)$, we have $\vdash B \to f \to . A \to \neg B$. But $B \to f$ is equivalent to $\neg B$, so $\vdash \neg B \to . A \to \neg B$. Similarly, $\vdash \neg A \to . A \to \neg B$. Hence in OR, $\vdash (\neg A \vee \neg B) \to . A \to \neg B$, whence $\vdash \neg(A \to \neg B) \to (A \mathbin{\&} B)$. But also in OR, $\vdash A \to . A \to \neg B \to \neg B$, so that $\vdash A \to . B \to \neg(A \to \neg B)$. Thus $\vdash A \to . B \to (A \mathbin{\&} B)$. Conjoining to the antecedent and disjoining to the final consequent, $\vdash A \mathbin{\&} (B \vee C) \to . B \to .(A \mathbin{\&} B) \vee C$; call this (1). Now for any C, $\vdash C \to . t \to C$ hence $\vdash t \to . C \to C$. Hence

$\vdash t \to . C \to C$. Hence if A is a secondary equation, $A \to . C \to C$. Conjoining to the antecedent etc. as before,

$\vdash A \& (B \vee C) \to . C \to (A \& B) \vee C$; call this (2).

Using both (1) and (2) and the properties of OR,

$\vdash A \& (B \vee C) \to . (B \vee C) \to . (A \& B) \vee C$; so

$\vdash A \& (B \vee C) \to (A \& B) \vee C$. But in OR, $\vdash A \to . A \to B \to . A \to B$. Hence $\vdash A \& (B \vee C) \to (A \& B) \vee C$. This is the desired result when A and B are secondary equations. If B and C are secondary unequations, then $\neg B, \neg C$ are secondary equations. Applying what we have just proved, $\vdash \neg C \& (\neg B \vee \neg A) \to (\neg C \& \neg B) \vee \neg A$. Rearranging using commutation and De Morgan's laws (all in OR), $\vdash \neg(A \& B) \& \neg C \to \neg A \vee \neg(B \& C)$. De Morgan again gives $\vdash \neg((A \& B) \vee C) \to \neg(A \& (B \vee C))$, and contraposition gives $\vdash A \& (B \vee C) \to (A \& B) \vee C$ as required.

Now we can state the main result of this section.

Proposition 2. In any inconsistent extension of $OR^{\#}$, if A,B,C are all extensional formulae (containing no $\to$s), then $\vdash A \& (B \vee C) \to (A \& B) \vee C$.

Proof: We need the following facts: if any of $\vdash A \to B$, $\vdash B \to A$, $\vdash B \to C$, $\vdash C \to B$ or $\vdash A \to C$ holds, then distribution does too. The proof of these is standard and easy to verify. Now, it is also true that in any inconsistent extension of $R^{\#}$, $\vdash t \to f$ (see [2] Proposition 21). In fact, the proof of this does not use distribution, so it holds also in $OR^{\#}$. Hence we have that any one of the following five conditions suffices for $A \& (B \vee C) \to (A \& B) \vee C$ to hold in any inconsistent extension of $R^{\#}$: (1) A a secondary equation and B a secondary unequation (2) B a secondary equation and A a secondary unequation (3) B a secondary equation and C a secondary unequation (4) C a secondary equation and B a secondary unequation (5) A a secondary equation and C a secondary unequation. To these we can add two more sufficient conditions, following from the lemma: (6) A and B both secondary equations (7) B and C both secondary unequations. Now it is also a fact that every extensional sentence is either a secondary equation or a secondary unequation (see [2] Proposition 21; we note that this does not hold for the intensional sentences of R so the proof here does not generalize). But now it is easy to see that it follows from (1) - (4) that it is impossible for distribution to fail for extensional sentences A, B, C unless they are all secondary equations, or all secondary unequations. And that is impossible by (6) and (7). This conclusion reinforces those of [2], that inconsistent arithmetics are surprisingly stable, at least in their extensional parts, and that inconsistent mathematics is a rich and interesting area.

REFERENCES

1. J. Michael Dunn, Quantum Mathematics , edited by P. Asquith and R. Giere, Philosophy of Science Association, in PSA 80, vol. 2, East Lansing, Michigan, 1982.
2. Robert K. Meyer and Chris Mortensen, Inconsistent Models for Relevant Arithmetics, Journal of Symbolic Logic, (forthcoming).
3. Alan R. Anderson and Nuel D. Belnap Jr., Entailment, Princeton, Princeton University Press, 1975.
4. Robert K. Meyer, R_I-the Bounds of Finitude , Zeitschrift für Mathematische Logik und Gründlagen der Mathematik, 14 (1968), 385-387.
5. Robert K. Meyer, The Consistency of Arithmetic, Australian National University, 1976.
6. Robert K. Meyer, Arithmetic Formulated Relevantly , Australian National University, 1976.

REMARKS ON SIGNIFICANCE AND MEANINGFUL APPLICABILITY

William N. Reinhardt/Department of Mathematics, University of Colorado, Boulder, Colorado

I am very pleased to be invited to contribute to this volume in honor of Newton da Costa. I believe the topic I have chosen is appropriate for two reasons. One, it is a first step in investigating a logical topic of philosophical interest. Two, while the theories and methods I use are completely classical, they give rise to some interesting non-classical systems and questions about these.

I have been interested for some time in pursuing Gödel's remark in [1944] that the paradoxes can perhaps be resolved by noting that some properties and relations (such as the Russell property, of not applying to itself) may not always be meaningfully applicable, and that the relation of meaningful applicability may itself be among these. It is possible to carry out this idea in a formal theory, but there are two difficulties. First, it is difficult to be sure one has a consistent theory, and second, it is extremely awkward because in classical theories there will generally be a property $\theta(x)$ and a term t such that the theory proves both $\theta(t)$ and that the property expressed by θ is <u>not</u> meaningfully applicable to the object denoted by t.

The first difficulty may be overcome (to be sure only to a very limited extent) using Kripke's suggestions in [75] regarding the notion of truth as a partial predicate (for one approach, see Maddy [83]). The second I have dealt with in [198?]. In brief, one can adopt the point of view that the formal theory is to be interpreted more strictly than is usual: rather than regarding provability of a sentence $\theta(t)$ as sufficient for meaningful assertability of the sentence, one requires that in addition one can prove the appropriate meaningful applicability.

Although Kripke's theory is not about properties and meaningful applicability, but rather about truth (and falsity) of sentences in a certain language, one may regard it as being about certain properties (namely those expressible in the language) and take meaningful applicability of the property expressed by θ to x to mean simply that the corresponding sentence $\theta(\overline{x})$ (where $\overline{x}$ denotes x) is either true or false.

Kripke points out that it is not quite natural to say that sentences such as the liar are not meaningful; rather they fail in some way to get truth values. We therefore use the term "significant" rather than "meaningful" for sentences which are meaningful in the strong sense intended here.

In this paper I will develop the theory of significance induced by the suggested definition in terms of truth and falsity far enough to obtain an alternative axiomatization in terms of truth and significance. This suggests the possibility of modifying the theory in T and F in the light of intuitions about S.

AXIOMS FOR TRUTH AND FALSITY

It is possible to axiomatize a significant fragment of the theory of the models given by Kripke's construction, or more precisely, of the models obtained from Kripke's by what he calls "closing off" the partial truth (and falsity) predicates. The following axioms were given by Feferman.

We add to ordinary first order Peano arithmetic two one-place predicates T and F (for true and false respectively). These predicates are allowed in the induction scheme, and the theory is formulated in classical logic. In addition, we have special axioms for T and F. These give the Kripke constraints on T and F as properties of sentences of the classical language of arithmetic with T and F adjoined. Thus the statement of the axioms requires the arithmetization of the syntax of this language. (For future reference we call it L_o). To present this we follow the Feferman [60] convention regarding primitive recursive functions f : by affixing a bar (Feferman uses a dot) to the notation for f, we obtain a notation for a term of arithmetic which represents the function f. (Thus, using the common practice of indicating a numeral for n by $\overline{n}$, we have $P \vdash \overline{f}(\overline{n}) = \overline{f(n)}$). In particular, we think of conjunction as an operation on formulas, and thus we have $P \vdash (\overline{\theta}\ \overline{\&}\ \overline{\phi}) = \overline{(\theta \ \& \ \phi)}$. The operation Sb of substitution, $Sb(\theta,x,n) = \theta(x / \overline{n})$ (the result of substituting a name for n for x in θ) deserves special mention. It will make many formulas more readable to use $[\theta]^x_y$ for an arithmetical term with x free and y not free, corresponding to this operation. (This notation is adopted from George Bealer's in [82]). Specifically, if θ has only x,y free, $[\theta]^x_y$ is $\overline{Sb}(\overline{\theta},\overline{x},x)$ and corresponds to $\theta(x / \overline{n})$. Similarly, $[\theta]^{x,y}$ would correspond to $\theta(x / \overline{n},\ y / \overline{m})$ and $[\theta]_{x,y}$ to θ itself. If no subscripts or superscripts are explicitly indicated, we suppose that $[\theta]$ has exactly the same free variables that θ does. (Thus $[\theta(x)]$ is a term, which when x is assigned the value n, will denote $\theta(\overline{n})$.) This notation is convenient to work with because we have in Peano arithmetic,

$$[\neg\theta] = \overline{\neg}[\theta] \ ,$$

$$[\theta \ \& \ \phi] = [\theta]\ \overline{\&}\ [\phi] ,$$

etc., corresponding to

$$(\neg\theta)(\overline{n}) = \neg(\theta(\overline{n}))$$

$$(\theta \ \& \ \phi)(\overline{n}) = \theta(\overline{n}) \ \& \ \phi(\overline{n}) \ ,$$

etc. Note that $[\theta(x)(x / \overline{n})]$ and $[\theta(x)](x / \overline{n})$ are distinct terms; we do have in Peano arithmetic

$$[\theta(x)(x / \overline{n})] = [\theta(x)](x / \overline{n}) \ .$$

We do <u>not</u>, however, have

$$[\theta(x)(x / \overline{f}(x))] = [\theta(x)](x / \overline{f}(x)) \ ;$$

this means that in certain cases the notation $[\theta(\overline{f}x)]$ can be misleading, so that caution in required in using it. In the Kripke-Feferman theory we do have

$$T[\theta(x)(x / \overline{f}x)] \leftrightarrow T([\theta(x)](x / \overline{f}x))$$

(and similarly for F), so that the notation $T[\theta(\overline{f}x)]$ is to that extent unambiguous.

We follow a similar convention for predicate to the one for functions. For example, if $Sent_L$ is the sentences of L, $\overline{Sent}_L$ is an arithmetical formula expressing 'x is a sentence of L'. Although one could allow T and F to apply to predicates not expressible in L_o, the Kripke-Feferman theory, for definiteness, supposes that the universalizations of

$$(L_o) \qquad T(x) \rightarrow \overline{Sent}_{L_o}(x)$$

$$F(x) \rightarrow \overline{Sent}_{L_o}(x)$$

are axioms. (We shall shortly modify this slightly). Finally, we let Tr_a be an arithmetical predicate expressing 'is a true atomic sentence of Peano arithmetic.'

We may now state the axioms for T and F. These are the universal closures (properly relativized to $\overline{Sent}_{L_o}$ in 3b and 4b) of:

1.a $\quad \overline{A}(x) \rightarrow (T(x) \leftrightarrow Tr_a(x))$

b $\quad A(x) \rightarrow (F(x) \leftrightarrow \neg Tr_a(x))$

where A is the set of atomic sentences of arithmetic

2.a $\quad T(\overline{\neg} x) \leftrightarrow F(x)$

b $\quad F(\overline{\neg} x) \leftrightarrow T(x)$

3.a $\quad T(x \mathbin{\overline{\&}} y) \leftrightarrow T(x) \;\&\; T(y)$

b $\quad F(x \mathbin{\overline{\&}} y) \leftrightarrow F(x) \text{ or } F(y)$

4.a $\quad T(\overline{\forall} ux) \leftrightarrow \forall y T(\overline{Sb}(x,u,y))$

b $\quad F(\overline{\forall} ux) \leftrightarrow \exists y F(\overline{Sb}(x,u,y))$

5.a $\quad T[T(x)] \leftrightarrow T(x)$

b $\quad F[T(x)] \leftrightarrow F(x)$

6.a $\quad T[F(x)] \leftrightarrow F(x)$

b $\quad F[F(x)] \leftrightarrow T(x)$

7. $\quad \neg(T(x) \;\&\; F(x))$

The resulting theory is easily shown to be consistent by Kripke's method.

We note that if we give the classical definitions for "or" and $\exists$, then these axioms yield (for x,y relativized to sentences of L_o) axiom 8:

8.

$T(x \ \overline{or} \ y) \leftrightarrow T(x) \ or \ T(y)$

$F(x \ \overline{or} \ y) \leftrightarrow F(x) \ \& \ F(y)$

$T(\overline{\exists} ux) \leftrightarrow \exists y T(\overline{Sb}(x,u,y))$

$F(\overline{\exists} ux) \leftrightarrow \forall y F(\overline{Sb}(x,u,y))$

Another useful observation is that

$F(x) \leftrightarrow T(\overline{\neg} x)$

and

$F(x) \leftrightarrow (T(x) \ or \ F(x)) \ \& \neg T(x)$.

It is easy to prove by induction on A that

$T[A] \rightarrow A$

$F[A] \rightarrow \neg A$

and hence that, assuming (T[A] or F[A]),

$T[A] \leftrightarrow A$.

We note that axiom 5 gives truth conditions for sentences $T(\overline{n})$. If we wish the convenience of terms more general than $\overline{n}$ we need a form of 5 allowing for these, and similarly for 6. For our purpose, the universal closure of

$T[T(t)] \leftrightarrow T(t)$

$F[T(t)] \leftrightarrow F(t)$

(for t any primitive recursive arithmetical term) will be adequate, and similarly for axiom 6.

PROPERTIES OF SIGNIFICANCE INDUCED BY THE AXIOMS FOR TRUTH AND FALSITY

We now define S(x) to be (T(x) or F(x)); we think of this as expressing the significance of x in the sense that x successfully achieves a truth value. One may think of this as saying that x expresses a genuine proposition. At least, if x does express a proposition (of the classical sort) we expect it to get a truth value (that of the proposition); and if it does succeed in getting a truth value, then it seems to express the proposition that its truth value is "true" .

The notion of "significance" or "expresses a proposition" seems however to be of independent interest and we would like to examine the theory of significance induced by the theory with T,F. In addition we wish to see what axioms on S and T are needed to yield those in T and F. To facilitate this comparison we consider the language L with three one-place predicates T, F, and S. The defining axiom for S is

$$S(x) \leftrightarrow T(x) \text{ or } F(x) . \tag{1}$$

In addition we must add the truth and falsity conditions for the new atomic formulas:

$$T[S(x)] \leftrightarrow T[T(x) \text{ or } F(x)]$$

$$F[S(x)] \leftrightarrow F[T(x) \text{ or } F(x)] . \tag{2}$$

The theory KF is like the Kripke-Feferman theory we have described, except that the condition (L_o) is replaced by

(L) $T(x) \rightarrow \overline{Sent}_L(x)$

$F(x) \rightarrow \overline{Sent}_L(x)$,

references to L_o in the axioms (namely in the induction schema and in 3b, 4b) are replaced by references to L, and of course we add 1) and 2).

Notice that because of condition 2), this is not simply a definitional expansion of the Kripke-Feferman theory we first described. Nevertheless it appears to have the same intuitive content, and the method of Kripke does show it is consistent.[1] It is now not difficult to check that we get the following properties of S, which we split into groups.

Group 1:

1.1 $S(x) \rightarrow S(\overline{\neg} x)$

1.2 $S(x) \ \& \ S(y) \rightarrow S(x \ \overline{\&} \ y)$

1.3 $S(x) \rightarrow S[T(x)]$

1.4 $S(x) \rightarrow S[S(x)]$

1.5 $\forall y S(\overline{Sb}(x,u,y)) \rightarrow S(\overline{\forall} ux)$

1.6 $\overline{A}(x) \rightarrow S(x)$

1.7 $T(x) \rightarrow S(x)$

This group gives sufficient conditions for significance. The next group gives the semantic consequences of significance, namely the appropriate truth conditions.

Group 2:

2.1 $S(\overline{\neg}x) \to (T(\overline{\neg}x) \leftrightarrow \neg T(x))$

2.2 $S(x \mathbin{\overline{\&}} y) \to (T(x \mathbin{\overline{\&}} y) \leftrightarrow (T(x) \mathbin{\&} T(y))$

2.3 $S(\overline{\forall}ux) \to (T(\overline{\forall}ux) \leftrightarrow \forall y T(\overline{Sb}(x,u,y))$

2.4 $S(x) \mathbin{\&} \bar{A}(x) \to (T(x) \leftrightarrow Tr_a(x))$

2.5 $S[T(x)] \to (T[T(x)] \leftrightarrow T(x))$

2.6 $S[S(x)] \to (T[S(x)] \leftrightarrow S(x))$

The axioms in both of the preceeding two groups seem extremely natural. We remark that in the presence of the condition

$$T(x) \to T(x \mathbin{\overline{or}} y) , \qquad (1)$$

these yield

$$T(x) \to S(x \mathbin{\overline{or}} y)$$

This may strike one as unnatural: it allows the possibility that

$$\neg S(y) \mathbin{\&} S(x \mathbin{\overline{or}} y),$$

which may seem excessively liberal. We have not yet assumed (1), however.

We remark that the antecedents in Group 2 could be strengthened (and the axioms correspondingly weakened) by requiring $S(x)$ (and $S(y)$). For example, 2.1 would become

$$S(\overline{\neg}x) \mathbin{\&} S(x) \to (T(\overline{\neg}x) \leftrightarrow \neg T(x)).$$

The next group of axioms give necessary conditions for significance which render these redundant except in the case of 2.2 and 2.3.

Group 3:

3.1 $S(\neg x) \to S(x)$

3.2 $S(x \mathbin{\overline{\&}} y) \to (Tx \mathbin{\&} Ty)$ or $(Sx \mathbin{\&} \neg Tx)$ or $(Sy \mathbin{\&} \neg Ty)$

3.3 $S(\overline{\forall}ux) \to \forall y T(\overline{Sb}(x,u,y))$ or $\exists y(S(\overline{Sb}(x,u,y) \mathbin{\&} \neg T(\overline{Sb}(x,u,y)))$

3.4 $S[Tx] \to S(x)$

3.5 $S[Sx] \to S(x)$

We observe that on the basis of 3.1 and earlier axioms, the two candidates for defining $F(x)$, namely $S(x) \mathbin{\&} \neg T(x)$ and $T(\overline{\neg}x)$, become classically equivalent:

$$S(x) \rightarrow S(\overline{\neg x}) \qquad 1.1$$

$$\rightarrow T(\overline{\neg x}) \leftrightarrow \neg T(x) \qquad 2.1$$

$$S(x) \,\&\, \neg T(x) \rightarrow T(\overline{\neg x}) \;;$$

and

$$T(\overline{\neg x}) \rightarrow S(\overline{\neg x}) \qquad 1.7$$

$$\rightarrow S(x) \,\&\, (T(\overline{\neg x}) \leftrightarrow \neg T(x)) \qquad 3.1,\ 2.1$$

$$\rightarrow S(x) \,\&\, \neg T(x).$$

We must of course express this definition by conditions analogous to 1) and 2) above:

1S) $F(x) \leftrightarrow S(x) \,\&\, \neg T(x)$

2S) $T[F(x)] \leftrightarrow T[S(x) \,\&\, \neg T(x)]$

$S[F(x)] \leftrightarrow S[S(x) \,\&\, \neg T(x)]$.

We note that these conditions are indeed provable in KF, as the reader can easily check.

Thus 3.2 and 3.3 may be written

3.2' $S(x \,\&\, y) \rightarrow (Tx \,\&\, Ty)$ or $F(x)$ or $F(y)$

3.3' $S(\overline{\forall} ux) \rightarrow \forall y\, T(\overline{Sb}(x,u,y))$ or $\exists y\, F(\overline{Sb}(x,u,y))$

FROM T AND S TO T AND F

We are now able to check some of the axioms of KF using our conditions in S and T, and the definition of F. Axiom 1 is immediate from 1.6, 2.4. Axioms 7 is trivial from the definition of F as $(S(x) \,\&\, \neg T(x))$. Axiom 2, 5a and 6a are obtained as follows.

2a; to see $T(\overline{\neg x}) \leftrightarrow F(x)$.

This is just the alternative definition of F we already checked.

2b; to see $F(\overline{\neg x}) \leftrightarrow T(x)$

$$F(\overline{\neg x}) \rightarrow S(\overline{\neg x}) \,\&\, \neg T(\overline{\neg x})$$

$$\rightarrow S(x) \,\&\, \neg\neg T(x) \qquad 2.1, 3.1$$

$$\rightarrow T(x) \;;$$

	$T(x) \to S(x)$	1.7
	$\to S(\overline{\neg} x)$	1.1
	$\to \neg T(\overline{\neg} x)$	2.1 ■
5a;	to see $T[T(x)] \leftrightarrow T(x)$.	
	$T[T(x)] \to S[T(x)]$	1.7
	$\to T[T(x)] \leftrightarrow T(x)$	2.5
	$\to T(x)$;	
	$T(x) \to S(x)$	1.7
	$\to S[T(x)]$	1.3
	$\to T[T(x)]$	2.5 ■
5b;	to see $F[T(x)] \leftrightarrow F(x)$.	
	$F[T(x)] \to S[T(x)] \,\&\, \neg T[T(x)]$	(1S)
	$\to S(x) \,\&\, \neg T(x)$	3.4, 2.5
	$\to F(x)$;	
	$F(x) \to S(x) \,\&\, \neg T(x)$	(1S)
	$\to S[Tx] \,\&\, \neg T[T(x)]$	1.3, 2.5
	$\to F[T(x)]$	(1S) ■
6a;	to see $T[F(x)] \leftrightarrow F(x)$.	
	$T[F(x)] \to S[F(x)]$	1.7
	$\to S[S(x) \,\&\, \neg T(x)]$	(2S)
	$\to T[S(x)] \,\&\, T[\neg T(x)]$	2.2
	$\to S(x) \,\&\, \neg T[T(x)]$	1.7, 2.6, 2.5
	$\to S(x) \,\&\, \neg T[T(x)]$	5a
	$\to F(x)$;	(1S)
	$F(x) \to S(x) \,\&\, \neg T(x)$	(1S)
	$\to S[S(x)] \,\&\, S[T(x)]$	1.4, 1.3
	$\to T[S(x)] \,\&\, S[\neg T(x)]$	2.6, 1.1

$$\rightarrow T[S(x) \ \& \neg T(x)] \qquad 1.2,\ 2.2$$

$$\rightarrow T[F(x)] \qquad (2S)\blacksquare$$

6b; to see $F[F(x)] \leftrightarrow T(x)$.

$$F[F(x)] \leftrightarrow S[F(x)] \ \& \neg T[F(x)] \qquad (1S)$$

$$\leftrightarrow S[S(x) \ \& \neg T(x)] \ \& \neg F(x) \qquad (2S),6a$$

$$\rightarrow (T[S(x)] \ \& \ T[\neg T(x)]) \ \text{or} \qquad 3.2$$

$$(S[S(x)] \ \& \neg T[S(x)]) \ \text{or}$$

$$(S[\neg T(x)] \& \neg T[\neg T(x)])$$

$$\rightarrow (S(x) \ \& \neg T(x)) \ \text{or} \qquad 1.2,\ 2.1,\ 5a$$

$$(S(x) \ \& \neg S(x)) \ \text{or} \qquad 3.5,\ 2.6$$

$$(\neg\neg T[T(x)]) \qquad 2.1$$

Recalling that we still have $\neg F(x)$, the first two disjuncts cancel, so

$$\rightarrow T(x) \qquad 5a;$$

$$T(x) \rightarrow \neg F(x) \qquad \text{tautology, (1S)}$$

Thus it will suffice by (1) to get $S[S(x) \ \& \neg T(x)]$

$$T(x) \rightarrow S(x)$$

$$\rightarrow S[S(x)] \ \& \ S[\neg T(x)] \qquad 1.4,\ 1.3,\ 1.1$$

$$\rightarrow S[S(x) \ \& \neg T(x)] \qquad 1.2$$

$$\rightarrow F[F(x)] \qquad \text{by (1)} \blacksquare$$

We observe that to obtain axioms 5 and 6 we have used all of the listed properties except the quantifier conditions.

Recall that we have also checked that the two possible definitions of F are classically equivalent. We now observe that in our present setting, two concepts will not be equivalent in a very interesting way unless they are significant for the same arguments. The two possible definitions of F are so equivalent, but this equivalence uses almost all the group 3 axioms.

Remark. From the S-axioms of groups 1-3 we get

$$S[S(x) \ \& \neg T(x)] \leftrightarrow S[T(\bar{\neg} x)].$$

Proof. $S[S(x) \& \neg T(x)] \to (T[S(x)] \& T[\neg T(x)]$ or $\quad$ 3.2

$(S[S(x)] \& \neg T[Sx])$ or

$(S[\neg T(x)] \& \neg T[\neg T(x)])$

$\to S(x)$ or $\quad$ 1.7, 2.6 or 1.7, 3.1, 3.4

$S(x)$ or $\quad$ 3.5

$S(x)$ $\quad$ 3.1,3.4

$\to S(\bar{\neg} x)$ $\quad$ 1.1

At this point we need the versions of our axioms which allow terms in formulas involving iterations of S, T (namely 1.3, 1.4, 2.5, 2.6, 3.4, 3.5). We could avoid this by considering the term -free version of $T(\bar{\neg} x)$, namely $\forall y(\neg(x,y) \to T(y))$, and supposing

$$S[\forall y(\bar{\neg}(x,y) \to T(y)] \leftrightarrow \forall y(\bar{\neg}(x,y) \to S[T(y)])$$

$$\leftrightarrow \exists y(\bar{\neg}(x,y) \& S[T(y)]).$$

For simplicity we proceed with terms.

$S(\bar{\neg} x) \to S[T(\bar{\neg} x)]$ $\quad$ 1.3

For the reverse direction,

$S[T(\bar{\neg} x) \to S(\bar{\neg} x)$ $\quad$ 3.4

$\to S(x)$

$\to S[S(x)] \& S[\neg T(x)]$ $\quad$ 1.4; 1.3, 1.1

$\to S[S(x) \& \neg T(x)]$ $\quad$ 1.2 ■

Note that $(S[\theta] \leftrightarrow S[\phi]) \& (\theta \leftrightarrow \phi) \to (T[\theta] \leftrightarrow T(\phi))$, that is, S-equivalence and classical equivalence imply T-equivalence.

In order to check axioms 3 and 4, we need one more group of axioms. These are the converses of 3.2 and 3.3

Group 4:

4.1 $(Sx \& \neg T(x))$ or $(Sy \& \neg T(y)) \to S(x \bar{\&} y)$

4.2 $(\exists y(S(\overline{Sb}(x,u,y)) \& \neg T(\overline{Sb}(x,u,y)) \to S(\bar{\forall} ux)$

We now check axioms 3 and 4

3a; to see $T(x \bar{\&} y) \leftrightarrow T(x) \& T(y)$.

$T(x \& y) \to S(x \bar{\&} y)$ $\quad$ 1.7

$$\rightarrow (T(x) \mathbin{\&} T(y))\ ; \qquad 2.2$$

$$T(x) \mathbin{\&} T(y) \rightarrow S(x) \mathbin{\&} S(y) \qquad 1.7$$

$$\rightarrow S(x \mathbin{\overline{\&}} y) \qquad 1.2$$

$$\rightarrow T(x \mathbin{\overline{\&}} y) \qquad 2.2 \blacksquare$$

3b; to see $F(x \mathbin{\overline{\&}} y) \leftrightarrow F(x)$ or $F(y)$.

$$F(x \mathbin{\overline{\&}} y) \rightarrow S(x \mathbin{\overline{\&}} y) \mathbin{\&} \neg T(x \mathbin{\overline{\&}} y) \qquad (1S)$$

$$\rightarrow \neg(T(x) \mathbin{\&} T(y)) \mathbin{\&} \qquad 2.2$$

$$(S(x) \mathbin{\&} \neg T(x)) \text{ or } \qquad 3.2$$

$$(S(y) \mathbin{\&} \neg T(y))$$

$$\rightarrow F(x) \text{ or } F(y)\ ; \qquad (1S)$$

$$F(x) \rightarrow Sx \mathbin{\&} \neg Tx \qquad (1S)$$

$$\rightarrow S(x \mathbin{\overline{\&}} y) \mathbin{\&} \neg T(x \mathbin{\overline{\&}} y) \qquad 4.1,\ 2.2$$

$$\rightarrow F(x \mathbin{\overline{\&}} y) \qquad (1S)$$

Similarly for $F(y)$, so we are done. ■

The proof of axiom 4 is exactly similar and is omitted.

KF is essentially axiomatized in terms of T and F; S is essentially a defined notion. We now have a version which is essentially axiomatized in terms of T and S, taking F to be essentially defined. We shall call this theory KS; specifically, it has Peano arithmetic (allowing formulas of L in the induction scheme), the axioms of groups 1-4 (with the proper relativization to $Sent_L$ in 4), the "definitions" (1S), (2S), and the restriction axiom

(LS) $$T(x) \rightarrow \overline{Sent_L}(x)$$

$$S(x) \rightarrow \overline{Sent_L}(x)\ .$$

We now have

Theorem. KF and KS are alternative axiomatizations of the same theory.

Proof. We have checked axioms 1-7 of KF, so we need only check (L), (1), and (2).

(L): $$F(x) \rightarrow S(x) \mathbin{\&} \neg T(x) \qquad (1S)$$

$$\rightarrow \overline{Sent_L}(x) \qquad (LS) \blacksquare$$

(1): $T(x) \text{ or } F(x) \leftrightarrow T(x) \text{ or } S(x) \,\&\, \neg T(x))$ (1S)

$\leftrightarrow S(x)$ 1.7, tautology ■

(2a): to see $T[S(x)] \leftrightarrow T[T(x) \text{ or } F(x)]$.

$T[T(x) \text{ or } F(x)] \leftrightarrow T[T(x)] \text{ or } T(F(x)]$

by Lemma 8 for L

$\leftrightarrow T(x) \text{ or } F(x)$ 5a, 6a

$\leftrightarrow S(x)$

$\leftrightarrow S[S(x)] \,\&\, S(x)$ 1.4, 3.5

$\leftrightarrow T[S(x)]$ 1.7, 2.6 ■

(2b): to see $F[S(x)] \leftrightarrow F[T(x) \text{ or } F(x)]$.

$F[T(x) \text{ or } F(x)] \leftrightarrow F[T(x) \,\&\, F[F(x)]$

by Lemma 8

$\leftrightarrow Fx \,\&\, Tx$ 5b, 6b

$\leftrightarrow$ nix , 7

while

$F[S(x)] \leftrightarrow S[S(x)] \,\&\, \neg T\, S(x)]$ (1S)

$\leftrightarrow S(x) \,\&\, \neg T[S(x)]$ 1.4, 3.5

$\leftrightarrow S(x) \,\&\, \neg S(x)$ 1.4, 2.6, 1.7

$\leftrightarrow$ nix

Thus the two are equivalent. This completes the proof of the theorem.

QUESTIONS

1. a) Is there an axiomatization of $\{\sigma \mid KF \vdash T[\sigma]\}$ which is natural and formulated entirely within the domain of significant sentences, $\{\sigma \mid KF \vdash S[\sigma]\}$.
   b) Similarly for the relation $\Gamma \vdash_S \sigma$ defined by $KF + \{T[\gamma] \mid \gamma \in \Gamma\} \vdash T[\sigma]$.

2. If $KF \vdash T[\sigma]$, is there a proof $p = \langle p_o, p_1, \ldots, p_n \rangle$, $p_n = \sigma$ such that for each i, $KF \vdash T[p_i]$?

We remark that $\{\sigma \mid KF \vdash S[\sigma]\}$ is closed under negation, but that

$\{\sigma \mid KF \vdash \sigma \rightarrow S[\sigma]\} = \{\sigma \mid KF \vdash \sigma \leftrightarrow T[\sigma]\}$

is not.

MODIFICATIONS OF THE T , F AXIOMS

It would be natural to relativize the quantifiers in axioms 5b and 6b to $Sent_L$ or even to S. We note that in the presence of $T(x) \rightarrow \overline{Sent}_L(x)$, $T(x) \rightarrow S(x)$ and the analogues for F, relativization does not affect most parts of 5,6.

$T(x) \rightarrow T[T(x)]$

$F(x) \rightarrow T[F(x)]$

$T(x) \rightarrow F[F(x)]$

$F(x) \rightarrow F[T(x)]$,

while we clearly want

$T[F(x)] \rightarrow F(x)$

$T[T(x)] \rightarrow T(x)$

unrelativized. In the case of

$F[F(x)] \rightarrow T(x)$

$F[T(x)] \rightarrow F(x)$,

however, the relativizations are both natural and make a difference. They amount to the unrelativized

5c $F[T(x)] \rightarrow F(x)$ or $\neg Sent(x)$

6c $F[F(x)] \rightarrow T(x)$ or $\neg Sent(x)$

5d $F[T(x)] \rightarrow F(x)$ or $\neg S(x)$

6d $F[F(x)] \rightarrow T(x)$ or $\neg S(x)$

These conditions are still adequate to prove the schemes

$T[\sigma] \rightarrow \sigma$

$F[\sigma] \rightarrow \neg\sigma$

What about the converses of c, d ? In the case of c, the Kripke method yields models. In the case of d, the converses are highly suspect because of the modified liar:

$\lambda \leftrightarrow \neg S[\lambda]$ or $F[\lambda]$

$S[\lambda] \to (\lambda \leftrightarrow F[\lambda])$	def of λ
$\to (\lambda \leftrightarrow T[\lambda])$	truth schema
$\to$ contradiction	
$\neg S[\lambda]$	
$F[T[\lambda]]$ & $F[F[\lambda]]$	proposed converse
$F[T[\lambda]$ or $F[\lambda]]$	or
$F[S[\lambda]]$	def of S
$T[\neg S[\lambda]]$	$\neg$
$T[\lambda]$	def of λ, or
$S[\lambda]$	
contradiction	

One might, however, consider

5e $\quad F[T(x)] \leftrightarrow F(x)$ or $(\neg S(x)$ & $S[S(x)])$

6e $\quad F[F(x)] \leftrightarrow T(x)$ or $(\neg S(x)$ & $S[S(x)])$

Assuming that $S(x) \leftrightarrow T(x)$ or $F(x)$ etc., the conditions e are strangely circular, but the obvious modified liar

$\lambda \leftrightarrow (F[\lambda]$ or $(\neg S[\lambda]$ & $S[S[\lambda]]))$

$\leftrightarrow F[T[\lambda]]$

seems to lead only to the conclusions $\neg\lambda$, $\neg F[\lambda]$, $\neg S[\lambda]$, $\neg S[S[\lambda]]$, which appear to contain no contradiction.

REFERENCES

1. Bealer, George. Quality & Concept, The Clarendon Press, Oxford, 1982.
2. Feferman, Solomon. Arithmetization of mathematics in a general setting, Fundamenta Mathematica 49 (1960), 35-92.
3. Gödel, Kurt. Russell's mathematical logic, in the philosophy of Bertrand Russell,Ed. Paul A. Schilpp,the Library of Living Philosophers,Evanston, Ill. (New York: The Tudor Publishing Company, 1944) pp. 125-153.

Reprinted in Philosophy of Mathematics, Ed. Paul Benacerraf and Hilary Putnam, Prentice-Hall, Englewood Cliffs, New Jersey, 1964, pp. 211-232.

3. Kripke, Saul. Outline of a theory of truth, J. of Phil. 78 (1975), 690-716.

4. Maddy, Penelope. Proper classes, J. of Symbolic Logic 48 (1983), pp. 113-139.

5. Reinhardt, William. Some remarks on extending and interpreting theories with partial predicates for truth. Submitted to the Journal of Philosophical logic, 1982.

NOTE

1. Indeed, we can interpret KF in the theory without S. To do this we apply the recursion theorem to find a primitive recursive term $\overline{I}$ such that the corresponding function I is a mapping from the language with S to that without S satisfying

$$(\neg\theta)^I = \neg(\theta^I)$$
$$(\theta\&\phi)^I = (\theta^I) \& (\phi^I)$$
$$(\forall u\theta)^I = \forall u(\theta^I)$$
$$\alpha^I = \alpha$$

for α atomic arithmetical, and

$$(T(x))^I = T(\overline{I}(x))$$
$$(F(x))^I = F(\overline{I}(x))$$
$$(S(x))^I = T(\overline{I}(x)) \text{ or } F(\overline{I}(x))$$

for all other atomic formulas. It is now not difficult to check that I is a suitable interpretation.

CHARACTERIZATIONS OF ELEMENTARY INTERPRETATIONS IN CATEGORY THEORY

A. M. Sette/Department of Mathematics, Universidade Estadual de Campinas, Campinas, São Paulo, Brazil

L. W. Szczerba*/Department of Mathematics, Iowa State University, Ames, Iowa

Abstract. We are concerned with interpretations of elementary theories as defined in Szczerba [12]. It will be proved that such interpretations are functors between categories whose objects are structures and whose morphisms are isomorphisms between structures. The necessary and sufficient condition for a functor to be an interpretation will be provided.

1. INTRODUCTION

Two classical theorems on definability theory provide, almost, an algebraical characterization of the notion definability. One of them is known by the name of Padoa's method of proving nondefinability (see Padoa [8] and Tarski [15]). The other is Beth's theorem (Beth [1]). To state exactly what "almost" means, in this context, we shall state both theorems in terms of category theory.

Let St_σ be a class of structures of signature σ. We may make this class into a category by choosing model theoretic isomorphisms as morphisms. The resulting category is denoted by St_σ^{iso} .

*Current Affiliation: Instytut Matematyki, Uniwersytet Warszawski, Warsaw, Poland

Let $\mathbb{U}$ be a structure, the operation consisting of adjoining to $\mathbb{U}$ a relation defined by a given formula is called definitional extension. (See below for exact definitions of the notions involved).

Padoa's theorem may be expressed as follows:

Definitional extensions are functors $\mathcal{St}_\sigma^{iso}$ to $\mathcal{St}_{\sigma^\frown\langle n\rangle}^{iso}$ for some σ and n.

It is easy to see that the Padoa's theorem establishes a necessary condition for definitional extensions. The sufficient condition is provided by Beth's theorem

If Γ is a functor from St_σ^{iso} to $St_{\sigma^\frown\langle n\rangle}^{iso}$ satisfying

(i) for any structure $\mathbb{U}$ of signature σ, $\mathbb{U}$ is the reduct of $\Gamma\mathbb{U}$ by the last relation:

(ii) there is a set of sentences Φ such that $\Gamma\mathbb{U}$ is the only expansion of $\mathbb{U}$ to a model of Φ; then Γ is definitional extension.

The condition (ii) involving notions of sentence and model, is responsible for the use of "almost" at the beginning of this section.

It is possible to get rid of the condition (ii) but at the expense of extending the category $\mathcal{St}_\sigma^{iso}$ by including some partial isomorphisms. The notion of the partial isomorphism and their hierarchy is due to Ehrenfeucht [3] and Fraïse [4]. (The precise definition of a partial isomorphism of order k will be given below). Now we shall state only that the category with structures of signature σ as object and partial isomorphism of order k as morphism will be denoted by $\mathcal{St}_\sigma^k$. The common generalization of Padoa's and Beth's theorems is the following (see theorem 7):

Γ is a definitional extension if and only if for some natural numbers k and n the diagram

$$\begin{array}{ccc} \mathcal{St}_\sigma^k & \xrightarrow{\ \Gamma\ } & \mathcal{St}_{\sigma^\frown\langle n\rangle}^0 \\ & \searrow\!\!\searrow \quad \swarrow\!\!\swarrow & \\ & \mathcal{Set} & \end{array}$$

commutes.

The symbol $\mathcal{Set}$ denotes the category of sets with partial functions as morphisms, and $\Rightarrow$ denotes the forgetful functor.

The above theorem may be generalized still further to give a necessary and sufficient condition for Γ to be an interpretation as defined in Szczerba [12] (see also Lindenbaum - Tarski [5]). The generalization is given below (theorems 7.3 and 7.4).

Main results of this paper, obtained on the basis of some ideas of Sette and Sette [9], have been reported in Sette and Szczerba [10].

2. NOTATIONS

For a function f we denote by Dom f the domain of f and by Rg f its range. Let $X \subseteq \mathrm{Dom}\, f$ and $Y \subseteq \mathrm{Rg}\, f$. The image (coimage) of X (of Y) will be denoted by $\vec{f}X$ (respectively by $\overleftarrow{f}Y$). The value of function f for the argument x will be denoted by fx. Sometimes when x will be given by a complicated formula, we shall add, for clarity, parenthesis around the argument.

The composition of two functions f and g will be denoted by fg. It is understood that $(fg)x = f(gx)$. Let F and G be two sets of functions. By FG we mean the set of all compositions of functions from F and G:

$$FG = \{fg : f \in F \text{ and } g \in G\}.$$

Instead of FF we shall write F^2, instead of FFF,F^3 etc. Following Mostowski [6] by F^{ω} we shall denote the set of all finite compositions of elements of the set F. Similar notation will be applied for individual functions: $f^2 = ff$, $f^3 = fff$ etc. We shall write f^{ω} instead of $\{f\}^{\omega}$.

A function a defined for all natural numbers less than p is called sequence. The number p is called length of the sequence a and denoted by $\ell h\ a$.

If a is a sequence of length α, and $\beta < \alpha$ then to denote the β-th element of the sequence a we write a_β rather than $a(\beta)$ or $a\beta$. The finite sequence a of length n will be denoted also by $\langle a_i \rangle_{i<n}$ or more explicitely by $\langle a_o, a_1, \ldots, a_{n-1} \rangle$.

Let a and b be two finite sequences of length m and n respectively. Concatenation or juxtaposition of a and b will be denoted by $a^\frown b$. Thus

$$a^\frown b = \langle a_o, \ldots, a_{m-1} \rangle^\frown \langle b_o, \ldots, b_{n-1} \rangle = \langle a_o, \ldots, a_{m-1}, b_o, \ldots, b_{n-1} \rangle$$

2.1. Definition. The relation $\preccurlyeq$ holds for finite sequence of positive natural numbers and is the least reflexive and transitive relation such that for every σ and τ being appropriate sequences, and any positive natural numbers m, n,

(i) $\tau^\frown\sigma \preccurlyeq \tau^\frown\langle m\rangle^\frown\sigma$

(ii) $\tau^\frown\langle m\rangle^\frown\sigma \preccurlyeq \tau^\frown\langle m+n\rangle^\frown\sigma$

(iii) $\tau^\frown\langle m+n\rangle^\frown\sigma \preccurlyeq \tau^\frown\langle m,n\rangle^\frown\sigma$

To be sure that $\preccurlyeq$ is an order we have to prove that if $\tau \preccurlyeq \sigma$ and $\sigma \preccurlyeq \tau$ then $\sigma = \tau$. Suppose therefore that there is a finite sequence (of length k) of finite sequences of positive natural numbers such that

$$\Sigma_{k-1} = \Sigma_0 = \tau ,$$

for some $p < k$ $\Sigma_p = \sigma$ and for each $i < k$ we have as applications of (i)-(iii)

$$\Sigma_i \preccurlyeq \Sigma_{i+1} . \qquad (1)$$

It is easy to see that in the rules (i)-(iii) the sequence on the left is not longer than the sequence on the right. In fact in case of (i) and (iii) they are actually shorter. Thus the applications of rules (i) and (iii) cannot occur among (1). Similarly the sum of terms of sequences on the left hand side of the rule (ii) is less than the sum of terms on the right hand side. Thus the application (ii) also cannot occur among (1). Therefore (1) is due to the reflexivity rule, thus $\sigma = \tau$.

2.2. Lemma. Suppose

$$\sigma^\frown\langle p\rangle \preccurlyeq \tau^\frown\langle q\rangle$$

Then

(i) $\sigma^\frown\langle p\rangle \preccurlyeq \tau$

or

(ii) $\sigma \preccurlyeq \tau$ and $p \leq q$.

<u>Proof</u>. Since $\sigma^\frown\langle p\rangle \preccurlyeq \tau^\frown\langle q\rangle$ then either $\sigma^\frown\langle p\rangle = \tau^\frown\langle q\rangle$, in which case we have $\sigma \preccurlyeq \tau$ and $p \leq q$, or else there is a sequence Σ of sequences of positive natural numbers such that

$$\Sigma_0 = \sigma^\frown\langle p\rangle$$

$$\Sigma_{\ell h\Sigma-1} = \tau^\frown\langle q\rangle$$

and for every $i < \ell h - 1$

$$\Sigma_i \preccurlyeq \Sigma_{i+1}$$

by an application of one of the rules (ii)-(iii) from the definition 2.1.

Suppose $\ell h\Sigma = 2$. Then

$$\sigma^\frown\langle p\rangle \preccurlyeq \tau^\frown\langle q\rangle$$

itself is an application of the rules.

Case 1. $\sigma^\frown\langle p\rangle = \sigma'^\frown\tau'$ and $\tau^\frown\langle q\rangle = \sigma'^\frown\langle m\rangle^\frown\tau'$

then either τ' is empty, and we obtain (i) or τ' is not empty: $\tau' = \tau''^\frown\langle r\rangle$. Thus we have $p = r = q$ and

$$\sigma = \sigma'^\frown\tau'' \preccurlyeq \tau = \sigma'^\frown\langle m\rangle^\frown\tau'' .$$

Case 2. $\sigma^\frown\langle p\rangle = \sigma'^\frown\langle m\rangle^\frown\tau'$ and $\tau^\frown\langle q\rangle = \sigma'^\frown\langle m+n\rangle^\frown\tau'$

Case 3. $\sigma^\frown\langle p\rangle = \sigma'\langle m+n\rangle^\frown\tau'$ and $\tau^\frown\langle q\rangle = \sigma'^\frown\langle m,n\rangle^\frown\tau'$.

The proofs in these two cases are analogous to that in the case 1.

Now let $\ell h\Sigma > 2$. Then $\Sigma_1 = \sigma'^\frown\langle r\rangle$ and using induction hypothesis we have the following two possibilities:

$$\sigma^\frown\langle p\rangle \preccurlyeq \sigma' \tag{2}$$

or

$$\sigma \preccurlyeq \sigma' \text{ and } p \le r \tag{3}$$

and

$$\sigma'^\frown\langle r\rangle \preccurlyeq \tau \tag{4}$$

or

$$\sigma' \preccurlyeq \tau \text{ and } r \le q \tag{5}$$

In case we have (2) and (4) we get

$$\sigma^\frown\langle p\rangle \preccurlyeq \sigma' \preccurlyeq \sigma'^\frown\langle r\rangle \preccurlyeq \tau$$

i.e. we get (i) by transitivisty of $\preccurlyeq$.

In case of (2) and (5) we get

$$\sigma^\frown\langle p\rangle \preccurlyeq \sigma' \preccurlyeq \tau$$

i.e. again we get (i).

Suppose (3) and (4). Thus

$$\sigma^\frown\langle p\rangle \preccurlyeq \sigma'^\frown\langle r\rangle \preccurlyeq \tau$$

by definition 2.1 (ii), and we get (i) again.

Finally let (3) and (5). Therefore have

$$\sigma \preccurlyeq \sigma' \preccurlyeq \tau$$

and

$$p \leq r \leq q$$

and we get (ii) by transitivity of $\preccurlyeq$ and $\leq$. ⊣

By a structure we mean a pair $\mathbb{U} = \langle U,P\rangle$. The first element of the pair is a set called the universe and is denoted often by $|\mathbb{U}|$. The second element is a sequence of relations over U. We say that σ is a signature of a structure $\mathbb{U}$ if $\ell hP = \ell h\sigma$ and for every $i < \ell h\sigma$ we have

$$P_i \subseteq U^{\sigma_i}$$

i.e. P_i is σ_i-ary relation over U. Instead of $\langle U,\langle P_o,\ldots,P_n\rangle\rangle$ we shall often write $\langle U;P_o,\ldots,P_n\rangle$. The class of all structures of signature σ will be denoted by St_σ.

Let $P \subseteq U^n$ be any n-ary relation on U and let $V \subseteq U$. The relation

$$P\restriction V = P \cap V^n = \{\langle a_o,\ldots,a_{n-1}\rangle : \langle a_o,\ldots,a_{n-1}\rangle \in P \cap a_o \in V \cap \ldots \cap a_{n-1} \in V\}$$

is called the restriction of P to the set V. Let $\approx$ be any equivalence relation on U. The equivalence class of an element a is denoted by $[a]_{/\approx}$. Suppose moreover that $\approx$ is a congruence with respect to P, i.e. for any $a_o,\ldots,a_{n-1}, b_o,\ldots,b_{n-1} \in U$ we have

$$a_o \approx b_o \wedge \ldots \wedge a_{n-1} \approx b_{n-1} \wedge P(a_o,\ldots,a_{n-1}) \rightarrow P(b_o,\ldots,b_{n-1}).$$

By the quotient relation of P we mean the relation $P_{/\approx}$ on equivalence classes of elements of U:

$$P/_{\approx}([a_o]_{/\approx},\ldots,[a_{n-1}]_{/\approx}) \leftrightarrow P(a_o,\ldots,a_{n-1}).$$

We assume that the proof of the soundness of this definition is known.

Let $\mathbb{U} = \langle U, P_o, \ldots, P_m \rangle$ be a structure of signature σ i.e. $\mathbb{U} \in St_\sigma$ and let $V \subseteq U$. The structure $\mathbb{U}\restriction V = \langle V, P_o\restriction V, \ldots, P_m\restriction V\rangle$ is called the restriction of $\mathbb{U}$ to V. Let $\approx$ be a congruence in $\mathbb{U}$ i.e. congruence with respect to all relations $P_o, \ldots, P_m$. The structure $\mathbb{U}_{/\approx} = \langle U_{/\approx}, R_{o/\approx}, \ldots, R_{m/\approx} \rangle$ is called the quotient structure by the congruence $\approx$.

To describe structures of signature σ internally, we use the elementary language of signature σ, i.e. the set of all formulas built up in usual way from the logical symbols $\exists$, $\vee$, $\neg$ (for existential quantifier, alternative and negation), and the infinite sequence of variables $v_o, v_1, \ldots$, equality sign $=$, and predicate symbols $P_o, \ldots, P_n$ where $n = \ell h\sigma - 1$. The symbol P_i is σ_i-ary. The set of all formulas with exactly $v_o, \ldots, v_{k-1}$ occuring free will be denoted by L^k_σ. In particular L^o_σ is the set of formulas without free variables, also called sentences. The symbols, $\forall, \wedge, \rightarrow, \leftrightarrow$ (universal quantifier, conjuction, implication, and equivalence) will also be used, but to denote abbreviations of appropriate formulas, not as symbols of logical operations. We shall also use the following abbreviations. Let $X = \{\phi_o, \ldots, \phi_n\}$ be a finite set of formulas. We shall use generalized alternative and conjuction

$$\vee X = \bigvee_{i=0}^{n} \phi_i = \phi_o \vee \ldots \vee \phi_n$$

$$\wedge X = \bigwedge_{i=0}^{n} \phi_i = \phi_o \wedge \ldots \wedge \phi_n$$

We shall also use occasionally the symbols $\exists, \forall, \neg, \vee, \wedge$ in metamathematical expressions. We hope it will not introduce any misunderstanding.

We assume that the reader is familiar with the notion of satisfaction and understands the expression $\mathbb{U} \models \phi[a]$ where a is a finite sequence of elements of $\mathbb{U}$. In case $\phi \in L^o_\sigma$ we often write $\mathbb{U} \models \phi$. The formula $\phi \in L^k_\sigma$ defines in $\mathbb{U} \in St_\sigma$ the k-ary relation

$$\phi\mathbb{U} = \{\langle a_o, \ldots, a_{k-1}\rangle : \mathbb{U} \models \phi[a_o, \ldots, a_k]\}.$$

Next we introduce a classification of formulas which is a modification of the prenex normal form classification. With any finite sequence τ of natural numbers we shall associate a set $L_\sigma(\tau)$ of formulas:

2.3. Definition

(i) if $\phi \in L_\sigma$ is quantifier-free then for any τ we have $\phi \in L_\sigma(\tau)$

(ii) if $\phi \in L_\sigma(\tau)$ then $(\neg\phi) \in L_\sigma(\tau)$

(iii) if $\phi, \psi \in L_\sigma(\tau)$ then $(\phi \vee \psi) \in L_\sigma(\tau)$

(iv) if $\phi \in L_\sigma(\tau)$ then $(\exists v_{i_o} \ldots \exists v_{i_p} \phi) \in L_\sigma(\tau^\frown\langle q\rangle)$ for any $q > p$

(v) $L_\sigma(\tau)$ are the least sets satisfying the conditions (i)-(iv).

Thus $L_\sigma(\emptyset)$ is the set of all quantifier-free formulas. Note that

$$(\exists v_o v_o = v_o) \in L_\sigma(\langle 1 \rangle) \quad \text{and} \quad (\forall v_o \exists v_1 v_o = v_1) \in L_\sigma(\langle 1,1 \rangle).$$

The next lemma is easy to prove by induction with respect to the length of formula ϕ and the number of applications of rules from definition 2.1.

2.4. Lemma

If $\tau \geqslant \tau'$ then $L_\sigma(\tau') \subseteq L_\sigma(\tau)$.

When $\phi \in L_\sigma(\tau)$ we say that ϕ is of type τ. It is not difficult to see that $L_\sigma(\tau)$ is closed under forming finite conjuctions and alternatives, as well as implications, equivalence and negation. Quantification however changes the type of a formula.

We shall combine the above notation with the one concerning free variables. By $\phi \in L^n_\sigma(\tau)$ we shall mean that ϕ is of type τ and has exactly variables $v_o,\ldots,v_{n-1}$ occuring free.

Sometimes it is convenient to use a simpler classification.

2.5. Definition

(i) if ϕ is a quantifier-free formula then $dpt\,\phi = 0$

(ii) $dpt(\neg\phi) = dpt\,\phi$

(iii) $dpt(\phi \vee \psi) = \max\{dpt\,\phi,\ dpt\,\psi\}$

(iv) $dpt(\exists v\phi) = dpt\,\phi + 1$

We say that $dpt\,\phi$ is the depth of the formula ϕ. If is easy to see that if

$$\phi \in L_\sigma(\tau) \quad \text{then} \quad dpt\,\phi \leq \sum_{i=1}^{\ell h\,\tau-1} \tau_i .$$

The only categories we are going to discuss are categories of structures. Such a cetegory will be always understood as a pair $\langle Ob, Mor\rangle$. The class Ob is the class of all structures of some signature σ, i.e. the class St_σ. The class Mor is a function mapping pairs of elements of Ob onto a set of partial transformations of one of them into another. More precisely, let $\mathbb{U}, \mathbb{B} \in St_\sigma$ then $Mor(\mathbb{U},\mathbb{B})$ will always be a set of triples of the form $\langle \mathbb{U},\mathbb{B}, f\rangle$ where f is a partial function from $|\mathbb{U}|$ into $|\mathbb{B}|$. We shall treat morphisms as functions, and we are going to apply them to elements of structures, e.g. if $a \in |\mathbb{U}|$ then we put $\langle \mathbb{U}, \mathbb{B}, f\rangle a = fa$. If f is an identity on $|\mathbb{U}|$ then the morphism $\langle \mathbb{U}, \mathbb{U}, f\rangle$ will be denoted by $id_{\mathbb{U}}$, or simply by id.

Composition of $\langle \mathbb{U},\mathbb{B}, f\rangle$ and $\langle \mathbb{B},\mathbb{C}, g\rangle$ is $\langle \mathbb{U},\mathbb{C}, gf\rangle$, and therefore $\langle \mathbb{U}, \mathbb{B}, f\rangle^{-1} = \langle \mathbb{U},\mathbb{B}, f^{-1}\rangle$. By $Dom\langle \mathbb{U},\mathbb{B}, f\rangle$ we mean Dom f.

A functor Φ from a category $\langle Ob_1, Mor_1\rangle$ into a category $\langle Ob_2, Mor_2\rangle$ is a funtion defined on the class $Ob_1 \cup Mor_1$ satisfying the following conditions:

(a) if Φ is defined for $\mathbb{U} \in Ob_1$ then $\Phi\mathbb{U} \in Ob_2$

(b) if Φ is defined for $\mathbb{U},\mathbb{B} \in Ob_1$ and $f \in Mor_1(\mathbb{U},\mathbb{B})$ then $\Phi f \in Mor_2(\Phi\mathbb{U},\Phi\mathbb{B})$

(c) if $f \in Mor_1(\mathbb{U},\mathbb{B})$ and $g \in Mor_1(\mathbb{B},\mathbb{C})$ then $\Phi(fg) = (\Phi f)(\Phi g)$.

It is important to observe that classes Ob and Mor are always disjoint.

We shall not distinguish between functor defined in a category and its restriction to a subcategory. This should not cause any misunderstanding.

As usual when using terminology of category theory, we are going to use diagrams. The fact the Γ is a functor from category K_1 into category K_2. We shall express as usual by

$$K_1 \xrightarrow{\Gamma} K_2 .$$

The very important examples of functors are imbedding and forgetful functors. Functor Γ is an imbedding iff K_1 is a subcategory of K_2 and Γ is an identity functor for both objects and morphisms of K_1. Since an imbedding Γ, if it exists, is uniquely determined by K_1 and K_2 instead of

$$K_1 \xrightarrow{\Gamma} K_2$$

we shall write simply

$$K_1 \longrightarrow K_2 .$$

Function Γ is a forgetful functor if it maps any structure $\mathbb{U} \in \mathcal{St}_\sigma$ onto its universe $|\mathbb{U}|$ and a morphism $\langle \mathbb{U}, \mathbb{B}, f \rangle$ onto to triple $\langle |\mathbb{U}|, |\mathbb{B}|, f \rangle$. The values of this functor are always in the category $\mathcal{Set} = \langle \text{Set}, \text{Pr f} \rangle$ where Set is the class of all sets and $\text{Pr f} \langle A,B \rangle$ is the set of all partial functions defined in A and with values in B. Again a forgetful functor is uniquely determined by the category of structures on which it is defined, and therefore instead of

$$K \xrightarrow{\Gamma} \mathcal{Set}$$

where Γ is a forgetful functor we shall write

$$K \Longrightarrow \mathcal{Set}.$$

3. PARTIAL ISOMORPHISMS

As one of the major tools in this paper we shall use the notion of partial isomorphism introduced by Fraïsse [4] and indirectly by Ehrenfeucht [3].

By an isomorphism from a structure $\mathbb{U} = \langle U,P \rangle$ to a structure $\mathbb{B} = \langle V,Q \rangle$ we mean a one-to-one transformation f from $\mathbb{U}$ onto $\mathbb{B}$ preserving all relations (in a strong sense) i.e. for any $i < \ell h$ and any $a_o, \ldots, a_{\sigma_i - 1} \in |\mathbb{U}|$ we have

$$P_i(a_o, \ldots, a_{\sigma_i - 1}) \text{ iff } Q_i(fa_o, \ldots, fa_{\sigma_i - 1}) . \tag{1}$$

The set of all isomorphisms from $\mathbb{U}$ to $\mathbb{B}$ will be denoted by $\mathcal{Iso}(\mathbb{U}, \mathbb{B})$. If the domain (or range) of f is not necessarily all of $|\mathbb{U}|$ (or $|\mathbb{B}|$), but condition (1) holds for elements for which f is defined, we say that f is a partial isomorphism. Defining partial isomorphisms we introduce an hierarchy of partial isomorphisms parametrizes with finite sequences of natual numbers.

3.1. Definition

$$Iso^{\emptyset}(\mathbb{U},\mathbb{B}) = \{\langle \mathbb{U}, \mathbb{B}, f\rangle : \exists A \subseteq |\mathbb{U}|\ \exists B \subseteq |\mathbb{B}|\ \langle \mathbb{U} \upharpoonright A, \mathbb{B} \upharpoonright B, f\rangle \in$$

$$\in Iso(\mathbb{U} \upharpoonright A, \mathbb{B} \upharpoonright B)\}$$

$$Iso^{\tau^\frown\langle p\rangle}(\mathbb{U},\mathbb{B}) = \{\langle \mathbb{U}, \mathbb{B}, f\rangle : \forall X\, \exists f' \in Iso^{\tau}(\mathbb{U},\mathbb{B}),\ f \subseteq f' \wedge$$

$$\wedge (\bar{\bar{X}} \leq p \wedge X \subseteq |\mathbb{U}| \rightarrow X \subseteq \mathrm{Dom}\, f') \wedge (\bar{\bar{X}} \leq p \wedge X \subseteq |\mathbb{B}| \rightarrow X \subseteq \mathrm{Rg}\, f')\}$$

or in plain words: $f \in Iso^{\tau^\frown\langle p\rangle}(\mathbb{U},\mathbb{B})$ iff whenever we choose p elements of $|\mathbb{U}|$ (or of $|\mathbb{B}|$) it is possible to extend f to a τ-isomorphism f' defined for the chosen elements or having these elements as values.

In case $f \in Iso^{\tau}(\mathbb{U},\mathbb{B})$ we say that f is τ-isomorphism from $\mathbb{U}$ to $\mathbb{B}$.

Sometimes it is convenient to use a simpler classification using the notion of an isomorphism of depth p:

3.2. Definition

$$Iso^{p}(\mathbb{U},\mathbb{B}) = \cup \{Iso^{\tau}(\mathbb{U},\mathbb{B}) : \sum_{i=0}^{\ell h\, \tau - 1} \tau_i = p\}.$$

In other words if $f \in Iso^{\tau}(\mathbb{U},\mathbb{B})$ and p is less than or equal to the sum of terms of the sequence τ then $f \in Iso^{p}(\mathbb{U},\mathbb{B})$.

Whenever $f \subseteq f'$, we sahll also write $\langle \mathbb{U}, \mathbb{B}, f\rangle \subseteq \langle \mathbb{U}, \mathbb{B}, f'\rangle$. In fact we shall assume that when $\langle \mathbb{U}, \mathbb{B}, f\rangle \subseteq \langle \mathbb{U}', \mathbb{B}'\, f'\rangle$ then $\mathbb{U} = \mathbb{U}'$, $\mathbb{B} = \mathbb{B}'$ and $f \subseteq f'$. The letters f,g (with possible primes) will be used to denote both partial isomorphisms and their underlying functions. It should not introduce any misunderstanding.

Now we shall prove some simple properties of the set $Iso^{\tau}(\mathbb{U},\mathbb{B})$.

3.3. Theorem

(i) if $f \subseteq f' \in Iso^{\tau}(\mathbb{U},\mathbb{B})$ then $f \subseteq Iso^{\tau}(\mathbb{U},\mathbb{B})$

(ii) $Iso(\mathbb{U},\mathbb{B}) \subseteq Iso^{\tau}(\mathbb{U},\mathbb{B}) \subseteq Iso^{\emptyset}(\mathbb{U},\mathbb{B})$

(iii) if $\sigma \preccurlyeq \tau$ then $Iso^{\tau}(\mathbb{U},\mathbb{B}) \subseteq Iso^{\sigma}(\mathbb{U},\mathbb{B})$

(iv) if $p \leq q$ then $Iso^{q}(\mathbb{U},\mathbb{B}) \subseteq Iso^{p}(\mathbb{U},\mathbb{B})$

Proof. Of all three properties is by induction. In case of (i) and (ii) the induction is with respect to the length of τ. It is pretty simple and therefore is left to the reader. The induction in case of (iii) is a little more complicated.

1. If $\sigma = \emptyset$ then $Iso^{\tau}(\mathbb{U},\mathbb{B}) \subseteq Iso^{\emptyset}(\mathbb{U},\mathbb{B}) = Iso^{\sigma}(\mathbb{U},\mathbb{B})$
2. If $\tau = \emptyset$ then $\sigma = \emptyset$ and the case reduces to 1.
3. If $\sigma = \sigma'^\frown\langle p\rangle$ and $\tau = \tau'^\frown\langle q\rangle$. Then by lemma 2.2 we have either $\sigma \lessapprox \tau'$ or $\sigma \lessapprox \tau'$ and $p \leq q$.

Suppose $\sigma \lessapprox \tau'$ and $f \in Iso^{\tau}(\mathbb{U},\mathbb{B})$. Then there is some f' such that $f \subseteq f' \in Iso^{\tau'}(\mathbb{U},\mathbb{B})$. By induction hypotheses $Iso^{\tau'}(\mathbb{U},\mathbb{B}) \subseteq Iso^{\sigma}(\mathbb{U},\mathbb{B})$ thus $f \in Iso^{\sigma}(\mathbb{U},\mathbb{B})$

Suppose now $\sigma' \lessapprox \tau'$ and $p \leq q$. Let $f \in Iso^{\tau}(\mathbb{U},\mathbb{B})$ then for every X with $\overline{\overline{X}} \leq q$ there is f' with $f \subseteq f' \in Iso^{\tau}(\mathbb{U},\mathbb{B})$, and by induction hypotheses, $Iso^{\tau'}(\mathbb{U},\mathbb{B}) \subseteq Iso^{\sigma'}(\mathbb{U},\mathbb{B})$. Thus $f \subseteq Iso^{\sigma}(\mathbb{U},\mathbb{B})$.

It is obvious that $Iso(\mathbb{U},\mathbb{B}) \subseteq Iso^{\tau}(\mathbb{U},\mathbb{B})$.

Part (iv) follows easily from (iii).

Structures of signature σ and partial isomorphisms of type τ form a category denoted by $St^{\tau}_{\sigma} = \langle St_{\sigma}, Iso^{\tau}\rangle$. Similarly we introduce $St^{iso}_{\sigma} = \langle St_{\sigma}\ Iso\rangle$ and $St^{p}_{\sigma} = \langle St_{\sigma}, Iso^{p}\rangle$.

3.4. Theorem

St^{τ}_{σ} is a category i.e.

(i) $id_{\mathbb{U}} \in Iso^{\tau}(\mathbb{U},\mathbb{U})$

(ii) if $f \in Iso^{\tau}(\mathbb{U},\mathbb{B})$ then $f^{-1}\ Iso^{\tau}(\mathbb{B},\mathbb{U})$

(iii) if $f \in Iso^{\tau}(\mathbb{U},\mathbb{B})$ and $g \in Iso^{\tau}(\mathbb{B},\mathbb{C})$ then $fg \in Iso^{\tau}(\mathbb{U},\mathbb{C})$.

Proof. Part (i) is a simple consequence of 3.5(ii). Let us turn to (ii). In case $\tau = \emptyset$ the theorem is obvious. Suppose $\tau = \tau'^\frown\langle p\rangle$, and let X be a subset of $|\mathbb{U}|$ (or $|\mathbb{B}|$ of power not bigger than p. Since $f \in Iso^{\tau}(\mathbb{U},\mathbb{B})$ then there is $g \in Iso^{\tau'}(\mathbb{U},\mathbb{B})$ such that $f \subseteq g$ and $X \subseteq \mathrm{Dom}\ g$ (respectively $X \subseteq \mathrm{Rg}\ g$). By the induction hypothesis we have $f^{-1} \subseteq g^{-1}$ and $X \subseteq \mathrm{Rg}\ g$ (or $X \subseteq \mathrm{Dom}\ g$). Thus $f^{-1} \in Iso^{\tau}(\mathbb{B},\mathbb{U})$.

Finally we prove (iii). Again if $\tau = \emptyset$, the theorem is obvious. Let $\tau = \tau'^\frown\langle p\rangle$, and let X be a subset of $|\mathbb{U}|$ of power not exceeding p. Since $f \in \mathit{Iso}^{\tau}(\mathbb{U},\mathbb{B})$ there is $f' \in \mathit{Iso}^{\tau'}(\mathbb{U},\mathbb{B})$ extending f such that $X \subseteq \mathrm{Dom}\, f'$ but $\vec{f}'X$ is of power not exceeding p. Therefore there is $g' \in \mathit{Iso}^{\tau'}(\mathbb{B},\mathbb{B})$ extending g, such that $f'X \subseteq \mathrm{Dom}\, g'$ and therefore $X \subseteq \mathrm{Dom}(f',g')$. By the induction hypothesis $fg \in \mathit{Iso}^{\tau}(\mathbb{U},\mathbb{C})$ and since $fg \subseteq f'g'$ we get $fg \in \mathit{Iso}^{\tau}(\mathbb{U},\mathbb{B})$. Suppose now $X \subseteq |\mathbb{C}|$ and $\overline{\overline{X}} \leq p$. We proceed analogously. Since $g \in \mathit{Iso}^{\tau}(\mathbb{B},\mathbb{C})$ there is $g' \in \mathit{Iso}^{\tau'}(\mathbb{B},\mathbb{C})$ such that $g \subseteq g'$ and $X \subseteq \mathrm{Rg}\, g'$. Of course $\overleftarrow{g}',X$ is of power less than or equal p and thus there is $f' \in \mathit{Iso}^{\tau'}(\mathbb{U},\mathbb{B})$ such that $f \subseteq f'$ and $\overleftarrow{g}'X \subseteq \mathrm{Rg}\, f'$. Thus $X \subseteq \mathrm{Rg}(f'g')$ and by the induction hypothesis $f'g' \in \mathit{Iso}^{\tau'}(\mathbb{U},\mathbb{C})$. Thus $fg \in \mathit{Iso}^{\tau}(\mathbb{U},\mathbb{B})$.

It is worthwhile to notice that all morphisms of $\mathit{St}_{\sigma}^{\tau}$ are reversible (cf. 3.6(ii)).

Similarly we may prove

3.5. Theorem

$\mathit{St}_{\sigma}^{iso}$ is a category.

3.6. Theorem

St_{σ}^{p} is a category.

Moreover the next theorem follows easily from 3.3(iii) and (iv).

3.7. Theorem

(i) $\mathit{St}_{\sigma}^{iso}$ is a subcategory of $\mathit{St}_{\sigma}^{\tau}$

(ii) if $\tau \preccurlyeq \tau'$ then $\mathit{St}_{\sigma}^{\tau'}$ is a subcategory of $\mathit{St}_{\sigma}^{\tau}$

(iii) $\mathit{St}_{\sigma}^{iso}$ is a subcategory of St_{σ}^{p}

(iv) if $p \leq q$ then St_{σ}^{q} is a subcategory of St_{σ}^{p}.

4. INTERPRETATIONS

We shall discuss elementary interpretations as defined in Szczerba[11].

The difference will be that in the quoted paper interpretations are functions from class of structures St_σ to a class St_τ for some σ and τ . In this paper the notion of interpretation will be much more involved. It will be defined by use of generators and therefore it will be convenient to define a notion of pre-interpretation. Later on we shall single out class of interpretations as a subclass of pre-interpretations.

4.1. Definition

By a pre-interpretation from signature σ into signature τ we mean any function Γ defined in the union of St_σ, L_τ and $\cup\{\mathrm{Mor}(\mathbb{U},\mathbb{B});\ \mathbb{U},\mathbb{B}\in St_\sigma\}$ such that

(i) if $\mathbb{U}\in St_\sigma$ and Γ is defined for $\mathbb{U}$ then $\Gamma\mathbb{U}\in St_\tau$

(ii) if $\phi\in L_\tau$ and Γ is defined for ϕ then $\Gamma\phi\in L_\sigma$

(iii) if $f\in\mathrm{Mor}(\mathbb{U},\mathbb{B})$ and Γ is defined for $\mathbb{U},\mathbb{B}$ then $\Gamma f\in\mathrm{Mor}(\Gamma\mathbb{U},\Gamma\mathbb{B})$

To have this definition be sound we have to make sure that the set of formulas L_τ and set of morphisms $\cup\{\mathrm{Mor}(\mathbb{U},\mathbb{B}) : \mathbb{U},\mathbb{B}\in St_\sigma\}$ are disjoint. Unlike the case of the definition of a functor the classes need not be automatically disjoint. In fact all elements of St_σ, i.e. all structures of signature σ, are pairs: set and sequence of relations over this set, and all morphisms are triples: structure, structure and function from universe of the first structure to the universe of the second structure. Since we treat both triples and pairs as sequences, i.e. as functions defined on an initial segment of natural numbers, we may say that no triple is a pair, and therefore no structure is a morphism and conversely. A formula is a sequence of symbols. There is no reason its length has to be greater than three. In order to have a class of formulas disjoint with the classes of structures and morphisms we assume that no symbol is a structure, nor a sequence of relations defined over another symbol. Such restrictions seem to reasonable.

As an example of an pre-interpretation from signature σ to singature σ we may give an identity interpretation: it is a function mapping every structure of signature σ onto itself, every formula of signature σ onto itself and every morphism of category $St_\sigma^\emptyset$ onto itself. The identity interpretation will be denoted by $\iota\delta_\sigma$. The subscript σ will be usually ommited.

4.2. Definition

Given two pre-interpretations Γ_1 and Γ_2 from signature σ to signature τ we say that Γ_1 and Γ_2 are isomorphic, in symbols $(\Gamma_1 \cong \Gamma_2)$, iff

(i) if $\phi \in L_\tau$ then $\Gamma_1\phi = \Gamma_2\phi$

(ii) if $\mathbb{U}, \mathbb{B} \in St_\sigma$ and $\mathbb{U} \cong \mathbb{B}$ then $\Gamma_1\mathbb{U} \cong \Gamma_2\mathbb{B}$

It is easy to see that the relation of isomorphism among pre-interpretations is an equivalence relation i.e. it is reflexive symmetric and transitive. Moreover if Γ_1 and Γ_2 are isomorphic pre-interpretations from signature σ to signature τ, and Δ_1, Δ_2 are two isomorphic pre-interpretations from signature τ to signature η then $\Gamma_1\Delta_1 \cong \Gamma_2\Delta_2$.

We say the pre-interpretation Γ belongs up to isomorphism to a class X (in symbols $\Gamma \tilde{\in} X$) iff there is $\Gamma_1 \cong \Gamma$ such that $\Gamma_1 \in X$.

We shall give the precise definition of interpretation at the end of this section. We shall give first some examples of interpretations. In fact these examples will constitute system of generators of interpretations. It will be therefore convenient to introduce some terminology: Let Γ be an interpretation. When saying construction we shall refer to the restriction of Γ to a class of structures and when saying translation - to the restriction of Γ to a set of formulas. We did not find it necessary to introduce a special term for restriction of Γ to morphisms.

4.3. Definition

By the identity interpretation on signature σ we mean the function $\iota\delta_\sigma$ mapping $\mathbb{U} \in St_\sigma$ onto $\mathbb{U}$, $\phi \in L_\sigma$ onto ϕ and any morphism $f \in Iso^{\emptyset}(\mathbb{U},\mathbb{U})$ onto f.

As the signature is always easily determined from the argument - we shall usually omit it writing simply $\iota\delta$ instead of $\iota\delta_\sigma$.

4.4. Definitional Extensions

4.4.1. Definition

Let $\phi \in L_{\sigma}^{n}$ and $\mathbb{U} = \langle A; P_o, \ldots, P_{m-1} \rangle$. We put

$$D_{\phi}\mathbb{U} = \langle A; P_o, \ldots, P_{m-1}, \phi^{\mathbb{U}} \rangle$$

Let further $\langle \mathbb{U}, \mathbb{B}, f \rangle \in \mathit{Iso}^{\emptyset}(\mathbb{U}, \mathbb{B})$ then we put

$$D_{\phi}\langle \mathbb{U}, \mathbb{B}, f \rangle = \langle D_{\phi}\mathbb{U}, D_{\phi}\mathbb{B}, f \rangle$$

Finally let $\psi, \eta, \xi \in L_{\sigma^\frown\langle n \rangle}$ then

$$D_{\phi}(v_i = v_j) = (v_i = v_i)$$

$$D_{\phi}(P_i(v_{j_o}, \ldots, v_{j_{\sigma_i - 1}})) = (P_i(v_{j_o}, \ldots, v_{j_{\sigma_i - 1}})) \quad (\text{for } i < m)$$

$$D_{\phi}(P_m(v_{j_o}, \ldots, v_{j_{n-1}})) = \left(\phi \begin{pmatrix} v_o, \ldots, v_{n-1} \\ v_{j_o}, \ldots, v_{j_{n-1}} \end{pmatrix} \right)$$

$$D_{\phi}(\neg \psi) = (\neg D_{\phi}\psi)$$

$$D_{\phi}(\xi \vee \eta) = (D_{\phi}\xi \vee D_{\phi}\eta)$$

$$D_{\phi}(\exists v_{k\psi}) = (\exists v_k D_{\phi} \psi)$$

(The expression $\phi \begin{pmatrix} v_o, \ldots, v_{n-1} \\ v_{j_o}, \ldots, v_{j_{n-1}} \end{pmatrix}$ means the result of substituting v_{j_o} for v_o, v_{j_1} for v_1, etc.

The function D_{ϕ} will be called a definitional extension. It is used whenever a mathematical theory is enriched by a defined notion. The class of definitional extensions is denoted by D.

The theorem we are about to prove has been called in [11] "interpretation lemma".

4.4.2. Theorem

If $\phi \in L^n_\sigma$, $\mathbb{U} \in St_\sigma$, $\psi \in L^m_{\sigma^\frown\langle n\rangle}$ and $a_o,\ldots,a_{m-1}$ are elements of $|\mathbb{U}|$ then

$$D_\phi \mathbb{U} \models \psi[a_o,\ldots,a_{m-1}] \quad \text{iff} \quad \models D_\phi\psi[a_o,\ldots,a_{m-1}]$$

Proof is by easy induction on the length of the formula ψ. As an example we give a proof in case

$$\psi = (P_{\ell h\sigma}(v_o,\ldots,v_{n-1})) \tag{1}$$

We proceed as follows:

$$D_\phi \mathbb{U} \models (P_{\ell h\sigma}(v_o,\ldots,v_{n-1}))[a_o,\ldots,a_{n-1}]$$

by the definition of satisfaction and definition of D_ϕ holds iff

$$\phi^{\mathbb{U}}(a_o,\ldots,a_{n-1})$$

i.e. iff

$$\mathbb{U} \models \phi[a_o,\ldots,a_{n-1}]$$

thus by definition of D_ϕ condition (1) is equivalent to

$$\mathbb{U} \models D_\phi(P_{\ell h\sigma}(v_o,\ldots,v_{n-1}))[a_o,\ldots,a_{n-1}]$$ ⊣

Before we prove that definitional extension is a functor we need thc following

4.4.3. Lemma

Let $\phi \in L^n_\sigma(\tau)$ and $f \in Iso^\tau(\mathbb{U},\mathbb{B})$ then

$$\mathbb{U} \models \phi[a_o,\ldots,a_{n-1}] \quad \text{iff} \quad \mathbb{B} \models \phi[fa_o,\ldots,fa_{n-1}].$$

Proof. In case τ is empty the lemma is obvious. Let now $\phi \in L_\sigma(\tau^\frown\langle m\rangle)$ and $f \in Iso^{\tau^\frown\langle m\rangle}(\mathbb{U},\mathbb{B})$. Therefore there is $\psi \in L^{n+m}_\sigma(\tau)$ such that

$$\psi = (\exists v_n \ldots \exists v_{n+m-1}\psi).$$

Thus we have

$\mathbb{U} \models \phi[a_o,\ldots,a_{n-1}]$ iff there are $a_n,\ldots,a_{n+m-1}$ such that

$$\mathbb{U} \models \psi[a_o,\ldots,a_{n+m-1}] \;. \tag{1}$$

By definition of $(\tau^\frown\langle m\rangle)$-isomorphism there is $f' \in Iso^{\tau}(\mathbb{U},\mathbb{B})$ an extension of f such that $a_n,\dots,a_{n+m-1}$ belong to the domain of f'. By induction hypothesis we have

$$\mathbb{U} \models \psi[a_o,\dots,a_{n+m-1}] \text{ iff } \mathbb{B} \models \psi[f'a_o,\dots,f'a_{n+m-1}]. \tag{2}$$

Therefore by definition of satisfaction and the fact that f' is an extension of f, we get the proof. ⊣

Let us turn now the proof that definitional extension is a functor.

4.4.4. Theorem

Let $\phi \in L_\sigma(\tau)$. Then D_ϕ is a functor from St_σ^{iso} into $St_{\sigma^\frown\langle n\rangle}^{iso}$ as well as from $St_\sigma^{\tau'}$ into $St_{\sigma^\frown\langle n\rangle}^{\emptyset}$ for any τ' such that $\tau \leqslant \tau'$.

Proof. To prove the second part of the theorem it will suffice to prove

If $f \in Iso^{\tau'}(\mathbb{U},\mathbb{B})$ then $D_\phi f \in Iso^{\emptyset}(D_\phi \mathbb{U}, D_\phi \mathbb{B})$. (1)

In fact, it will suffice to show that $D_\phi f$ maps $\phi^{\mathbb{U}}$, the additional relation in $D_\phi\mathbb{U}$, onto $\phi^{\mathbb{B}}$, the additional relation in $D_\phi \mathbb{B}$. This is however obvious because of lemma 4.3.3.

Suppose now $f \in Iso(\mathbb{U},\mathbb{B})$. Thus f is a one-to-one function from $|\mathbb{U}|$ to $|\mathbb{B}|$ belonging to $Iso^{\tau}(\mathbb{U},\mathbb{B})$. Therefore by (1) $D_\phi f \in Iso^{\emptyset}(D_\phi\mathbb{U}, D_\phi\mathbb{B})$ and, being one-to-one has to be an isomorphism. ⊣

Now let us consider a composition of definitional extensions

4.4.5. Definition

If $\phi_o,\dots,\phi_n \in L_\sigma$ then we say that the composition

$$D_{\phi_o} \dots D_{\phi_n}$$

is independent.

4.4.6. Theorem

Any composition of definitional extensions is equal to an independent composition of definitional extensions.

Proof. It is enough to prove the theorem for the composition of two definitional extensions D_{ϕ_1}, D_{ϕ_0}. Suppose $\phi_0 \in L^n_\sigma$. Then $\phi_1 \in L_{\sigma^\frown\langle n\rangle}$. Let $\psi_1 = D_{\phi_1}\phi_1$ be the result of eliminating the defined term. By theorem 4.3.4 it is easy to see that $D_{\phi_1} D_{\phi_0} = D_{\phi_1} D_{\psi_0}$. ⊣

4.4.7. Theorem

Let $\phi_o,\ldots,\phi_n \in L_\sigma(\tau)$ then the composition $\Gamma = D_\phi \ldots D_{\phi_n}$ is a functor from St^τ_σ into $St^\emptyset_{\sigma'}$, and from St^{iso}_σ into $St^{iso}_{\sigma'}$ for some σ', an extension of σ.

Proof. It suffices to prove the theorem in the case of the composition of two indepedent definitional extensions. Let $\phi \in L^m_\sigma$ and $\psi \in L^n_\sigma$. By lemma 4.5.3 if $f \in Iso^\tau(\mathbb{U},\mathbb{B})$ then $D_\phi f = f \in Iso^\emptyset(\mathbb{U},\mathbb{B})$ and

$D_\phi f = f \in Iso^\emptyset(\mathbb{U},\mathbb{B})$

thus in particular f preserves $\phi^{\mathbb{U}}$ and $\psi^{\mathbb{B}}$ restricted to the field of f. Thus $f \in Iso^0(D_\phi D_\psi \mathbb{U}, D_\phi D_\psi \mathbb{B})$. ⊣

As immediate consequences of 4.3.7 we get

4.4.8. Corollary

Any composition of a finite number of definitional extensions is a functor from St^τ_σ into $St^\emptyset_{\sigma'}$, for some σ, σ' and τ.

4.4.9. Corollary

Any composition of a finite number of definitional extensions is a functor from St^k_σ into $St^0_{\sigma'}$, for some σ, σ' and k.

The next example of interpretation is

4.5. Reduction

4.5.1. Definition

Let $\mathbb{U} = \langle A;P_o,\ldots,P_m\rangle \in St_\sigma, \phi,\psi \in L_\tau$, $\sigma = \langle k\rangle^\frown\tau$ and $\langle \mathbb{U}, \mathbb{B}, f\rangle \in Iso^\emptyset(\mathbb{U},\mathbb{B})$. We put

$\Box_\sigma \mathbb{U} = \langle A; P_1, \ldots, P_m \rangle$

$\langle \Box_\sigma, \mathbb{U}, \mathbb{B}, f \rangle = (\Box_\sigma \mathbb{U}, \Box_\sigma \mathbb{B})$

$\Box_\sigma (v_i = v_j)\ (v_i = v_j)$

$\Box_\sigma (P_i(v_{j_o}, \ldots, v_{j_{\sigma_i - 1}})) = (P_{i+1}(v_{j_o}, \ldots, v_{j_{\sigma_i - 1}}))$

$\Box_\sigma (\neg \phi) = (\neg \Box_\sigma \phi)$

$\Box_\sigma (\phi \vee \psi) = (\Box_\sigma \phi \vee \Box_\sigma \psi)$

$\Box_\sigma (\exists v_i \phi) = (\exists v_i \Box_\sigma \phi)$.

The function $\Box_\sigma$ will be called reduction. We have defined an entire class of reductions: one for every type σ. This class will be denoted by R. In most cases we shall omit the subscript, since it will be easy to reconstruct it from the context. Applications of reductions are seemingly scarce. They occur only as a part of some more general class of interpretations, which we are going to discuss a little later.

Proving properties of reduction, we first turn to interpretation lemma

4.5.2. Theorem

If $\mathbb{U} \in St_{\langle k \rangle ^\frown \sigma}$ and $\phi \in L^n_\phi$ then

$\Box\mathbb{U} \models \psi[a_o, \ldots, a_{n-1}]$ iff $\models \Box\phi[a_o, \ldots, a_{n-1}]$

Proof is obvious. ⊣

It is obvious that reduction is a functor. In fact

4.5.3. Theorem

Reduction is a functor from the category $St^\tau_{\langle k \rangle ^\frown \sigma}$ into St^τ_σ as well as from $St^{iso}_{\langle k \rangle ^\frown \sigma}$ into St^{iso}_σ.

It is easy to see that composition of extensions, and reductions need not be commutative. In fact if σ is of the length m and the symbol P_m actually occurs in $\phi \in L_\sigma$ then $D_\phi \Box$ is not defined for structures of signature σ while $\Box D_\phi$ is. However

4.5.4. Theorem

$$D_{\phi}\Box \subseteq \Box D_{\Box\phi}$$

4.6 Q-Interpretations

The class of all compositions of a finite number of definitional extensions and reductions is called Q-interpretation and is denoted by Q. Because of theorem 4.5.4 it is easy to see that

4.6.1. Theorem

For every $\Gamma \in Q$ there is an extension $\Gamma' \in R^{\underset{\smile}{\omega}} Q^{\underset{\smile}{\omega}}$.

In other words: for every Q-interpretation Γ there is Γ' such that $\Gamma \subseteq \Gamma'$ and Γ' consists of a finite number of definitional extensions performed first, and a finite number of reductions performed thereafter.

Examples of application of Q-interpretations are numerous. We shall give here just two. Let $\mathbb{U} = (A;\cup,\cap)$ be a lattice, and let

$$\phi = (v_o = v_o \cap v_1) .$$

Then the interpretation $\Box^2 D_{\phi}$ makes so the lattice $\mathbb{U}$ is a partially ordered set $\mathbb{B} = \langle A,\leq \rangle$ in which any two elements have a least upper bound and greater lower bound. Let $\mathbb{B} = \langle B;\leq \rangle$ be a partially order set with l.u.b. and g.l.b. for any two elements and let

$$\psi_1 = (v_o \leq v_1 \wedge v_o \leq v_2 \wedge \forall v_3(v_o \leq v_3 \rightarrow (v_1 \leq v_3 \vee v_2 \leq v_3)))$$

$$\psi_2 = (v_1 \leq v_o \wedge v_2 \leq v_o \wedge \forall v_3(v_3 \leq v_1 \wedge v_3 \leq v_2) \rightarrow v_3 \leq v_o))).$$

Then

$$\Box D_{\psi_2} D_{\psi_1} \mathbb{B}$$

is a lattice. The existence of the above teo interpretations allows algebraists to treat the theory of lattices interchangeably as the theory of certain partial orders or the theory of some algebras with two binary operations.

By examining theorems 4.4.6, 4.4.7, 4.5.3 and 4.5.4 it is easy to prove the following

4.6.2. Theorem

Let $\phi_i \in L_\sigma(\tau)$ $(i = 0,1,\ldots,n)$ then $\Box^k D_{\phi_o} \ldots D_{\phi_n}$ is a functor from St_σ^τ into $St_{\sigma'}^\emptyset$, as well as from $St_{\sigma'}^{iso}$ into $St_{\sigma'}^{iso}$ for some σ'.

A particular case of Q-interpretation is permutation of relations or, in other words, renaming:

4.6.3. Definition

Let f be a permutation of the natural numbers less than n, σ - a signature of length n, $\mathbb{U} = \langle A; P_o, \ldots, P_{n-1} \rangle \in St_\sigma$, $\mathbb{B} \in St_\tau$ $\langle \mathbb{U}, \mathbb{B}, g \rangle \in Iso^\emptyset(\mathbb{U}, \mathbb{B})$ and $\phi, \psi \in L_{\langle \sigma_{fi} \rangle_{i<n}}$. We put

$$L_f \mathbb{U} = \langle A; P_{f_o}, \ldots, P_{f(n-1)} \rangle$$

$$L_f \langle \mathbb{U}, \mathbb{B}, g \rangle = \langle L_f \mathbb{U}, L_f \mathbb{B}, g \rangle$$

$$L_f(v_i = v_j) = (v_i = v_j)$$

$$L_f(P_i(v_{j_o}, \ldots, v_{j_{\sigma_i - 1}})) = (P_{f^{-1}(i)}(v_{j_o}, \ldots, v_{j_{\sigma_j - 1}}))$$

$$L_f(\neg \phi) = (\neg L_f \phi)$$

$$L_f(\phi \vee \psi) = (L_f \phi \vee L_f \psi)$$

$$L_f(\exists v_i \phi) = (\exists v_i L_f \phi)$$

To prove that renaming is a Q-interpretation it is sufficient to observe that it may be expressed as a composition of definitional extensions with atomic formulas taken in the proper order and then reduction of proper number of relations.

Let $\ell h \sigma = n$ and let

$$f(i) = \begin{cases} 0 & \text{if } i = n \\ i+1 & \text{otherwise} \end{cases}$$

Let further $\phi = (v_o = v_o)$. Then

$$\iota\delta = \Box L_f D_\phi .$$

Therefore we have proved

4.6.4. Theorem

$\iota\delta \in Q$

4.7. Restriction

The third example of interpretation is restriction denoted by the symbol $\upharpoonright$.

4.7.1. Definition

Let $\mathbb{U} = \langle A, P_o, \ldots, P_m \rangle$ and $\mathbb{B} = \langle B; Q_o, \ldots, Q_m \rangle$ be two structures of signature $\tau = \langle 1 \rangle^\frown \sigma$. Let further $\langle \mathbb{U}, \mathbb{B}, f \rangle \in \mathit{Iso}^{\emptyset}(\mathbb{U}, \mathbb{B})$ and $\phi\ \psi \in L_\sigma$. We put

$$\upharpoonright_\tau \mathbb{U} = \langle P_o; P_1 \cap P_o^{\sigma_o}, \ldots, P_m \cap P_o^{\sigma_{m-1}} \rangle$$

$$\upharpoonright_\tau \langle \mathbb{U}, \mathbb{B}, f \rangle\rangle = \langle \upharpoonright_\tau \mathbb{U}, \upharpoonright_\tau \mathbb{B}, f \cap (P_o \times Q_o \rangle$$

$$\upharpoonright_\tau (v_i = v_j) = ((v_i = v_j) \wedge P_o(v_i) \wedge P_o(v_j))$$

$$\upharpoonright_\tau P_i(v_{j_o}, \ldots, v_{j_{\sigma_i - 1}})) = (P_{i+1}(v_{j_o}, \ldots, v_{j\sigma_i - 1}) \bigwedge_{k<\sigma_i} P_o(v_{j_k})) \text{ for } i < m$$

$$\upharpoonright_\tau (\neg \phi) = (\neg \upharpoonright_\tau \phi)$$

$$\upharpoonright_\tau (\phi \vee \psi) = (\upharpoonright_\tau \phi \vee \upharpoonright_\tau \psi)$$

$$\upharpoonright_\tau (\exists v_k \phi) = (\exists v_k (\upharpoonright_\tau \phi \wedge P(v_k)) \ .$$

The construction of restriction is thus a usual restriction to the first relation which therefore has to be unary. In fact we may write using traditional meaning of restriction:

$$\upharpoonright_\tau \mathbb{U} = (\Box\, \mathbb{U}) \upharpoonright P_o \ .$$

We assume that restriction is defined only for structures with unary first relation. Again as in the case of reduct restriction depends on the signature τ. We shall normally omit this subscript. The class of all restrictions will be denoted by T.

By an induction with respect to the length of the formula ϕ we prove

4.7.2. Theorem

If $\mathbb{U} = \langle A; P_o, \ldots, P_m \rangle \in \mathit{St}_{\langle 1 \rangle^\frown \sigma}$ and $\phi \in L^n_\sigma$ then

$\upharpoonright \mathbb{U} \models \phi[a_o,\ldots,a_{n-1}]$ iff $\mathbb{U} \models \upharpoonright\phi[a_o,\ldots,a_{n-1}]$ and $P_i(a_i)$ for $i < n$. A little more difficult is the proof of the next theorem.

4.7.3. Theorem

The restriction is a functor from $St^{\tau}_{\langle 1\rangle^\frown\sigma}$ into St^{τ}_{σ} as well as from $St^{iso}_{\langle 1\rangle^\frown\sigma}$ into St^{iso}_{σ}

Proof. Let $f \in Iso^{\tau}(\mathbb{U},\mathbb{B})$ where $\mathbb{U},\mathbb{B} \in St_{\langle 1\rangle^\frown\sigma}$. We have to prove that $\upharpoonright f$ $Iso^{\tau}(\upharpoonright\mathbb{U}, \upharpoonright\mathbb{B})$. We shall prove it by induction with respect to the length of τ. Let therefore $\tau = \emptyset$, and let $a \in P_o \cap \mathrm{Dom}\, f$. Since f is a partial isomorphism, it has to map elements of Q_o and therefore $fa \in Q_o$. Thus $\langle a,fa\rangle \in P_o \upharpoonright Q_o$ and therefore $a \in \mathrm{Dom} \upharpoonright f$. The fact that $\upharpoonright f$ maps P_i on elements of Q_i ($i = 1,\ldots,\ell h\sigma$) is obvious.

Let now $\tau = \langle m\rangle^\frown\tau'$, and let $a_o,\ldots,a_m \in P_o$. Since $f \in Iso^{\langle m\rangle^\frown\tau}(\mathbb{U},\mathbb{B})$ there is $f' \in Iso^{\tau}(\mathbb{U},\mathbb{B})$ such that $f \subseteq f'$, and $a_o,\ldots,a_m \in \mathrm{Dom}\, f'$. Therefore $\upharpoonright f \subseteq \upharpoonright f'$, $a_o,\ldots,a_m \in \mathrm{Dom} \upharpoonright f'$, and, by induction hypothesis $\upharpoonright f'$ $Iso^{\tau}(\upharpoonright\mathbb{U}, \upharpoonright\mathbb{B})$.

—|

Now we turn to the problem of commutativity of restriction with other interpretations.

4.7.4. Theorem

(i) there is $\Gamma \in Q$ such that $\upharpoonright\upharpoonright = \upharpoonright\Gamma$

(ii) if $\phi \in L_\sigma$ then $D_\phi \upharpoonright = \upharpoonright D_{\upharpoonright\phi}$

(iii) there is $\Gamma \in Q$ such that $\square\upharpoonright = \upharpoonright\square\,\Gamma$

Proof. (i) Suppose that $\upharpoonright\upharpoonright$ is defined for structures of signature $\langle 1,1\rangle^\frown\sigma$ and let $\ell h\sigma = n$. We put

$$\phi = (P_o(v_o) \wedge P_1(v_o)).$$

Let f map numbers less than $n+1$ onto themselves such that

$f(i) = i + 1$ (for $i < n$)

$f(n) = 0$.

Then $\upharpoonright\upharpoonright = \upharpoonright L_f \square\square\, D_\phi$

(ii) $D_\phi \upharpoonright = \upharpoonright D_{\upharpoonright\phi}$ follows from 4.7.2

(iii) Let f be a function such that

$f(0) = 1$

$f(1) = 0$

$f(n) = n$ for $n > 1$

Since the remaining L_f belongs to Q, the theorem follows from the equality

$$\Box \upharpoonright = \upharpoonright \Box L_f$$

⊣

Remark. The proof of the theorem 4.6.4(i) is one of the exceptions when it is essential to notice that for different signatures we have different restriction interpretations.

4.7.5. Theorem

There is $\Gamma \in Q$ such that

$$\iota\delta_\sigma = \upharpoonright\Gamma$$

Proof. Let

$$f(n) = \begin{cases} n+1 & \text{for } n < \ell h\sigma \\ 0 & \text{for } n = \ell h\sigma \end{cases}$$

and let

$$\phi = (v_o = v_o).$$

Thus we have

$$\iota\delta_\sigma = \upharpoonright L_f D_\phi .$$

Theorem 4.7.5 may be reformulated to state

$$\iota\delta \in TQ.$$

4.8. Quotient Interpretation

The next example of interpretation is based on the notion of quotient structure.

4.8.1. Definition

Let $\mathbb{U} = \langle A; P_o, \ldots, P_m \rangle \in St_\tau$ and let P_o be a congruence in $\mathbb{U}$, let $\mathbb{B} = \langle B; Q_o, \ldots, Q_m \rangle \in St_\tau$ and let Q_o be a congruence in $\mathbb{B}$, let further $f \in Iso^{\emptyset}(\mathbb{U}, \mathbb{B})$ and $\phi, \psi \in L_\sigma$. If $\tau = \langle 2 \rangle ^\frown \sigma$ we put

$$/_\tau \mathbb{U} = \langle A; P_1, \ldots, P_m \rangle /_{P_o}$$
$$(/_\tau f)[a]/_{P_o} = [fa]/_{Q_o}$$
$$/_\tau (v_i = v_j) = (P_o(v_i, v_j))$$
$$/_\tau (\neg \phi) = \neg /_\tau \phi$$
$$/_\tau (\phi \vee \psi) = /_\tau \phi \vee /_\tau \psi$$
$$/_\tau (\exists v_i \phi) = (\exists v_i /_\tau \phi)$$

The quotient construction is therefore the usual construction of taking quotient structure with respect to the first relation, which has to be a congruence. If the first relation is not a congruence the construction is not defined. The quotient translation consists in fact of the introduction of a new equation which in the original language was an extralogical symbol.

Mycielski [7] avoids the use of quotient constructions and translation by allowing non standard meaning for the equality symbol. Thus in his approach the equality need not be treated separately from other relations.

Just like the case of reduct and restriction we shall usually omit the subscript σ in the symbol $/_\sigma$. It will always be easy to reconstruct the signature the quotient refers to. The class of all quotient interpretations will be denoted by E.

4.8.2. Theorem

If $\mathbb{U} = \langle A; P_o, \ldots, P_n \rangle \in St_{\langle 2 \rangle ^\frown \sigma}$, $\phi \in L_\sigma$ and P_o is a congruence in $\mathbb{U}$ then for any $a_o, \ldots, a_{m-1} \in A$ we have

$$/\mathbb{U} \models \phi[[a_o]/_{P_o}, \ldots, [a_{m-1}]/_{P_o}] \quad \text{iff} \quad \mathbb{U} \models /\psi[a_o, \ldots, a_{m-1}]$$

Proof is by induction with respect to the length of the formula ϕ. The only interesting step in the proof is when ϕ is an atomic formula and

$\phi = (v_i = v_j)$,

but then we have $/\mathbb{U} \models (v_i = v_j)[[a_o]/_{P_o}, [a_1]/_{P_o}]$ iff $P_o(a_o,a_1)$ i.e. iff

$\mathbb{U} \models P_o(v_i,v_j)[a_o,a_1]$. ⊣

The quotient construction is so commonly used in algebra that there is hardly any need to give an example. We shall turn therefore directly to the problem of quotient interpretation being a functor.

4.8.3. Theorem

Quotient construction is a functor from the category $St^{\tau}_{\langle 2\rangle^\frown\sigma}$ into St^{τ}_{σ} and from $St^{iso}_{\langle 2\rangle^\frown\sigma}$ into St^{iso}_{σ} for every σ and τ .

Proof. (By induction with respect to the length of τ). We put $\mathbb{U} = \langle A;P_o,\ldots,P_m\rangle$ and $\mathbb{B} = \langle B,Q_o,\ldots,Q_m\rangle$. Let $\tau = \emptyset$ and $f \in Iso^{\tau}(\mathbb{U},\mathbb{B})$

$(f/P_o)[a_i[/_{P_o} = [fa_i]/_{Q_o}$ for $i = 0,\ldots,0_j$.

Suppose $a_o,\ldots,a_{\sigma_j-1} \in$ Dom f. We have

$P_j/P_o([a_o]/_{P_o},\ldots,[a\ _{-1}]/_{P_o})$ iff $P_j(a_o,\ldots,a_{\sigma_j-1})$ iff

$Q_j(fa_o,\ldots,fa_{\sigma_j-1})$ iff $Q_j/_{Q_o}([fa_o]/_{Q_o},\ldots,[fa_{\sigma_j-1}]/_{Q_o})$

Thus $/f \in Iso^{\emptyset}(/\mathbb{U},/\mathbb{B})$.

Suppose now $f \in Iso^{\tau^\frown(n)}(/\mathbb{U},/\mathbb{B})$, and $[a_o]/_{P_o},\ldots,[a_{n-1}]/_{P_o} \in A/_{P_o}$. Thus $a_o,\ldots,a_{n-1} \in A$. Thus there is $f' \in Iso^{\tau}(\mathbb{U},\mathbb{B})$ such that $a_o,\ldots,a_{n-1} \in$ Dom f'. By induction hypothesis $/f' \in Iso^{\tau}(/\mathbb{U}, /\mathbb{B})$. Of course $/f \subseteq /f'$. Analogously we may prove that if $[b_o]/_{Q_o},\ldots,[b_{n-1}]/_{Q_n} \in B/_{Q_o}$ then there is $f'' \in Iso^{\tau}(\mathbb{U},\mathbb{B})$ such that $f \subseteq f''$. Thus $/f \subseteq /f''$ and $/f'' \in Iso^{\tau}(/\mathbb{U},/\mathbb{B})$. Finally therefore $/f \in Iso^{\tau^\frown(n)}(/\mathbb{U}, /\mathbb{B})$.

If $f \in Iso(\mathbb{U},\mathbb{B})$ then f is a one-to-one mapping of A onto B. Since P_o is an equivalence relation thus /f is a one-to-one mapping of $A/_{P_o}$ onto $B/_{Q_o}$. Moreover $f \in Iso^{\emptyset}(\mathbb{U},\mathbb{B})$ and thus $/f \in Iso^{\emptyset}(\mathbb{U},\mathbb{B})$.

Therefore $/f \in \mathcal{Iso}(/\mathbb{U}, /\mathbb{B})$ ⊣

Finally we shall discuss commutativity laws for quotient interpretations.

4.8.4. Theorem

(i) $// \cong /\Box$

(ii) $D_\phi/ = /D/_\phi$

(iii) there is $\Gamma \in Q$ such that $/ = /\Gamma$

(iv) there is $\Gamma \in Q$ such that $\upharpoonleft / \cong / \upharpoonleft \Gamma$

Proof. Let $\mathbb{U} = \langle A; P_o, P_1, \dots, P_n \rangle$.

(i) The required isomorphism is

$$g([[a]/_{P_o}]/_{P_1/P_o}) = [a]/_{P_1}$$

(ii) is obvious.

To prove (iii) and (iv) we need the following function

$$f(n) = \begin{cases} 1 & \text{if } n = 0 \\ 0 & \text{if } n = 1 \\ n & \text{otherwise} \end{cases}$$

(iii) We have

$$\Box / = / \Box L_f$$

(iv) We have

$$\upharpoonleft / \cong / \upharpoonleft L_f .$$

In fact the required isomorphism is

$$h([a]/_{P_o}) = [a]/_{P_o \cap P_1^2}$$ ⊣

4.8.5. Theorem

These is $\Gamma \in Q$ such that

$$\iota\delta_\sigma \cong /\Gamma$$

Proof. Let

$$f(n) = \begin{cases} n+1 & \text{for } n < \ell h\sigma \\ 0 & \text{for } n = \ell h\sigma \end{cases}$$

and let

$$\phi = (v_o = v_1).$$

We have

$$\iota\delta \cong /L_f D_\phi .$$

In fact for $\mathbb{U} \in St_\sigma$ the required isomorphism would be

$$ga = [a]/_{=} = \{\{a\} : a \in |\mathbb{U}|\}$$

⊣

We may reformulate the theorem 4.8.5 to state

$$\iota\delta \,\tilde{\in}\, EQ.$$

4.9. Cartesian Interpretation

The last example of interpretation is that of Cartesian interpretation.

Let P be an m-n-ary relation over A i.e. let $P \subseteq A^{m-n}$ then $P^{(n)} = \{\langle\langle a_{oo},\ldots,a_{o(n-1)}\rangle,\ldots,\langle a_{(m-1)0},\ldots,a_{(m-1)(n-1)}\rangle\rangle : P(a_{oo},\ldots,a_{(m-1)(n-1)})\}$ is an m-ary relation over A^n.

4.9.1. Definition

Let $\mathbb{U} = \langle A; P_o,\ldots,P_{m-1}\rangle \in St_{\langle n\cdot\sigma_i\rangle_{i\in m}}$, $f : A \to B$ and $\phi,\psi \in L_\sigma$. We put

$$C^{(n)}\mathbb{U} = \langle A^n; P_o^{(n)},\ldots,P_{m-1}^{(n)}\rangle,$$

$$C^{(n)}f\,\langle\langle a_o,\ldots,a_{n-1}\rangle\rangle = \langle fa_o,\ldots,fa_{n-1}\rangle \text{ for all } a_i \in A\ (i=0,\ldots,n-1),$$

$$C^{(n)}(v_i = v_j) = (v_{i\cdot n} = v_{j\cdot n}) \wedge (v_{i\cdot n+1} = v_{j\cdot n+1}) \wedge \ldots \wedge (v_{i(n+1)-1} = v_{j(n+1)-1})$$

$$C^{(n)}P_i(v_{j_o},\ldots,v_{j\sigma_i-1}) = P_{i-1}(v_{j_o}{}^\frown \ldots {}^\frown v_{j\sigma_i-1})$$

(where $v_k = \langle v_{kn},\ldots,v_{k(n+1)-1}\rangle$

$$C^{(n)}\neg\phi = \neg C^{(n)}\phi,$$

$$C^{(n)}(\phi \vee \psi) = C^{(n)}\phi \vee C^{(n)}\psi \ ,$$

$$C^{(n)}(\exists v_i \phi) = \exists v_{i \cdot n} \cdots \exists v_{i(n+1)-1} C^{(n)}\phi \ .$$

The Cartesian interpretation is in fact just another way of looking at the same structure. The n-th Cartesian interpretation is defined only for structures with all terms of their signatures divisible by n. The class of all Cartesian interpretations will be denoted by C.

The best known application of Cartesian interpretation is a Cartesian plane. The construction of a model of Euclidean plane geometry from the field of real numbers consists of defining relations between numbers e.g. the 6-ary relation given by the condition

$$(v_o - v_2)(v_o - v_4) + (v_1 - v_3)(v_1 - v_5) = 0.$$

Then apllying Cartesian interpretation of order 2 we get the relation "points (v_o,v_1), (v_2,v_3) and (v_4,v_3) and (v_4,v_5) form a triangle with right angle at (v_o,v_1)". Boffa [2] has proved that the use of Cartesian interpretation cannot be avoided in this case.

As usual we start with interpretation lemma:

4.9.2. Theorem

If $\mathbb{U} \in St_{\langle n \cdot \sigma_i \rangle_i k}$ and $\phi \in L^m_\sigma$ then

$$C^{(n)} \mathbb{U} \models \phi[\langle a_{oo},\ldots,a_{o(n-1)} \rangle,\ldots,\langle a_{(m-1)o},\ldots,a_{(m-1)(n-1)} \rangle]$$

iff $\mathbb{U} = C^{(n)}\phi[a_{oo},\ldots,a_{(m-1)(n-1)}]$ for all $a_{oo},\ldots,a_{(m-1)(n-1)} \in A$.

Proof is by straight forward induction with respect to the length of the formula ϕ using only the definition of satisfaction and Cartesian interpretation. ⊣

4.9.3. Theorem

Cartesian construction of order n is a functor from the category $St^\tau_{\langle n \cdot \sigma_i \rangle_{i \in m}}$ into St^τ_σ and from $St^{iso}_{\langle n \cdot \sigma_i \rangle_{i \in m}}$ into St^{iso}_σ where $m+1 = \ell h \sigma$.

Proof. (By induction with respect to the length of τ). Let

$\mathbb{U} = \langle A;P_o,\ldots,P_k\rangle$ and $\mathbb{B} = \langle A;P_o,\ldots,P_k\rangle$. Suppose moreover that $\mathbb{U},\mathbb{B} \in St_{\langle n\cdot\sigma_i\rangle_{i\in m}}$, where $m+1 = \ell h\sigma$.

Let $\tau = \emptyset$ and $f \in Iso^{\emptyset}(\mathbb{U},\mathbb{B})$ then

$$C^{(n)}f\langle a_o,\ldots,a_{n-1}\rangle = \langle fa_o,\ldots,fa_{n-1}\rangle .$$

Suppose $a_{oo},\ldots,a_{(\sigma_i-1(n-1)} \in$ Dom f. Then

$$C^{(n)}P_i\langle\langle a_{oo},\ldots,a_{o(n-1)}\rangle,\ldots,\langle a_{(\sigma_i-1)0},\ldots,a_{(\sigma_i-1)(n-1)}\rangle\rangle$$

iff $P_i\langle a_{oo},\ldots,a_{(\sigma_i-1)(n-1)}\rangle$

iff $Q_i\langle fa_{oo},\ldots,fa_{(\sigma_i-1)(n-1)}\rangle$

iff $C^{(n)}Q_i\langle\langle C^{(n)}\ fa_{oo},\ldots,a_{o(n-1)}\rangle,\ldots,C^{(n)}f\langle a_{(\sigma_i-1)0},\ldots,{}_{(\sigma_i-1)n-1)}\rangle\rangle$.

Thus $C^{(n)}f \in Iso^{\emptyset}(C^{(n)}\mathbb{U},C^{(n)}\mathbb{B})$.

Suppose now $f \in Iso^{\tau^\frown\langle m\rangle}(\mathbb{U},\mathbb{B})$ and

$$a_o = \langle a_{oo},\ldots,a_{o(n-1)}\rangle,\ldots,a_{m-1} = \langle a_{(m-1)0},\ldots,a_{(n-1)(-1)}\rangle \in C^{(n)}A$$

Thus $a_{oo},\ldots,a_{(m-1)(n-1)} \in A$. There is $f' \in Iso^{\tau}(\mathbb{U},\mathbb{B})$ such that $a_{oo},\ldots,a_{(m-1)(n-1)} \in Df'$. By induction hypothesis ${}^{(n)}f' \in Iso^{\tau}(C^{(n)}\mathbb{U},C^{(n)}\mathbb{B}$. Since it is obvious that $C^{(n)}f \subseteq C^{(n)}f'$, and $a_o,\ldots,a_{m-1} \in DC^{(n)}f'$, as well as nice for $b_o,\ldots,b_{m-1} \in C^{(n)}B$ we may give an analogous proof, therefore $C^{(n)}f' \in Iso^{\tau^\frown(m)}(C^{(n)}\mathbb{U},C^{(n)}\mathbb{B})$. ⊣

Finally we discuss commutativity laws for Cartesian interpretation.

4.9.4. Theorem

(i) $C^{(n)}C^{(m)} \cong C^{(n\cdot m)}$

(ii) $D_\phi C^{(n)} = C^{(n)}D_{(C^{(n)}\phi)}$

(iii) if $\mathbb{U} \in$ Dom $C^{(n)}$ $\Box\mathbb{U} = \Box C^{(n)}\mathbb{U}$

(iv) there is $\Gamma \in Q$ such that $C^{(n)}\restriction\ = \restriction C^{(n)}\Gamma$

(v) there is $\Gamma \in Q$ such that $C^{(n)}/ = /C^{(n)}\Gamma$

Proof of (i). It is easy to see that both the composition of n-th and m-th Cartesian interpretation and the (m·n)-th Cartesian interpretation are defined for $\mathbb{U}$ iff all terms of signature of $\mathbb{U}$ are divisible by $m \cdot n$. If this is the case then the mapping of $(|\mathbb{U}|^n)^m$ which assigns to

$$\langle\langle a_{oo},\ldots,a_{o(m-1)}\rangle,\ldots,\langle a_{(n-1)0},\ldots,a_{(n-1)(m-1)}\rangle\rangle$$

the $n \cdot m$-tuple

$$\langle a_{oo},\ldots,a_{(m+1)(n-1)}\rangle$$

is an isomorphism of $C^{(n)}(C^{(m)}\mathbb{U})$ onto $C^{(nm)}\mathbb{U}$.

(ii) Let $\mathbb{U} \in St_{\langle n\sigma_i\rangle_{i\in m}}$ and $\phi \in L_\sigma$ then

$$D_\phi C^{(n)}\mathbb{U} = C^{(n)}D_{(C^{(n)}\phi)}$$

(iii) Obvious

(iv) Let $\mathbb{U} \in St_{\langle 1\rangle^\frown\sigma}$, $\ell h\sigma = m+1$, $f(i) = i+1$ for $i=1,\ldots,m$ and $f(m+1) = 1$.

Finally let

$$\phi = P_o(v_o) \wedge P_o(v_1) \wedge \ldots \wedge P_o(v_{n-1}).$$

Then

$$C^{(n)} \restriction \mathbb{U} = \restriction C^{(n)} \; L_f D\mathbb{U}$$

(v) Let $\mathbb{U} = \langle A;P_o,\ldots,P_m\rangle \in St_{\langle 2\rangle^\frown\sigma}$ and let P_o be a congruence in $\mathbb{U}$.

Let f be as in the proof of (iv) and let

$$\phi = P_o(v_o v_1) \wedge \ldots \wedge P_o(v_{2n-2)}v_{2n-1}).$$

Then the function g: such that

$$g(\langle [a_o]/_{P_o},\ldots,[a_{n-1}]/_{P_o}\rangle) = [\langle a_o,\ldots,a_{n-1}\rangle]/_P$$

(where $P = \phi C^{(n)}\mathbb{U}$) is an isomorphism from $C^{(n)}/\mathbb{U}$ onto $/C^{(n)}\square L_f D_\phi \mathbb{U}$. ⊣

4.9.5. Theorem

$$C^{(1)} \cong \iota\delta$$

Proof. Let $\mathbb{U} \in St_\sigma$. The required isomorphism is

$$f\langle a\rangle = a .$$

⊣

We may reformulate this theorem to state

$\iota\delta \tilde{\in} C$.

4.10. General Notion of Interpretation

We conclude this section with the general definition of interpretation.

4.10.1. Definition

By an interpretation we mean a composition of any finite number of definitional extensions, reducts, restrictions, quotient interpretations and Cartesian interpretations.

In this definition we have excluded interpretations with parameters as described in Zsczerba [12]. Thus we shall discuss only paraterfree interpretations. In fact the following discussion will depend very heavily on this limitation.

Examining the commutativity theorems (4.4.4, 4.6.4, 4.7.4 and 4.8.4) as well as theorems 4.5.4, 4.6.5, 4.7.5 and 4.8.5 we prove

4.10.2. Theorem

For every interpretation Γ there is an interpretation $\Gamma' \in Q$ such that for some natural number n we have

$$\Gamma \;\tilde{\equiv}\; / \restriction C^{(n)}\Gamma'.$$

⊣

The intuitive meaning of this theorem is that all interpretations belong essentially to ETCQ. The word "essentially" is used not only because we may be forced to use an isomorphic interpretation instead of an original one, but also because the interpretation $/ \restriction C^{(n)}\Gamma'$ may be defined for some more structures than Γ.

Because of the theorem 4.9.2 we shall concentrate below on the interpretations in standard form i.e. on the interpretation of the form

$[/ \restriction C^{(n)}\Gamma$

where $\Gamma \in Q$. In fact this was the definition of interpretation given in Szczerba [11].

Let us turn now to interpretation lemmas. Examining theorems 4.4.2,

4.5.2, 4.7.2, 4.8.2 and 4.9.2 we get

4.10.3. Theorem

If Γ is an interpretation defined for structure $\mathbb{U}$ and sentence ϕ then

$$\Gamma\mathbb{U} \models \phi \Leftrightarrow \mathbb{U} \models \Gamma\phi.$$

In case Γ is in standard form if is possible to extend this theorem to formulas:

4.10.4. Theorem

Let $\phi_i \in L_\sigma$ (for $i < k$), let $\phi \in L^m_{\sigma'}$, $a_o, \ldots, a_{o(n-1)}, \ldots, a_{(m-1)(n-1)} \in |\mathbb{U}|$ and let an interpretation

$$\Gamma = /\mathbb{U}^{(n)} \square^P D_{\phi_o} \ldots D_{\phi_k}$$

be defined for $\mathbb{U} = \langle A; P_o, \ldots, P_q \rangle$ then

$$\Gamma\mathbb{U} \models [[a_{oo}, \ldots, a_{o(n-1)}] / P_{p+1}, \ldots, [a_{(m-1)0}, \ldots, a_{(n-1)(\ -1)} / P_{p+1}]$$

$$\text{iff } \mathbb{U} \models \Gamma\phi\, [a_{oo}, \ldots, a_{(m-1)(n-1)}].$$

⊣

As an immediate consequence of theorem 4.4.4, 4.5.3, 4.7.3, 4.8.3 and 4.9.3 we get

4.10.5. Theorem

Any interpretation is a functor from category St^τ_σ into $St^\emptyset_\sigma$, as well as from St^{iso}_σ into $St^{iso}_{\sigma'}$ for some σ, σ' and τ.

⊣

In case Γ is in standard form we may be again more specific. In fact considering 4.6.2 instead of 4.4.4 and 4.5.3 in the above proof we get

4.10.6. Theorem

Let $\phi_i \in L^{\rho_i}_\sigma(\tau)$ and $\rho_i = n\rho'_i$ for $i \leq \ell h\rho = k$. Let $\sigma = \sigma''{}^\frown\sigma'''$ and $\sigma'''_i = n\sigma'_i$ for $i < \ell h\sigma - m$, where $m = \ell h\sigma''$. Let

$$\Gamma = /\restriction C^{(n)}\, {}^m D_{\phi_0} \ldots D_{\phi_{k-1}}$$

then Γ is a functor from St^{τ}_{σ} into $St_{\sigma'\wedge\rho'}$ as well as from St^{iso}_{σ} into $St^{iso}_{\sigma'\wedge\rho'}$.

5. AUTOMORPHIC SEQUENCES

5.1. Definition

Let a and b be two sequences of the same length, we put

$f^a_b = \{\langle a_i, b_i \rangle : i < \ell h\, a = \ell h\, b\}$.

It is esay to see

5.2. Theorem

If a,b are without repetitions i.e. if $a_i = a_j$ implies $i = j$, and $b_i = b_i$ implies $i = j$ for all $i,j < \ell h\, a = \ell h\, b$ then:

(i) $(f^a_b)^{-1} = f^b_a$

(ii) $f^b_c f^a_b = f^a_b$

(iii) f^a_b is a function

$\dashv$

Let a and b be sequences of elements of the same structures $\mathbb{U}$. Suppose $\ell h\, a = \ell h\, b$. We say that a and b are σ-isomorphic if $f^a_b \in \mathit{Iso}^{\sigma}(\mathbb{U},\mathbb{U})$.

5.3. Definition

$a \overset{\sigma}{\cong} b$ iff $f^a_b \in \mathit{Iso}^{\sigma}(\mathbb{U},\mathbb{U})$

The relation $\overset{\sigma}{\cong}$ is an equivalence relation:

5.4. Theorem

(i) $a \overset{\sigma}{\cong} a$

(ii) if $a \overset{\sigma}{\cong} b$ then $b \overset{\sigma}{\cong} a$

(iii) if $a \overset{\sigma}{\cong} b$ and $b \overset{\sigma}{\cong} c$ then $a \overset{\sigma}{\cong} c$

Proof. (i) The function f^a_a is a subset of identity and thus by 3.4(i) and 3.3(ii) we have $f^a_a \in Iso^{\sigma}(\mathbb{U},\mathbb{U})$ for every σ.

(ii) Since $a \overset{\sigma}{\cong} b$ then $f^a_b \in Iso^{\sigma}(\mathbb{U},\mathbb{U})$. Thus by 3.4(ii) we have $f^b_a \in Iso^{\sigma}(\mathbb{U},\mathbb{U})$ and therefore $b \overset{\sigma}{\cong} a$.

(iii) By assumptions we have $f^a_b \in Iso^{\sigma}(\mathbb{U},\mathbb{U})$ and therefore $a \overset{\sigma}{\cong} c$. ⊣

Now we shall turn to the theorem which is the crucial point of this paper: we already know that the relation $\overset{\tau}{\cong}$ is an equivalence relation on a class $\mathbb{U}(p) = \{\langle A \cdot P^\frown a \rangle : a \in A^p\}$; here $\mathbb{U} = \langle A;P \rangle$. We shall prove that there is only a finite number of equivalence classes. To prove it need two lemmas:

5.5. Lemma

If $|\overline{\overline{\mathbb{U}}}| \geq \ell h\ a + p$ then $f^a_b \in Iso^{\tau^\frown \langle p \rangle}(\mathbb{U},\mathbb{U})$ iff the following two conditions hold

(i) for every $a' \in |\mathbb{U}|^p$ there is $b' \in |\mathbb{U}|^p$ such that $f^{a^\frown a'}_{b^\frown b'} \in Iso^{\tau}(\mathbb{U},\mathbb{U})$

(ii) for every $b' \in |\mathbb{U}|^p$ there is $a' \in |\mathbb{U}|^p$ such that $f^{a^\frown a'}_{b^\frown b'} \in Iso^{\tau}(\mathbb{U},\mathbb{U})$

Proof. Suppose $f^a_a \in Iso^{\tau^\frown p}(\mathbb{U},\mathbb{U})$ and $a' \in |\mathbb{U}|^p$. Thus $Rg\ a' \leq p'$ and therefore there is $f \in Iso^{\tau}(\mathbb{U},\mathbb{U})$ such that $f^a_b \subseteq f$ and $Rg\ a' \subseteq Dom\ f$. Let $b' = (f(a'_i))_{i<p}$. Then $f^{a^\frown a'}_{b^\frown b'} \subseteq f$ and therefore by 3.3(i) we get $f^{a^\frown a'}_{b^\frown b'} \in Iso^{\tau}(\mathbb{U},\mathbb{U})$. Analogously we prove that for every $b' \in |\mathbb{U}|^p$ there is $a' \in |\mathbb{U}|^p$ such that $f^{a^\frown a'}_{b^\frown b'} \in Iso^{\tau}(\mathbb{U},\mathbb{U})$.

Suppose now the right hand side of the equivalence holds and suppose $X \subseteq |\mathbb{U}|$ and moreover $\overline{\overline{X}} \leq p$. Without loss of generality we may assume that $\overline{\overline{X}} = p$ and therefore we may order it in a sequence a of length p. By hypothesis we have some $b \in |\mathbb{U}|^p$ such that

$$f^{a^\frown a'}_{b^\frown b'} \in Iso^{\tau}(\mathbb{U},\mathbb{U})$$

at the same time $f^a_b \subseteq f^{a^\frown a'}_{b^\frown b'}$ and $X \subseteq Dom\ f^{a^\frown a'}_{b^\frown b'}$. If $X \subseteq |\mathbb{U}|$ we proceed analogously and thus we get $f^a_b \in Iso^{\tau^\frown\langle p\rangle}(\mathbb{U},\mathbb{U})$. ⊣

5.6. Lemma

If $|\overline{\overline{\mathbb{U}}}| \geq \ell h\ a + \ell h\ b$ and

$\{[a^\frown a']/_\tau : a' \in |\mathbb{U}|^p\} = \{[b^\frown b']/_\tau : b \in |\mathbb{B}|^p\}$

then $a \overset{\tau^\frown\langle p\rangle}{\cong} b$.

Proof. Suppose both sets are equal. Then for any $a' \in |\mathbb{U}|^p$ there is $b' \in |\mathbb{U}|^p$ such that $a^\frown a' \cong b^\frown b'$ i.e. for every $a' \in |\mathbb{U}|^p$ there is $b' \in |\mathbb{U}|^p$ such that $f^{a^\frown a'}_{b^\frown b'} \in Iso^\tau(\mathbb{U},\mathbb{U})$. Similarly we may prove that for any $b' \in |\mathbb{U}|^p$ there is $a' \in |\mathbb{U}|^p$ such that $f^{a^\frown a'}_{b^\frown b'} \in Iso^\tau(\mathbb{U},\mathbb{U})$. Thus by 5.5 we have $f^a_b \in Iso^{\tau^\frown\langle p\rangle}(\mathbb{U},\mathbb{U})$ i.e. $a \overset{\tau^\frown\langle p\rangle}{\cong} b$

⊣

Intuitively the lemma 5.6 says that mapping

$f([a]/_{\overset{\tau}{\cong}}) = \{[a^\frown a']/_\tau : a' \in |\mathbb{U}|^p\}$

is an embedding of $St_\sigma(p)/_{\overset{\tau}{\cong}}$ where $q = \ell h\ a$.

5.7. Theorem

$\mathbb{U}(p)/_{\overset{\tau}{\cong}}$ is finite

Proof. Since by the definition of partial isomorphisms

$a \overset{\emptyset}{\cong} b$ iff $\mathbb{U} \restriction Rg\ a \cong \mathbb{U} \restriction Rg\ b$

thus $\mathbb{U}(p)/_{\overset{\emptyset}{\cong}}$ is finite: there is only a finite number of isomorphic types of structures of a given finite power.

Now let $\tau = \tau'^\frown\langle q\rangle$. If $|\overline{\overline{\mathbb{U}}}|$ is less or equal to $t = \sum_{i=0}^{\ell h\ \tau-1} \tau_i$ then $\mathbb{U}(p)/_{\overset{\tau}{\cong}}$ is equal again because there is only a finite number of isomorphism types of structures of power at most t. Suppose therefore that $|\bar{\bar{\mathbb{U}}}| > t$. Thus by lemma 5.6 $\mathbb{U}(p)/_{\underset{\cong}{\tau}}$ in embeddable is the class of subclasses of $\mathbb{U}(p+q)/_{\underset{\cong}{\tau'}}$. By the induction hypothesis $\mathbb{U}(p+q)/_{\underset{\cong}{\tau'}}$ is finite, thus its class of subclasses is finite and therefore $\mathbb{U}(p)/_{\underset{\cong}{\tau}}$ is finite. $\dashv$

Further on elements of $\mathbb{U}(p)/_{\underset{\cong}{\tau}}$ will be called τ-isomorphic types with p constants or simply τ-isomorphic types. Thus theorem 5.7 may be expressed: there are only finitely many τ-isomorphic types with p constants for every natural number p and every $\tau \in \omega^{\underset{\smile}{\omega}}$.

The immediate consequence of theorem 5.7 is

5.8. Corollary

$(A^p)/_{\underset{\cong}{\tau}}$ is finite.

6. DEFINING ISOMORPHIC TYPES

Throughout this section we fix a structure $\mathbb{U} = \langle A;P\rangle \in St_\sigma$. We shall prove that $[a]/_{\underset{\cong}{\tau}}$ is definable i.e. there is a formula $\phi_{[a]/_{\underset{\cong}{\tau}}}$ such that $b \in X$ iff $\mathbb{U} \models \phi_{[a]/_{\underset{\cong}{\tau}}}[b]$. Before giving the precise definition of $\phi_{[a]/_{\underset{\cong}{\tau}}}$ we shall describe the intuitions tied to it.

Let $X = [a]/_{\underset{\cong}{\tau}}$. In case $\tau = \emptyset$, for ϕ_X we take a full description of $\mathbb{U} \upharpoonright \mathrm{Rg}\, a$ i.e. the diagram of $\mathbb{U} \upharpoonright \mathrm{Rg}\, a$. Since $\mathrm{Rg}\, a$ is finite we get an elementary formula. In case $\tau = \tau'^\frown\langle q\rangle$ we get ϕ_X as follows: Take any $X' = (A^{p+q})/_{\underset{\cong}{\tau}}$. If it is possible to extend a by sequence a' of q new elements such that $a^\frown a' \in X'$, we take for $\psi_{X'}$ the formula "there is sequence $a' \in |A|^q$ such that $\phi_{X'}$" otherwise for $\psi_{X'}$ we take "there is no sequence $a' \in |A|^q$ such that $\phi_{X'}$". Since there is only a finite number of τ-isomorphic types with (p+q) constants (cf. corollary 5.8) we may take for ϕ_X the conjuction of all $\psi_{X'}$.

6.1. Definition

(i) Let $X = [a]/_{\overset{\emptyset}{\cong}}$. We put

$$\phi_X = (\bigwedge_{i=0}^{\ell h\,\sigma} (\bigwedge \{P_i(v_{f_0},\dots,v_{f(\sigma_i-1)}) : \mathbb{U} \models P_i(v_0,\dots,v_{\sigma_i-1})[a_{f_0},\dots,a_{f(\sigma_i-1)}]$$

and f maps $\{0,\dots,p-1\}$ into itself}

$$\wedge \{\neg P_i(v_{f_0},\dots,v_{f(\sigma_i-1)}) : \mathbb{U} \models \neg P_i(v_0,\dots,v_{\sigma_i-1})[a_{f_0},\dots,a_{f(\sigma_i-1)}]$$

and f maps $\{0,\dots,p-1\}$ into itself}))

(ii) Let $X = [a]/_{\overset{\tau^\frown\langle q\rangle}{\cong}}$. We put

$$\phi_X (\bigwedge \{\exists v_p,\dots,v_{p+q-1}\phi_{X'} : \text{there is } a' \in A^q \text{ such that } a^\frown a' \in X' \in (A^{p+q})/_{\overset{\tau}{\cong}}\}$$

$$\wedge \{\forall v_p,\dots,v_{p+q-1}\neg\phi_{X'} : \text{there is no } a' \in A^q \text{ such that } a^\frown a' \in X' \in (A^{p+q})/_{\tau}\}).$$

6.2. Theorem

The formula $\phi_{[a]/_{\overset{\tau}{\cong}}}$ defines the set $[a]/_{\overset{\tau}{\cong}}$ i.e. $a \overset{\tau}{\equiv} b$ iff $\mathbb{U} \models \phi_{[a]/_\tau}[b]$

Proof. Let $X = [a]/_{\overset{\tau}{\cong}}$. In case $\tau = \emptyset$ the theorem is obvious: ϕ_X is a complete description of $\mathbb{U} \upharpoonright Rg\ a$. Thus let us assume that $\tau = \tau'^\frown\langle q\rangle$ and for every $X' \in (A^{p+q})/_\tau$ the formula $\phi_{X'}$ defines X'. We have $a \overset{\tau}{\cong} b$ iff $f^a_b \in \mathit{Iso}^\tau(\mathbb{U},\mathbb{U})$. By the definition of τ-isomorphism this means that $a \overset{\tau}{\cong} b$ is equivalent to

$$\forall a' \in A^q \exists f' \in \mathit{Iso}^{\tau'}(\mathbb{U},\mathbb{U}) \quad f^a_b \subseteq f' \wedge Rg\ a' \subseteq Dom\ f'$$

$$\wedge\ \forall b' \in A^q \exists f' \in \mathit{Iso}^{\tau'}(\mathbb{U},\mathbb{U}) \quad f^a_b \in f' \wedge Rg\ b' \subseteq Dom\ f'$$

i.e. by the definition of τ'-equivalence $a \overset{\tau}{\cong} b$ is equivalent to

$$\forall a' \in A^q \, \exists b' \in A^q \quad a^\frown a' \overset{\tau}{\cong} b^\frown b' \quad \text{and} \quad b' \in A^q \, \exists a' \in A^q \quad a^\frown a' \overset{\tau}{\cong} b^\frown b' \, .$$

Thus by the induction hypothesis $a \overset{\tau}{\cong} b$ is equivalent to

$$\forall X' \subseteq (A^{p+q})/_{\overset{\tau}{\cong}} \quad \mathbb{U} \models \phi_{X'}([a] \Leftrightarrow \models \phi_{X'}[b])$$

i.e.

$$\mathbb{U} \models \phi_X[b] \, . \qquad \dashv$$

By inspection of definition 5.1 it is easy to prove

6.3. Theorem

If $X = [a]/_{\overset{\tau}{\cong}}$ then $\phi_X \in L_\sigma(\tau)$ ⊣

7. DIAGRAMS

Let $\Gamma \in Q$ i.e. let Γ be a composition of finite number of definitional extensions and reducts. Since all the components, i.e. restrictions and definitional extensions, change neither the universe of structures, nor the morphisms, it is easy to see that the diagram

$$\begin{array}{ccc} St_\sigma^\tau & \xrightarrow{\Gamma} & St_{\sigma'}^{\emptyset} \\ & \searrow \quad \swarrow & \\ & Set & \end{array} \qquad (1)$$

commutes. It turns out however that

7.1. Theorem

The diagram (1) commutes iff $\Gamma \in Q$

Proof. Let diagram (1) commute, and let $\mathbb{U} = \langle A;P \rangle \in St_{\sigma'}$ let moreover $\Gamma\mathbb{U} = \langle B;Q \rangle \in St_{\sigma'}$. We assume that $\ell h\, \sigma = n$ and $\ell h\, \sigma' = m$. Since diagram (1) commutes, it is easy to see that $A = B$. By the same reason any τ-isomorphism of $\mathbb{U}$ into itself is mapped onto a $\emptyset$-isomorphism of $\Gamma\mathbb{U}$ into itself, and moreover if $\langle \mathbb{U}, \mathbb{U}, f \rangle \in Iso^\tau(\mathbb{U},\mathbb{U})$ then

$\langle \Gamma \mathbb{U}, \Gamma \mathbb{U}, f \rangle \in Iso^{\emptyset}(\Gamma \mathbb{U}, \Gamma \mathbb{U})$. Thus for any $i \leq n$ and any $a \in A^{\sigma_i}$ we have

$$\text{if} \quad a \in Q_i \quad \text{then} \quad \{b: a \overset{\tau}{\cong} b\} \subseteq Q_i$$

and therefore Q_i is a sum of a finite number of equivalence classes with respect to $\overset{\tau}{\cong}$. Therefore

$$a \in Q_i \quad \text{iff} \quad \mathbb{U} \models \bigvee\{\phi_{[b]/\overset{\tau}{\cong}} : b \in Q_i\}[a]$$

Thus Q_i is definable by the formula

$$\eta_i = \{\phi_{[a]/\overset{\tau}{\cong}} : b \in Q_i\}.$$

It should be stressed that formula η_i is elementary since the set of formulas whose alternative we take is finite.

Therefore $\Gamma = \Box^n D_{\eta_m} \ldots D_{\eta_o} \in Q.$

$\dashv$

Any elementary interpretation Γ may be decomposed (by theorem 4.10.2): $\Gamma = /C^{(n)}\Gamma'$ where n is some natural number and $\Gamma' \in Q$. Thus by 7.1 we have

7.2. Pernambuco Theorem

$\Gamma \in ETCQ$ iff for some n, σ, τ, τ' and ρ the following diagram commutes

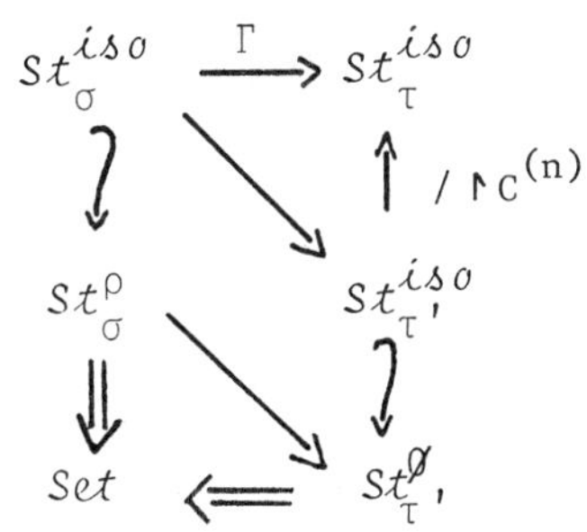

The Pernambuco theorem, after some modifications, may be used to estimate the complexity of interpretations. e.g. the results of Szczerba [13] have been obtained using the Pernambuco theorem, however the Pernanbuco theorem may be usually eliminated from the proofs just as it

has been done in Szczerba [13].

The mentioned modification consists of introducing a variable X ranging over functors. Then we modify the diagram as follows

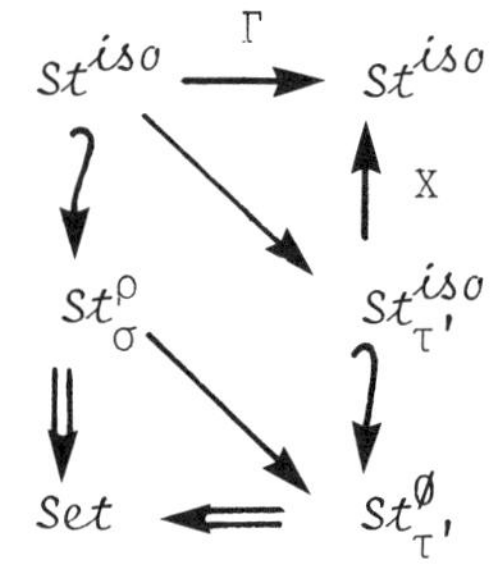

(2)

By a slight modification of the proof of theorem 7.2 we may prove that if diagram (2) commutes with $X = \upharpoonleft, /, C^{(n)}, /\upharpoonleft, \upharpoonleft C^{(n)}$, or $/C^{(n)}$, then $\Gamma \in$ TQ, EQ, CQ, ETQ, TCQ or ECQ, respectively.

The sequence ρ may be used to determine the types of the defending formulas in definitional extensions occuring in Γ.

Sometimes it is difficult to determine τ-isomorphisms but relatively easier to determine p-isomorphisms. Thus we may use the following collorary of theorem 7.2.

7.3. Corollary

$\Gamma \in$ ETCQ iff for some natural numbers n, p and some sequences σ, τ, τ' the following diagram commutes

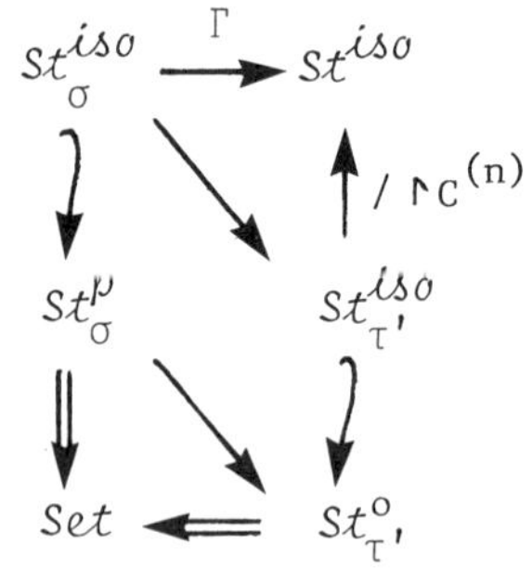

Of course collorary 7.3 may be modified and used just like theorem 7.2.

8. MEANS OF INTERPRETATION

We shall present now an application of the Pernambuco theorem: all the negative results of this section have been obtained with the help of the Pernambuco theorem. Then it was convenient to eliminate this theorem from the proofs.

8.1. Definition

Let X be a class of interpretations. We say that a theory T is X-interpretable in a theory S or in symbols

$$T \trianglelefteq_{X} S$$

iff there is an interpretation $\Gamma \in X$ such that $T = \overleftarrow{\Gamma} S$

It is possible to replace the condition $T = \overleftarrow{\Gamma} S$ by an equivalent one:

$$\mathrm{Mod}\ S \subseteq \mathrm{Dom}\ \overleftarrow{\Gamma} \quad \text{and} \quad T = Th(\overleftarrow{\Gamma}\ \mathrm{Mod}\ S).$$

It is obvious that

(1) if $X \subseteq Y$ and $T \trianglelefteq_{X} T'$ then $T \trianglelefteq_{Y} T'$

and therefore all the implications shown in fig. 1 are also obvious (an implication (1) is visualised in the figure as $X \to Y$). We shall prove that there are no more implications than the ones on fig. 1 (together with their compositions). It will suffice to prove that $\trianglelefteq_{CQ}$ is not included in $\trianglelefteq_{ETQ}$ as well as $\trianglelefteq_{EQ} \nsubseteq \trianglelefteq_{TCQ}$ and $\trianglelefteq_{TQ} \nsubseteq \trianglelefteq_{ECQ}$: that there are theories S, S', T, T' and U, U' such that

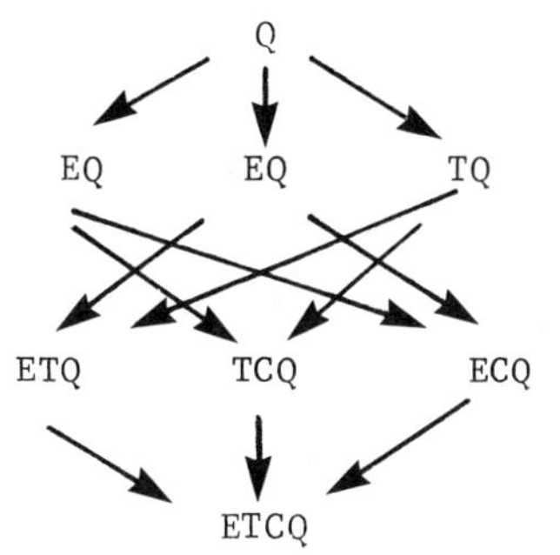

(2) $S \trianglelefteq_{CQ} S'$ but not $S \trianglelefteq_{ETQ} S$

(3) $T \trianglelefteq_{EQ} T'$ but not $T \trianglelefteq_{TCQ} T'$

(4) $U \trianglelefteq_{TQ} U'$ but not $U \trianglelefteq_{QCE} U'$

8.2 Example

Let S be the theory of equality in a four element set i.e. $S = Th\langle 4;\rangle$ since equality is a standard logical relation, and let S' be the theory of equality in a two element set: $S' = Th\langle 2;\rangle$. Furthermore $\phi = (v_o \wedge v_1 = v_3)$. Then

$$C^{(2)} \square D_\phi \langle 2;\rangle \cong \langle 4;\rangle$$

Thus

$$S \trianglelefteq_{CQ} S' .$$

On the other hand any Q-interpretation preserves cardinality of models and $\upharpoonright$ or $/$ may only make it smaller. Thus if $\mathbb{U} \models S$, $\Gamma \in ETQ$ then $\Gamma\mathbb{U}$ is not a model of S', i.e.

$$\neg S \trianglelefteq_{ETQ} S' .$$

This proves (2).

8.3 Example

Let T be the theory of equality in one-element set: $T = Th\langle 1;\rangle$ and let T' be the theory of equality in an infinite set i.e. $T' = Th\langle \omega;\rangle$. Let moreover

$$\phi = (v_o = v_o \wedge v_1 = v_1),$$

then for any nonempty set A

$$/D_\phi\langle A;\rangle \cong \langle 1;\rangle$$

and thus

$$T \trianglelefteq_{EQ} T' .$$

Suppose now $T \trianglelefteq_{TCQ} T'$. Thus there is a Q-interpretation Γ such that for some natural number n $\upharpoonright C^{(n)}\Gamma\langle\omega;\rangle$ has exactly one element. Therefore $C^{(n)}\Gamma\langle\omega;\rangle = \langle \omega^n;P_o\rangle$ where P_o is a singleton i.e. contains exactly one element. Therefore $\Gamma\langle\omega;\rangle = \langle\omega;Q_o\rangle$ where Q_o contains exactly one n-tuple $\langle a_o,\dots,a_{n-1}\rangle$. Let $b \neq a_i$ for all $i < n$. There is an automorphism f of $\langle \omega; \rangle$ mapping a_o onto b. Then f is not an automorphism of $\langle \omega; Q_o \rangle$ contrary to the fact that $f = \Gamma f$ should be an isomorphism of $\Gamma\langle \omega; \rangle$. This proves (3).

8.4 Example

Let $\mathcal{U}$ be the elementary theory of the successor relation in the set of integers, i.e. $\mathcal{U} = \mathit{Th}\ \mathrm{B}$ where $\mathrm{B} = \langle Z; S \rangle$, Z is the set of integers and for any $x \in z$ we have $S(x, x+1)$. For $\mathcal{U}'$ we have the elementary theory of the structure $\mathrm{B}' = \langle Z \cup \{\omega\}; Z, S \rangle$ i.e. of the structure B ammended by an extra element. It is essential that $\omega \notin Z$. Since $\upharpoonright \mathrm{B}' = \mathrm{B}$ then

$$\mathcal{U} \trianglelefteq_{TQ} \mathcal{U}' .$$

To prove (4) we need first to discuss partial automorphism of B'. As a useful convention we adopt

$$\omega + n = \omega,$$

$$\omega - n = \omega,$$

$$n - \omega = \omega,$$

$$\omega - \omega = \omega,$$

$$|\omega| = \omega,$$

where $n \in Z$.

8.4.1 Lemma

Let f be a partial function mapping a subset of $Z \cup \{\omega\}$. Suppose moreover that for any $a, b \in \mathrm{Dom}\ f$

$$\text{if } |a - b| < 2^k \text{ or } |fa - fb| < 2^k \text{ then } a - b = fa - fb$$

then $f \in \mathit{Iso}^k(\mathrm{B}', \mathrm{B}')$.

Proof. For $k = 0$ the theorem is obvious. Suppose the theorem is true for $k = p$. We shall prove it for $k = p+1$. We assume therefore that for every $a, b \in \mathrm{Dom}\ f$

$$\text{if } |a - b| < 2^{k+1} \text{ or } |fa - fb| < 2^k \text{ then } a - b = fa - fb \qquad (1)$$

Let $c \in Z$. Our objevtive is to find f', an extension of f such that $c \in \text{Dom } f'$ and $f' \in \mathit{Iso}^k(B', B')$.

If $c \in \text{Dom } f$ then we may take $f' = f$ and from the induction hypothesis get that $f' \in \mathit{Iso}^k(B', B')$.

Suppose now that $c \neq \omega$ and $c \notin \text{Dom } f$. Without any loss of generality we may assume that there are $x, y \in \text{Dom } f$ such that $x < c < y$. In fact the construction in the remaining cases is analogous. We may assume moreover that x and y are closest to c i.e. that for any $z \in \text{Dom } f$ we have either $z \leq x$ or $y \leq z$. We have of course $fx \neq fy$ and again without the loss of generality, we may assume that $fx < fy$.

Case 1. $|x - y| < 2^{k+1}$ then by (1) we have $|fx - fy| = |x - y|$ and therefore $x \leq y$ and $fx \leq fy$ we get $fy = fx + y - x$. We extend f to f' by putting

$$f'c = fx + c - x.$$

It is easy to check that f' satisfies the premise of the induction hypothesis and therefore $f' \in \mathit{Iso}^k(B', B')$.

Case 2. $|x - y| \geq 2^{k+1}$ and $|c - x| < 2^k$. Then $|c - y| \geq 2^k$ and by (1) we have $|fx + fy| \geq 2^{k+1}$. Thus $|(fx - c - x) - fy| \geq 2^k$ and therefore we extend f to f' by puting

$$f'c = fx + c - x$$

we get $f' \in \mathit{Iso}^k(B', B')$ by the induction hypothesis.

Case 3. $|x - y| \geq 2^{k+1}$ and $|c - y| < 2^k$. A discussion analogous to that in case 2 shows that the extension f' of f defined by putting

$$f'c = fy + c - y$$

belongs to $\mathit{Iso}^k(B', B')$.

Case 4. $|c - x| \geq 2^k$ and $|c - y| \geq 2^k$. Therefore $|x - y| \geq 2^k$ and by (1) we have $|fx - fy| \geq 2^{k+1}$. Thus there is c' such that $fx < c' < fy$ and $|c' - fx| > 2^k$ and $|c' - fy| > 2^k$. If we extend f to f' by putting

$$f'c = c',$$

we get that f' satisfies the premises of the induction hypothesis and therefore $f' \in \mathit{Iso}^k(B', B')$.

It remains to discuss the case when $c = \omega$. In this case we extend f to f' by the stipulation that

$$f'c = \omega$$

and again by the induction hypothesis we get $f' \in Iso^k(B', B')$.

Thus we have proved that for any $c \in Z$ there is a k-isomorphism f' extending f and defined for c. The proof that there is a k-isomorphism f'' extending f and taking c as its value, is analogous. Thus $f \in Iso^{k+1}(B', B')$. ⊣

Suppose now that $\mathcal{U}' \trianglelefteq_{ECQ} \mathcal{U}$. Thus there is an interpretation $\Gamma \in ECQ$ such that $\Gamma B'$ is at least elementarily equivalent with B . Since $\Gamma \in ECQ$, there is an interpretation $\Gamma \in Q$ such that for some n

$$\Gamma B' \cong (/C^n\Gamma')B' .$$

Let $C^n\Gamma' B' = B'' = \langle Z'', \approx, S'' \rangle$. Since $/$ is to be defined for B'', the relation $\approx$ has to be a congruence in B''. Elements of Z are n-tuples of $Z \cup \{\omega\}$. Thus one of these elements is $\Omega = \langle \omega, \ldots, \omega \rangle$. Let Z_o be the least subset of Z'' such that $\Omega \in Z_o$ and if $a \in Z_o$ then $S''(a,b)$ or $S''(b,a)$ imply $b \in Z_o$. Since $/B''$ is elementarily equivalent to B , therefore $Z_o/_\approx$ has to have infinite number of elements.

Note $\Gamma'' \in Q$. Thus for some k, Γ' is a functor $St^k_{\langle 2\rangle}$ into $St^0_{\langle 2\rangle}$. Discussing k-automorphisms of B' we shall prove that $Z_o/_x$ is finite thus getting a contradiction.

For further discussion we need the notion of diameter of n-tuple a

$$\delta a = \max\{|a_i - a_j| : i,j < n \wedge a_i \neq \omega \neq a_j\}.$$

We prove

8.4.2 Lemma

Let $a \in b_o$ and $|p| \geq 2^k + \delta a$ then $a \approx \langle a_i + p \rangle_{i<n}$.

Proof. If $a = \Omega$ the theorem is obvious, since $\omega + p = \omega$, and $\approx$ is reflexive. Suppose the theorem is true for some $a' \in Z_o$ and let $S''(a',a)$. We shall prove it is true for a.

By induction hypothesis we may assume that for all $i,j < n$ we have

$$|a_i - a'_j| \geq 2^k \ , \quad \text{and} \quad |a_i + p - a'_j| \geq 2^k$$

otherwise we could obtain an n-tuple equivalent to a' and removed far enough by adding to all its terms

$$q = p + \delta a' + a_o - a'_o \ .$$

Let $a'' = \langle a_i + p \rangle_{i<n}$. Then by lemma 8.3.1

$$f^{a'\ a}_{a'\ a''} \in \mathcal{Iso}^k(\ B', B\ ) \subseteq \mathcal{Iso}^0(\ B',\ B'),$$

and therefore from S"(a',a) we say infer S"(a',a) But $\Gamma\, B' = /\, B'' = \langle Z'', S'' \rangle /_{\approx}$ is elementarily equivalent to B and therefore $[a] /_{\approx} = [a''] /_{\approx}$ i.e. $a \approx a''$. ⊣

The above lemma contains some superfluous assumptions:

8.4.3 Lemma

Let $a \in Z_o$ then $a \approx \langle a_i + p \rangle_{i<n}$ for any $p \in Z$.

Proof. By lemma 8.3.2 we have

$$a \approx \langle a_i + q \rangle_{i<n} \approx \langle a_i + p \rangle_{i<n}$$

where $q \geq 2^k + \delta a + |p|$. Since $\approx$ is transitive this proves the lemma. ⊣

8.4.4 Lemma

Let $a \in Z_o$ then there is $b \approx a$ such that $\delta b \leq 2^k(n-1)$

Proof. The theorem is obvious for $a = \Omega$. Suppose therefore that the theorem is true for a' and S"(a',a). Suppose moreover that $a_{i_o} \leq a_{i_1} \leq \ldots \leq a_{i_{n-1}}$ and for $j < r$ we have $|a_{i_{j+1}} - a_{i_j}| \leq 2^k$ and $|a_{i_{p+1}} - a_{i_p}| > 2^k$. Then by lemma 8.3.1 we get

$$f^c_{c'} \in \mathcal{Iso}^k(\ B',\ B')$$

where $c = \langle a_{i_j} \rangle_{j<n-1}$ and $c' = \langle a_{i_j} \rangle_{j \leq p} \langle a_{i_j} + 2^k + a_{i_p} - a_{i_{p+1}} \rangle_{j=p+1,\ldots,n-1}$.

By an argument analogous to the one used in proof of lemma 8.3.2 we get that $a \approx f^c_{c'}a$. Repeating the above procedure at most (n-1)-times we get $b \approx a$ such that $\delta b \leq 2^k(n-1)$. ⊣

From lemmas 8.3.3 and 8.3.4 it follows immediately that for every $a \in Z_o$ there is $b \approx a$ such that $|b_i| \leq 2^k(n-1)$ for every $i < n$. Thus $Z_{o/\approx}$ has to be finite. Thus we have the required contradiction. This proves (4).

REFERENCES

1. E.W. Beth, On Padoa's method in the theory of definition, Indagationes Mathematicae XV (1953), pp. 330-339.
2. M. Boffa, Elimination des quantificateurs en algebre, Bulletin de la Societe Mathematique de Belgique, vol. XXXII, 1980, pp.107-130.
3. A. Ehrenfeucht, An application of games to the completeness problem for formalised theories, Fundamenta Mathematicae, vol. 49 (1960), pp. 129-141.
4. R. Fraïse , Sur quelques classifications des relations basees sur des isomorphismes restraines, Publications Scientifiques de l'Unicersite' d'Alger, Ser A (mathematiques), vol. 2 (1955).
5. A. Lindenbaum & A. Tarski, Uber die Beschreinheit der Ansdruckmitted deduktiver theories, Ergebnisse eines mathematiscken Kolloquiums fase 7 (1939-35) pp. 15-22 [English translation:] On the limitation of the means of expression of deductive theories [in:] Logic, Semantics, Metamathematics, Oxford 1956. Clarendon Press, pp. 384-392.
6. A. Mostowski, Constructible sets with applications, Warszawa 1969 PWN.
7. J. Mycielski, A lattice of interpretability types of theories, Journal of Symbolic Logic, vol. 42 (1977), pp. 297-305.
8. A. Padoa, Un nouveau système irreducible de postulats pour l'algebre, C.R. Deuxieme Congrès International des Mathematitians (Paris 1902), pp. 249-256.
9. A.M. Sette & J.S. Sette, Functionalization of first-order language with finitely many predicates, Mathematical Logic-Proceedings of the first Brasilian Conference. Eds: A.I. Arruda, N.C.A. da Costa, R. Chuaqui, New York, 1977, Marcel Dekker Inc. pp. 1-11.
10. A.M. Sette & L.W. Szczerba, Algebraic characterization of interpretability, Notas e Comunicações de Matemática Nº 94, Departamento de Matemática Universidade Federal de Pernambuco.
11. L.W.Szczerba, Elementary Means of creating entities [in:] Conference on Technology of Science, Towarzystwo Nankowe P Pxockie 1970.
12. L. Szczerba, Interpretability of elementay theories [in:] Foundations of Mathematics and Computability Theory. Reidel Publ. Co. 1977 Dordrecht, pp. 129-145.

13. L.W. Szczerba, Interpretations with parameters, Zeitschrift fur mathematische Logik und Grundlagen der Mathematik, Bd. 26 (1980), pp.35-39.

14. L.W. Szczerba, Parameters are essential in one dimensional geometry [to be published in] Geometria.

15. A. Tarski, Some methodological investigations on the definability of concepts. [in:] Logic, Semantic, Metamathematics, Clarendon Press, 1956, Oxford, pp. 296-319.

INDEX